Advances in
Elastomers and
Rubber Elasticity

Advances in Elastomers and Rubber Elasticity

Edited by
Joginder Lal
Formerly, The Goodyear Tire and Rubber Company
Akron, Ohio

and
James E. Mark
University of Cincinnati
Cincinnati, Ohio

SPRINGER SCIENCE+BUSINESS MEDIA, LLC

Library of Congress Cataloging in Publication Data

Advances in elastomers and rubber elasticity.

"Proceedings of a symposium on Advances in Elastomers and Rubber Elasticity,
sponsored by the Division of Polymer Chemistry, American Chemical Society, held
September 9–13, 1985, at the National American Chemical Society meeting in
Chicago, Illinois"—T.p. verso.
Includes bibliographies and index.
1. Elastomers—Congresses. 2. Elasticity—Congresses. I. Lal, Joginder, 1923–
II. Mark, James E., 1934–   . III. American Chemical Society. Division of Polymer
Chemistry. IV. American Chemical Society. Meeting (190th: 1985: Chicago, Ill.)
TS1925.A38 1986   678   86-25416
ISBN 978-1-4757-1438-8   ISBN 978-1-4757-1436-4 (eBook)
DOI 10.1007/978-1-4757-1436-4

Proceedings of a symposium on Advances in Elastomers and Rubber
Elasticity, sponsored by the Division of Polymer Chemistry,
American Chemical Society, held September 9–13, 1985, at the
National American Chemical Society meeting in Chicago, Illinois

The views, opinions, and/or findings contained in this symposium volume
are those of the author(s) and should not be construed as an official
Department of the Army position, policy, or decision, unless so designated
by other documentation. The same is true in regard to the other sponsors
of the symposium.

PREFACE

The present book is a sequel to "Elastomers and Rubber Elasticity,"
edited by J.E. Mark and J. Lal and published by the American Chemical
Society in 1982. It is also based on papers presented at an ACS Symposium,
sponsored by the Division of Polymer Chemistry, Inc., in this case one held
in Chicago in September of 1985. The keynote speaker was to have been Professor Paul J. Flory, and his untimely death just prior to the symposium
was a tremendous loss to all of polymer science, in particular to those interested in elastomeric materials. It is to his memory that this book is
dedicated.

There has been a great deal of progress in preparing and studying elastomers since the preceding symposium, which was in 1981. In the case of the
synthesis and curing of elastomers, much of the background necessary to an
appreciation of these advances is given in the first, introductory chapter.
More specific subjects include the control of microstructure in the anionic
polymerization of elastomers, polyurethane elastomers with monodisperse
sequence length distributions, hysteresis and heat build-up in polyurethanes,
polyurethane-polyurea copolymers, and new block copolymers with improved
high-temperature properties, and methods for curing some of them. Also
covered are the effects of microstructure on traction, fatigue resistance,
and rolling resistance, curing of elastomers that are difficult to cure,
polymer-bound antioxidants, and the chemical modification of elastomers (for
example the hydrogenation of polydienes and the epoxidation of natural
rubber).

A similarly wide range of subjects is covered in the area of rubberlike
elasticity. Primarily theoretical studies include the use of quantum statistical mechanical ideas in theories of entangled networks, calculations
of molecular deformation tensors and molecular orientation, a memory-lattice
theory of rubberlike elasticity, and strain-induced crystallization and its
effects on the stress and other mechanical properties. More experimentally
oriented topics include cross-linking reactions and the network imperfections they can introduce, model networks for estimating the possible importance of inter-chain entanglements, light scattering and small-angle neutron scattering in the study of network topology and chain extension, segmental orientation from fluorescence polarization, comparisons between
stress-strain relationships for elastomeric networks and polymer melts, and
the effects of chemical structure on the strength of elastomers (in tearing
and abrasive wear).

The vitality and importance of the subject of elastomers and rubber
elasticity are obvious from the above incomplete list of diverse topics,
but also from the very wide organizational and geographical distribution of
the authors. Various chapters in fact come from universities, industrial
laboratories, and research institutes in the United States, Canada, England,

France, Germany, the Netherlands, Italy, Turkey, and Japan.

Finally, it is a pleasure to acknowledge that the symposium on which this book is based received financial support from the following organizations: ACS Division of Polymer Chemistry, Inc., Petroleum Research Fund, U.S. Army Research Office, American Hoechst Corporation, Dow Chemical, U.S.A., E.I. du Pont de Nemours & Company, Firestone Tire & Rubber Company, GenCorp, BFGoodrich Company, Goodyear Tire & Rubber Company, Hercules Incorporated, Monsanto Company, Nippon Zeon of American, Inc., Phillips Petroleum, Polysar Limited, Shell Development Company, and Uniroyal Chemical Division.

Joginder Lal
Retired from
The Goodyear Tire & Rubber Co.
Akron, OH 44316

Lal Associates
855 Shullo Drive
Akron, OH 44313

James E. Mark
The University of Cincinnati
Cincinnati, OH 45221

CONTENTS

INTRODUCTION TO SYNTHESIS OF ELASTOMERS*

George Odian

College of Staten Island
City University of New York
Staten Island, New York 10301

ABSTRACT

Polymerization and crosslinking reactions used for the synthesis of elastomers are reviewed. Step and chain polymerizations are characterized in terms of the reaction variables which must be controlled to obtain an elastomer. Radical and ionic chain polymerizations are discussed as well as the structural variations possible through copolymerization and stereoregularity

INTRODUCTION

Elastomers or rubbers are polymers which can undergo very large, reversible deformations at relatively low stresses. Only a polymer molecule can undergo large deformations since it can respond to stress without bond rupture by extension from a random coil to an extended chain molecule through bond rotation. Only an amorphous polymer with low glass transition temperature and low secondary forces has the required chain flexibilty to meet these requirements. Crystalline polymers (e.g., polyethylene or isotactic polypropylene) or highly polar polymers (e.g., polyamides) do not have the necessary chain flexibility. Amorphous polymers with large bulky substituents (e.g., poly(methyl methacrylate) and polystyrene) are also too rigid to be elastomers. Elastomeric behavior is limited to those non-polar amorphous polymers with irregular structures (e.g., ethylene-propylene copolymers) and/or flexible chain units (e.g., polyisobutylene, polysiloxane, polysulfide,

---

*Adapted with permission from reference 14.

1,4-polyisoprene). Crosslinking of a polymer to form a network structure is an essential part of synthesizing an elastomeric product. The presence of a crosslinked network prevents polymer chains from irreversibly slipping past one another upon deformation. Either chemical or physical crosslinking can be used to produce the network structure. The various reactions for synthesizing polymers from monomers are discussed below together with a consideration of the processes used to achieve crosslinking.

POLYMERIZATION

The prime consideration in any polymerization reaction is the control of polymer molecular weight (MW). A polymer requires a minimum molecular weight in order to possess sufficient physical strength to be useful. This minimum MW may be as low as a few thousand but is generally considerbly higher. In any specific application, the desired MW is some compromise high MW which yields sufficient strength for the end-use application while retaining ease of processing the polymer into its final product shape. The need to control molecular weight places significant restrictions on any reaction used for polymer synthesis.

Step Polymerization

Polymerizations are classified as either chain or step polymerizations. The two processes differ in the time required for the growth of large-sized (i.e., polymer) molecules compared to the time for achieving high conversions. Step polymerizations typically involve two different bifunctional reactants (referred to as monomers), each containing a different functional group X or Y. Polymer growth occurs through the formation of xy functional units by reaction of X and Y groups

$$nX-X + nY-Y \longrightarrow (X-xy-Y)_n \qquad\qquad (1)$$

The synthesis of polysulfides (Eq. 2) and polyurethanes (Eq. 3) are examples of step polmerizations which yield elastomeric materials.

$$nCl-R-Cl + nNa_2S_m \xrightarrow{-NaCl} (R-S_m)_n \qquad\qquad (2)$$

$$nHO-R-OH + nOCN-R'-NCO \longrightarrow (O-R-OCO-NH-R'-NH-CO)_n \qquad (3)$$

The structural entity which repeats over and over again in the polymer, i.e., the structure within the parenthesis in Eq. 1 or 2 or 3, is referred to as the polymer repeat unit.

Step polymerization proceeds by the stepwise reaction between X and Y groups to form dimer, trimer, tetramer, pentamer, and so on until eventually polymer molecules are formed. As each larger-sized species is formed, it competes with the smaller-sized species for further reaction. Any two molecular species containing X and Y groups can react with each other throughout the reaction. This results in a final polymer with a distribution of molecular weights. The average size of the molecules increases slowly with conversion since the reaction rate constants are typically not high (of the order of 0.001-0.1 L/mol-s).

The average degree of polymerization DP, defined as the average number of monomer units linked together per molecule in the reaction mixture, is dependent on the fractional conversion p and stoichiometic ratio r (the molar ratio of X and Y groups defined such that $r \leqslant 1$) according to the expression

$$DP = (1 + r)/(1 + r - 2rp) \tag{4}$$

Achieving high molecular weights (DP $\sim$ 50, MW $\sim$ 5000) requires both high conversions and near stoichiometric amounts of X and Y groups. Table 1 shows the dependence of DP on r and p as calculated from Eq. 4 for

Table 1. Variation of DP with p and r

p	r	DP
0.995	1.000	200
0.990	1.000	100
0.990	0.995	80.1
0.990	0.990	66.8
0.990	0.980	50.0
0.980	1.000	50.0
0.980	0.995	44.5
0.980	0.990	40.0
0.980	0.980	33.4
0.970	1.000	33.0
0.970	0.990	28.7
0.970	0.970	22.7
0.950	1.000	20.0
0.900	1.000	10.0

selected values of r and p. Polymer synthesis is much more difficult than carrying out the corresponding small molecule reaction. Conversions such as 90 and 95%, considered outstanding for small molecule reactions, are of little value for polymer synthesis since DP values of only 10 and 20, respectively, are produced even when r = 1. (Some exceptions to this generalization are described below in the section on Physical Crosslinking.) Minimal conversions of 98% are typically required in polymerizations. The need for such high conversions dictates that only a small fraction of all known chemical reactions be used in polymer synthesis. Few reactions can be carried to these conversions. Achieving conversions of 98% and higher requires reactions generally devoid of side reactions. Reversible reactions must be capable of being driven to high conversion by displacement of the equilibrium. Polymerization must be performed in media in which the polymer does not precipitate prior to reaching the desired MW.

3

Reasonable reaction rates are needed to achieve the synthesis in a
reasonable time.

Step polymerizations proceed with second-order kinetics

$$1/[M] = 1/[M]_0 + kt \tag{5}$$

or $DP = 1 + [M]_0 kt$ (6)

where $[M]_0$ and $[M]$ are the concentrations of X(or Y) groups at times 0
and t, respectively. A characteristic of the second-order kinetics is
that it takes progressively longer and longer reaction times to achieve
each of the last few percent conversion. For example, it takes about as
long to go from 96% conversion to 98% conversion as it takes to reach
96% conversion from 0% conversion.

Molecular weight control at the desired level is achieved by the
simultaneous control of p and r. MW increases at any conversion as more
nearly stoichiometric amounts of X and Y groups are used.
Stoichiometric imbalance lowers DP since at some point in the reaction
all molecules in the polymerization system contain the same functional
group, e.g., Y if Y groups are in excess, and further increase in
molecular size does not occur since Y groups react only with X groups.
The stoichiometric ratio must be precisely controlled since a difference
of only a few tenths of a percent of excess of one reactant over the
other yields a significant difference in DP. Control of r requires the
use of high purity monomers. Reactant purity and stoichiometric balance
is much more critical in step polymerization compared to a small
molecule reaction. An impurity is generally carried along in the latter
and ends up as a minor impurity in the final product. The same level of
impurity in the polymerization can be disastrous -- yielding a lowered
DP which makes the polymer unsuitable for a specified application, i.e.,
0% yield of the desired product.

Chain Polymerization

Typical chain polymerizations are those of monomers containing the
carbon-carbon double bond, e.g., ethylene, isobutylene, isoprene,
styrene and acrylonitrile. Polymerization is initiated by radical,
cationic, anionic or Ziegler-Natta (coordination) initiators. All
monomers except 1-alkylethylenes, 1,1-dialkylethylenes and vinyl ethers
undergo radical polymerization. Ionic chain polymerizations are much
more selective than radical polymerizations. Cationic initiation is
limited to monomers containing electron-donating substituents, e.g.,
alkoxy, 1,1-dialkyl, phenyl and vinyl. Anionic initiation is limited to
monomers with electron-withdrawing substituents, e.g., CN, COOR, phenyl
and vinyl. 1-Alkylethylenes such as propylene are polymerized only by

4

Ziegler-Natta initiators. 1,2-Disubstituted ethylenes are not polymerized by any initiators due to steric hindrance although they can be copolymerized.

Radical Chain Polymerization. Radical chain polymerization involves initiation, propagation and termination. Consider the polymerization of ethylene. The most widely used method of initiation is the thermal homolysis of an initiator such as benzoyl peroxide

$$(\phi COO)_2 \longrightarrow 2 \ \phi COO\cdot \tag{7}$$

$$\phi COO\cdot + CH_2=CH_2 \longrightarrow \phi COOCH_2CH_2\cdot \tag{8}$$

Other initiators such as alkyl peroxides and hydroperoxides and azo compounds allow the generation of initiator radicals over a wide range of temperatures (50-200 ^{o}C). Redox systems (e.g., Fe^{2+} or N,N-dimethylaniline with a peroxide) extend this temperature range down to 0 ^{o}C and lower. Ultraviolet radiation, usually in the presence of an initiator or photosensitizer, is useful for coatings, imaging and printed cicuit board applications. Other techniques include ionizing radiation and electroinitiation.

Propagation of the radical center proceeds by the successive additions of large numbers of monomer molecules

$$\sim\sim\sim CH_2CH_2\cdot + CH_2=CH_2 \longrightarrow \sim\sim\sim CH_2CH_2CH_2CH_2\cdot \tag{9}$$

Termination occurs when two propagating radicals combine

$$2 \ \sim\sim\sim CH_2CH_2\cdot \longrightarrow \sim\sim\sim CH_2CH_2CH_2CH_2\sim\sim\sim \tag{10}$$

Disproportionation between propagating radicals also occurs but to a much lesser extent.

Propagation is favored over termination even though termination rate constants are larger than propagation rate constants (10^6-10^8 L/mol-s vs. 10^2-10^4 L/mol-s) because monomer concentrations are much larger than radical concentrations (0.1-10 M vs. 10^{-7}-10^{-9} M). Many hundreds and even thousands of monomer molecules add to a propagating radical in times of 10^{-1}-10 s. The achievement of high MW does not require high conversions. High MW polymer is produced almost immediately after the start of reaction in a chain polymerization, in contrast to step polymerization, and continues throughout the complete conversion range.

The polymerization rate R_p and degree of polymerization are given by

$$R_p = k_p[M](R_i/2k_t)^{1/2} \tag{11}$$

$$DP = 2k_p[M]/(k_tR_i)^{1/2} \tag{12}$$

where R_i is the rate of initiation. Successful application of radical polymerization requires that the initiator and initiator concentration be chosen to give the appropriate initiation rate to achieve the desired

5

DP at the desired polymerization termperature. One must also realize that higher polymerization rates achieved by higher initiation rates come at the expense of lower polymer molecular weights.

Polymer molecular weight is often lower than that described by Eq. 12 because of chain transfer (radical displacement) reactions of the type

$$\sim\sim CH_2CH_2\cdot \ + \ XA \ \longrightarrow \ \sim\sim CH_2CH_2X \ + \ A\cdot \tag{13}$$

where XA may be monomer, initiator, solvent or any other substance present in the reaction system. Chain transfer decreases the polymer DP by prematurely terminating growth of the propagating radical although the polymerization rate is usually not affected since A· is sufficiently reactive to reinitiate polymerization. Control of the polymer MW requires that one control the identities of the components of the reaction system to avoid chain transfer reactions which would decrease MW below the desired level. On the other hand, the deliberate addition of a strong chain transfer agent allows one to decrease MW to the desired level when the MW in a polymerization system is too high.

Radical polymerizations are performed in bulk, solution, suspension and emulsion. The reaction characteristics of emulsion polymerization are different than those of the other techniques. Radicals produced in an aqueous phase diffuse into colloidal particles (micelles) where propagation takes place with an alternating on-off mechanism. Immediate termination occurs whenever a radical enters a micelle containing a propagating radical since the micelle size is such that the presence of two radicals corresponds to an exceptionally high molar radical concentration. Successive radicals entering a micelle alternately terminate and re-initiate polymerization. The rate and degree of polymerization are given by

$$R_p = Nk_p[M]/2 \tag{14}$$

$$DP = Nk_p[M]/R_i \tag{15}$$

where N is the steady-state concentration of micelles. Emulsion polymerization has the unique feature that both R_p and DP can be increased by increasing N. For radical polymerizations in bulk, solution and suspension, increasing R_p by altering a reaction variable such as R_i or temperature almost always decreases DP. Emulsion polymerization is especially useful for low reactivity monomers such as 1,3-dienes. Bulk, solution and suspension polymerization typically yield low rates and low molecular weights. High R_p and DP are achieved in emulsion polymerization by using high concentrations of micelles.

Ionic Chain Polymerization. Ionic chain polymerizations take place at relatively low or moderate temperatures and in solvating media so that the ionic centers propagate to polymeric size prior to termination. Only solvents of low or moderate polarity, e.g., alkanes, chlorinated hydrocarbons, toluene, nitrobenzene and tetrahydrofuran, are employed. Highly polar solvents such as alcohols or ketones cannot be used since they inactivate ionic initiators and propagating centers by reaction or strong complexation.

Lewis acids such as $AlCl_3$ or BF_3 together with small concentrations of water or other proton source are most often used to initiate cationic chain polymerization. The two components of the initiating system form an initiator-coinitiator complex which donates a proton to monomer

$$AlCl_3 + H_2O \longrightarrow AlCl_3 \cdot H_2O \tag{16}$$

$$AlCl_3\ H_2O + (CH_3)_2C=CH_2 \longrightarrow (CH_3)_3C^+(AlCl_3OH^-) \tag{17}$$

Propagation proceeds by successive additions of monomer molecules to the carbenium ion center

$$\sim\!\!\sim\!\!\sim CH_2C^+(CH_3)_2(AlCl_3OH^-) + (CH_3)_2C=CH_2 \longrightarrow$$

$$\sim\!\!\sim\!\!\sim CH_2C^+(CH_3)_2CH_2C(CH_3)_2(AlCl_3OH^-) \tag{18}$$

Protonic acids such as hydrocholoric and sulfuric acids are far less suitable as initiators. Their conjugate bases, being strong nucleophiles, react rapidly with the carbenium ion center and prevent propagation. The advantage of the Lewis acid system is the low nucleophilicity of the anion ($AlCl_3OH^-$ in the example above).

The most important termination for many polymerizations is chain transfer to monomer in which a proton along with the counter-anion $AlCl_3OH^-$ is expelled from the propagating center with the formation of terminal unsaturation in the polymer

$$\sim\!\!\sim CH_2C^+(CH_3)_2(AlCl_3OH^-) \longrightarrow \sim\!\!\sim CH=C(CH_3)_2 + H^+(AlCl_3OH^-) \tag{19}$$

Chain transfer of the propagating carbenium ion to a negative fragment from the counter-ion (e.g., transfer of OH^- or Cl^- from $AlCl_3OH^-$) or solvent or some other component (e.g., water) of the reaction system are also important.

Anionic chain polymerization can be initiated by metal alkoxides, aryls and alkyls and electron-transfer from sodium naphthalene. Alkyllithiums are among the most useful, being employed commercially in the polymerization of 1,3-butadiene, isoprene, and styrene. Initiation involves addition of alkyl anion to monomer

$$R^-(Li^+) + CH_2=CH\phi \longrightarrow RCH_2-\bar{C}H\phi(Li^+) \tag{20}$$

The anionic chain polymerizations of polar monomers, such as methyl methacrylate, methyl vinyl ketone and acrylonitrile, often yield complex polymer structures due to nucleophilic reactions of the carbonyl and nitrile groups.

For many anionic polymerizations of nonpolar monomers such as styrene, isoprene and 1,3-butadiene, there are no effective termination reactions if moisture, oxygen and carbon dioxide are absent. Propagation proceeds with complete consumption of monomer and the propagating anionic centers remain intact as long as one employs solvents such as benzene, n-hexane and tetrahydrofuran which are inactive in transferring a proton to the propagating anion. These polymerizations, referred to as living polymerizations, are terminated when desired by the deliberate addition of a proton source such as water or alcohol.

Both cationic and anionic chain polymerization are very sensitive to changes in the reaction medium due to changes in the nature of the propagating centers. In the low to moderate polarity solvents used, two types of propagating species coexist -- the free ion and ion-pair. The ion-pair consists of the propagating center and its tightly held counter-ion. The free ion consists of the propagating center separated from the counter-ion by solvent. The ion-pair and free ion propagate concurrently and in equilibrium with each other. The ion-pair is the more plentiful species but the free ion is much more reactive, often by as much as three orders of magnitude. Changing the reaction solvent to a more polar solvent (e.g., from n-hexane to THF) results in large increases in both rate and degree of polymerization by increasing the concentration of free ion. The identity of the counter-ion also affects polymerization. Less tightly-held counter-ions yield more reactive ion-pairs.

Copolymerization. Chain copolymerization, the polymerization of a mixture of two monomers, yields a copolymer with two different repeat units distributed along the polymer chain

$$M_1 + M_2 \longrightarrow \sim\!\!\sim\!\!\sim M_1M_2M_2M_2M_1M_1M_2M_1M_1M_1M_1 \sim\!\!\sim\!\!\sim \tag{21}$$

Copolymerization has practical utility for changing the properties of a homopolymer in a desired direction. A number of commercially-important elastomers are copolymers. Butyl rubber is a copolymer of isobutylene with 1-2% isoprene. The isoprene units in the copolymer allow it to be crosslinked. Although polystyrene is far too rigid to be elastomeric, styrene-1,3-butadiene copolymers (SBR) are useful as elastomers. Polyethylene is a semi-crystalline plastic while ethylene-propylene copolymers and terpolymers of ethylene, propylene and a diene (e.g.,

8

hexa-1,4-diene, dicyclopentadiene, 2-ethylidenenorborn-5-ene) are elastomers (EPR and EPDM rubbers). Nitrile or NBR rubber is a copolymer of acrylonitrile and 1,3-butadiene. Vinylidene fluoride-chlorotrifluoroethylene, olefin-acrylic ester copolymers and 1,3-butadiene-styrene-vinyl pyridine terpolymer are examples of specialty elastomers.

The properties of a copolymer are dependent on the identities of the monomers and their relative proportions within the copolymer chain. The copolymer composition is determined by the feed composition and the relative reactivities of the monomers undergoing copolymerization. The latter is determined by the competition among four propagation reactions. There are two types of propagating centers -- those ending in M_1 and those ending in M_2 -- and each type of propagating center can react with either of two monomers, M_1 or M_2. The copolymer composition is usually different from the comonomer feed compostion, the difference depending on the interplay of the four propagation reactions.

Structural and Stereo-Isomerism. Stereo-isomerism is possible in the polymerization of a monosubstituted alkene such as propylene. Every other carbon in the polymer chain is a chiral center and the substituent on each chiral center can have either of two configurations. Two ordered or steroregular polmers are possible -- isotactic and syndiotactic -- where the substituent groups on successive chiral carbons have the same or opposite configurations, respectively. The unordered or atactic structure has a random distribution of equal numbers of the two configurations.

Both structural and stereo-isomerism are possible in the polymerization of conjugated dienes such as 1,3-butadiene. Polymerization can proceed by 1,2- and 1,4-reaction. Isotactic, syndiotactic and atactic polymers are possible for 1,2-polymerization analogous to the situation for a monosubstituted alkene. 1,4-Polymerization yields products in which the repeat units can be either cis(I) or trans (II)

     I   II

The polymer produced in a polymerization is highly dependent on the specific choice of reaction conditions (initiator, solvent, counter-ion and temperature). 1,4-Polymerization dominates over 1,2-polymerization and trans-1,4-polymerization over cis-1,4-polymerization for radical polymerization of conjugated dienes. These preferences are stronger at lower reaction temperatures. Higher temperatures yield more random placement of successive monomer units in the polymer chain. Cationic polymerizations proceed similarly but are not generally useful (except for Butyl rubber) because the products are usually low MW and extensively cyclized. 1,2-Polymerization is favored over 1,4-polymerization in anionic polymerizations in polar solvents (where the counter-ion is weakly coordinated with the propagating center). The Symposium paper by Bywater and Worsfold describes the large increase in the extent of 1,2-polymerization achieved upon the addition of a polar material to the reaction system.

The trends noted above are not exceptionally strong. Dramatically different results are observed for anionic polymerizations under conditions wherein there is strong coordination among the propagating center, counter-ion and monomer. Strong coordination results in very greatly increased preferences -- some in the same direction as for polymerization under non-coordination conditions and some in the opposite directions. The most remarkable results are obtained with the Ziegler-Natta catalysts obtained by combining a Group I-III metal derivative (e.g., AlR_3, AlR_2Cl, MgR_2) with a transition metal derivative (e.g., $TiCl_3$, $TiCl_4$, VCl_3). Polymerizations initiated by these catalysts yield polymers with high degrees of structural purity. Judicious choice of the components of the catalyst system allow one to synthesize either the cis-1,4-polymer or the trans-1,4-polymer or the 1,2-polymer each in high purity (>95% pure). For example, trans-1,4-polybutadiene is obtained using $TiCl_4/AlR_3$ or Co chelates/AlR_3, and 1,2-polybutadiene using $Ti(OR)_4/AlR_3$ or V(acetylacetonate)$_3$/AlR_3. The identity of the counter-ion determines the specific coordination among counter-ion, propagating center, and monomer which in turn determines how successive monomer molecules are allowed to enter the polymer chain in these coordination polymerizations.

The ability to synthesize each of the isomeric polymers in high purity is important since the different polymers have different properties. For example, cis-1-4-polyisoprene is an excellent elastomer over a large temperature range due to its very low degree of crystallinity and low glass transition and melting temperatures. About two billion pounds are used annually in the United State for such typical applications as tires, coated fabrics, molded objects, adhesives and rubber bands. Trans-1,4-polyisoprene is a much harder and less rubbery elastomer since it crystallizes to an appreciable extent and has higher glass transition and melting temperatures. Small amounts

10

(compared to the cis-isomer) are used in golf balls and electrical cable covering. Both cis- and trans-1,4-polyisoprenes are found in nature; the cis-isomer is the much more abundant. Both isomers have also been produced commercially using coordination polymerization processes. 1,4-Polybutadienes with high cis-1,4 and mixed cis-1,4/trans-1,4 contents are produced commercially using Ziegler-Natta or lithium catalysts. The high cis elastomer crystallizes on stretching whereas the mixed cis/trans elastomer shows no tendency to crystallize. The high cis elastomer has higher strength but poorer low temperature properties compared to the mixed cis/trans elastomer.

Coordination polymerization also produces high sterospecificity in the polymerization of alkenes. Isotactic and syndiotactic polymers can be obtained by appropriate choice of the catalyst components although such polymers are not useful as elastomers. However, Ziegler-Natta catalysts are used to produce EPR and EPDM rubbers. (Coordination polymerization is important for the synthesis of linear polyethylene and isotactic polypropylene which find extensive utility as plastics.) The Symposium paper by Su and Shih describes the synthesis of propylene-1-hexene block copolymers using several catalysts based on titanium and aluminum components.

Ring-Opening Polymerization

Ring-opening polymerization of cyclic monomers is another route to elastomers. These include the anionic polymerization of cyclotetrasiloxanes (Eq. 22), the polymerization of cyclopentene with

$$\left[\begin{array}{c} CH_3 \\ | \\ Si-O \\ | \\ CH_3 \end{array}\right]_n \tag{22}$$

$MoCl_5$, WCl_6 and vanadium catalysts (Eq. 23), and the thermal polymerization of hexachlorocyclotriphosphazene (Eq. 24).

$$\longrightarrow (CH_2CH=CHCH_2CH_2)_n \tag{23}$$

$$\longrightarrow (N=PCl_2)_n \tag{24}$$

CROSSLINKING

Crosslinking (also referred to as vulcanization or curing) is required for an elastomer to rapidly and completely recover from deformations. Crosslinking can be accomplished either by chemical reaction or physical means depending on the elastomer.

Chemical Crosslinking

The production of an elastomeric product such as an automobile tire is a two-step process. A linear polymer is obtained from nature or is synthesized by appropriate choice of reactant(s) and reaction conditions so that crosslinking cannot occur. The linear polymer is mixed with the reactant(s) which subsequently induce crosslinking, the mixture shaped into the desired product and then heated in a mold. Crosslinking takes place during the heating step.

The 1,4-polymers of 1,3-butadiene and isoprene and many of their copolymers (Butyl, SBR, NBR) and EPDM rubbers are usually vulcanized by heating with sulfur. Crosslinking involves both sulfur addition to double bonds and allylic substitution

$$\sim CH_2CH=CHCH_2 \sim \quad \xrightarrow{\text{Sulfur}} \quad \begin{array}{c} \sim CH_2\underset{|}{C}HCH_2CH_2 \sim \\ S_m \\ \sim \underset{|}{C}HCH=CHCH_2 \sim \end{array} \qquad (25)$$

The efficiency of crosslinking is increased by including accelerators (e.g., tetraalkylthiuram disulfide, zinc dialkyldithiocarbamate) and activators (e.g., ZnO plus stearic acid) which inhibit the formation of long polysulfide crosslinks, vicinal crosslinks and intramolecular cyclic sulfide structures. 1,4-Poly-1,3-dienes are also crosslinked by heating with p-nitrosobenzene, phenolic resins or maleimides for applications requiring improved thermal stability.

Polychloroprene, not efficiently crosslinked with sulfur since Cl deactivates the double bond, is vulcanized by heating with ZnO or MgO. Crosslinking occurs by Cl displacement with formation of ether and/or -OZnO- or -OMgO- crosslinks.

Ethylene-propylene and silicone rubbers are crosslinked by compounding with a peroxide such as dicumyl or di-t-butyl peroxide and then heating the mixture. Peroxy radicals abstract hydrogens from the polymer to form polymer radicals. Crosslinks form by coupling of the polymer radicals.

Elastomeric copolymers of vinylidene fluoride are crosslinked by heating with diamine and basic oxide. Crosslinking involves dehydrofluorination followed by addition of the diamine with the metal oxide acting as an acid neutralizer.

The synthesis of polysulfide elastomers involves the presence of a small amount of trichloroalkane with the dichloroalkane to form low molecular weight branched prepolymer. This branched polymer is reacted with sodium hydrosulfide and sodium sulfite followed by acidification to convert all end-groups to thiol groups. Continued polymerization to increase MW and crosslinking are accomplished by oxidative coupling of thiol end-groups with lead dioxide, p-quinone dioxime or other oxidizing agent

$$
HS\text{~}\text{}\text{~}SH \quad \underset{}{\overset{PbO_2}{\longrightarrow}} \qquad \begin{array}{c} \text{~}S\text{-}S\text{~}\text{~}S\text{-}S\text{~} \\ | \\ S \\ | \\ S \\ | \\ \text{~}S\text{-}S\text{~}\text{~}S\text{-}S\text{~} \end{array} \tag{26}
$$

Polyurethanes are crosslinked in different ways depending on the choice and stoichiometry of reactants and reaction conditions. For example, an isocyanate-terminated trifunctional prepolymer is synthesized from a polyol and diisocyanate. The polyol may be a polyester prepolymer synthesized from a triol or a hydroxyl-terminated polyether. Further polymerization and crosslinking of the isocyanate-terminated prepolymer is achieved by reaction with a diamine

$$
\begin{array}{c}
OCN\text{~}\text{}\text{~}NCO \\
NCO \\
+ \\
H_2N\text{-}R\text{-}NH_2
\end{array}
\quad \longrightarrow \quad
\begin{array}{c}
\text{~}NHCONHRNHCONH\text{~}\text{~}NHCONHRNHCONH\text{~} \\
| \\
NH \\
| \\
CO \\
| \\
NH \\
| \\
R \\
| \\
NH \\
| \\
CO \\
| \\
NH \\
\text{~}NHCONHRNHCONH\text{~}\text{~}NHCONHRNHCONH\text{~}
\end{array}
\tag{27}
$$

Crosslinking can also be achieved through the formation of urethane linkages by using a diol instead of a diamine.

Physical Crosslinking

Crosslinking occurs by a physical process in block copolymers which possess a microheterogeneous, two-phase network morphology. The two important types of block copolymers are the triblock, ABA, and multiblock (AB)$_n$ copolymers (III and IV, respectively) where A and B represent blocks of two different repeat units. The A blocks are hard

$$
\underline{}\boxed{A}\underline{}\text{---}B\text{---}\boxed{A}\underline{} \qquad III
$$

$$
\text{---}\boxed{A}\text{---}B\text{---}\boxed{A}\text{---}B\text{---}\boxed{A}\text{---}B\text{---}\boxed{A}\text{---}B\text{---} \qquad IV
$$

(e.g., polystyrene) and short while the B blocks are flexible (e.g., 1,4-polyisoprene) and long. Such block copolymers, referred to as thermoplastic elastomers, behave as elastomers at ambient temperatures but are thermoplastic at elevated temperatures where fabrication is accomplished.

The dual behavior of thermoplastic elastomers results from their microheterogeneous, two-phase morphology. The hard A blocks from different polymer chains aggregate to form rigid domains at ambient temperatures. The A blocks are hard if they are either crystalline or glassy. The rigid domains comprise a minor, discontinuous phase dispersed in the major, continuous phase composed of the rubbery B blocks from various polymer chains. The rigid domains act as "physical crosslinks" to hold together the soft, rubbery B blocks in a network structure. Physical crosslinking is thermally-reversible since heating above the crystalline melting or glass transition temperature of the A blocks softens the rigid domains and the polymer flows. Cooling reestablishes the rigid domains and the polymer again behaves as a crosslinked elastomer.

Thermoplastic elastomers have the important advantage over conventional elastomers that there is no need for the additional chemical crosslinking reaction and fabrication is achieved in the same way as for thermoplastics. However, only certain polymerization methods can be used to synthesis block copolymers -- primarily living anionic chain polymerization and certain step polymerizations. Triblock copolymers are produced by living anionic polymerization by sequential addition of different momomer charges to a living anionic system, for example, a styrene-isoprene-styrene is synthesized by the sequence

$$S \xrightarrow{RLi} RS_x^- \xrightarrow{D} RS_xD_y^- \xrightarrow{S} RS_xD_yS_z^- \xrightarrow{ROH} RS_xD_yS_zH \qquad (28)$$

Styrene (S) is polymerized to polystyryl anions using RLi and the diene (D) added. A second styrene charge is added after the diene is consumed. Finally a terminating agent (ROH) is added. The lengths of each of the three blocks are controlled by the ratios of the concentrations of the three monomer charges to initiator concentration. Coupling reactions have also been used to synthesize triblock copolymers, e.g., living polystyrene-polyisoprene anions are coupled with a dibromo compound

$$RS_mD_n^- \xrightarrow{BrR'Br} RS_mD_nR'D_nS_mR \qquad (29)$$

Step polymerization can be adopted to the synthesis of multiblock copolymers. An example is the polyether-polyurethane system produced by the reaction of a diisocyanate with a mixture of macro diol and

$$nHO(R-O)_bR-OH + naHO-R'-OH + (n+na)OCN-R''-NCO \longrightarrow$$

$$[O(R-O)_bR-O(CONH-R''-NHCO-O-R'-O)_aCONH-R''-NHCO]_n \qquad (30)$$

small-sized diol (Eq. 30). The macro diol, referred to as diol prepolymer, is synthesized by the ring-opening polymerization of a cyclic ether such as tetrahydrofuran or propylene oxide under conditions which yield hydroxyl groups at both chain ends. The length of the polyether blocks (the value of b) is determined by the monomer:initiator ratio in the polyether synthesis. The length of the polyurethane blocks (the value of a) is determined by the relative amounts of diol prepolymer and small sized diol. The polyether and polyurethane blocks function as the soft and hard segments, respectively, of the thermoplastic elastomer. A variation of this multiblock copolymer is the polyester-polyurethane system in which the diol prepolymer is produced by polyesterification.

The growing importance of thermoplastic elastomers is clearly evident from the papers presented at this Symposium. The majority of papers on elastomer synthesis deal with these types of block copolymers. The papers by Quirk and Tung describe the triblock copolymers based on p-methylstyrene-butadiene and styrene-α-methylstyrene-diene systems, respectively. Anionic block copolymerization of hexamethyl- and hexaphenylcyclotrisiloxane is discussed in the paper (not included in this book) by Ibemesi, Gvozdic, Keumin, Lynch and Meier.

REFERENCES

1. H. R. Allcock and F.W. Lampe, "Contemporary Polymer Chemistry," Prentice-Hall, Englewood Cliffs, 1981.
2. G. Alliger and I.J.C Sjothun, "Vulcanization of Elastomers," Reinhold, New York, 1964.
3. F. W. Billmeyer, Jr., "Textbook of Polymer Science," 3rd Edition, Wiley, New York, 1984.
4. J. Boor, Jr., "Ziegler-Natta Catalysts and Polymerization," Academic New York, 1979.
5. A. Y. Coran, "Vulcanization," in F. Eirich, ed., "Science and Technology of Rubber," Academic, New York, 1978, Chapt. 7.
6. H.-G. Elias, "Macromolecules," Vols. 1 and 2, Plenum, New York 1977
7. K. J. Ivin and T. Saegusa, eds., "Ring-Opening Polymerization," Elsevier, London, 1984.
8. J. P. Kennedy and E. G. M. Tornqvist, eds., "Polymer Chemistry of Synthetic Elastomers," Wiley, New York, 1968.
9. J. P. Kennedy and E. Marichal, "Carbocationic Polymerization," Wiley, New York, 1983.
10. R. W.. Lenz, "Organic Chemistry of Synthetic High Polymers," Wiley, New York, 1967.

11. H. F. Mark, J. Appl. Polym. Sci.: Appl. Polym. Symp. 39, 1 (1984).

12. M. Morton, "Anionic Polymerization: Principles and Practice," Academic, New York 1983.

13. G. Odian, "Principles of Polymerization," 2nd ed., Wiley, New York, 1981.

14. G. Odian, "Basic Concepts in Elastomer Synthesis," in J. E. Mark and J. Lal, eds., "Elastomers and Rubber Elasticity," ACS Symposium Series 193, American Chemical Society, Washington, D. C., 1982, Chapt. 1.

15. F. Rodriquez, "Principles of Polymer System," 2nd ed., McGraw-Hill, New York, 1982.

16. C.E. Schildknecht and I. Skeist, eds., "Polymerization Processes," Vol. 29 of "High Polymers," Wiley, New York, 1977.

17. R. B. Seymour and C.E. Carraher, Jr., "Polymer Chemistry," Dekker, New York, 1981.

18. D. H. Solomon, ed., "Step-growth Polymerizations," Dekker, New York 1972.

STRUCTURE AND PROPERTIES OF TIRE RUBBERS PREPARED BY ANIONIC POLYMERIZATION

S. L. Aggarwal, I. G. Hargis, R. A. Livigni, H. J. Fabris,
and L. F. Marker

GenCorp
Research Division
2990 Gilchrist Road
Akron, Ohio 44305

ABSTRACT

The two functional properties of tire tread rubbers of prime
importance today are low rolling resistance and high wet traction during
braking. Rolling resistance and wet traction are related to low tan δ and
Tg, respectively. In most of the conventional synthetic rubbers, high Tg
is accompanied with high tan δ. Studies on polybutadiene rubbers prepared
with anionic catalysts showed that the rubbers containing a high percent of
vinyl structure provide exceptions to this general rule. This work demonstrates that vinyl groups in vinyl BR's do not contribute significantly to
tan δ in the frequency/temperature regime corresponding to the rubber
plateau region. Guided by our results, a number of experimental rubber
compositions were developed for tire treads. These compositions have the
desirable combination of high traction and low rolling resistance.

INTRODUCTION

The rapid development of polymerization technology which occurred
after World War II led to a profusion of new synthetic rubbers. Several
of these synthetic rubbers or blends of these have achieved technological
importance for use in tires. Strategic questions for tire rubber research
to address during recent years have been: (1) Should we devote resources
to develop continually new synthetic rubbers, and (2) on what type of new
rubbers should we focus our efforts?

The approach that has been used in our laboratories for synthetic
rubber research for improved tire tread composition is shown in Figure 1.
The first step is to delineate that combination of functional properties,
or use properties of technological interest, that is not satisfied by the
presently available rubbers. In recent years, because of the interest in
fuel savings, technological interest developed in tire rubbers that show
the lowest dissipation of energy, that is, lowest rolling resistance,
highest traction during braking, and highest wear resistance. Several
studies have recently been reported on this topic.[1-3]

The second step in our approach is to relate such functional properties

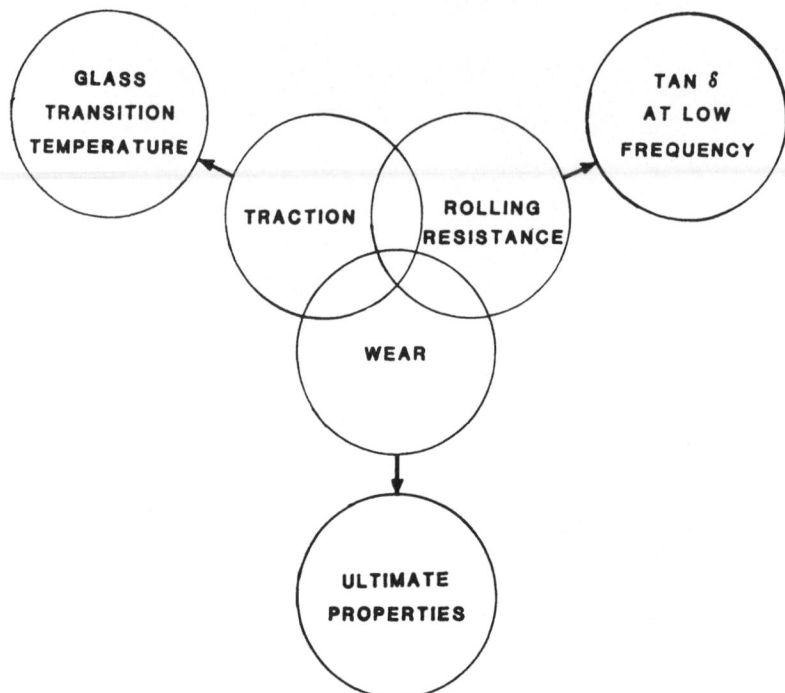

Fig. 1. Our approach for synthetic rubber research.

to fundamental properties; and then to devise and synthesize polymers of controlled structures. Thus, our approach has been to work back from the technologically desired combination of properties to fundamental properties and to relate these fundamental properties to the desired rubber structure.

The ultimate properties, such as tear strength and tensile strength, relate to wear. The relationship of the ultimate properties to the molecular structure of rubber vulcanizates has been studied and reported by several workers over the years.[4-6]

In this presentation we will discuss the research in our laboratory that led to tire rubbers that provide the combination of lowest rolling resistance and highest sliding traction. We concluded, from considerations that will be explained later, that rolling resistance, i.e., fractional energy dissipated in a rolling tire, is predominately related to loss tangent, i.e., tan δ at comparatively low frequency and at appropriate temperatures above the glass transition temperature. The sliding friction or traction is predominately dependent on the glass transition temperature of the rubber.

The deformation versus time of a footprint section of a typical passenger tire rolling at about 80 km/hr is shown schematically in Figure 2. The footprint of such a tire is about 1/10th of the circumference of the tire. The rotational frequency corresponding to 80 km/hr speed is ~ 12 Hz. Thus, during contact, the tread rubber will be deformed at an effective frequency equal to the rolling frequency times the ratio of the circumference of the tire to the footprint length, i.e., about 100-115 Hz. Thus the principal deformation frequencies of the tread rubber are in the range of 100-115 Hz. The deformation of a tire tread is a complex process.

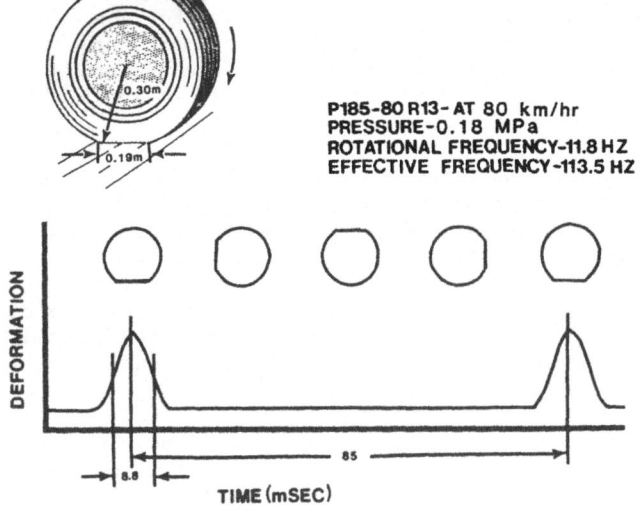

P185-80 R13- AT 80 km/hr
PRESSURE-0.18 MPa
ROTATIONAL FREQUENCY-11.8 HZ
EFFECTIVE FREQUENCY-113.5 HZ

Fig. 2.   Deformation of a tire during rolling.

DEFORMATION CAUSED BY FINE
TEXTURE OF ROAD ASPERITIES

EFFECTIVE FREQUENCY $\approx 10^5-10^6$
HZ FOR WET SURFACE

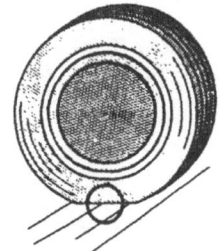

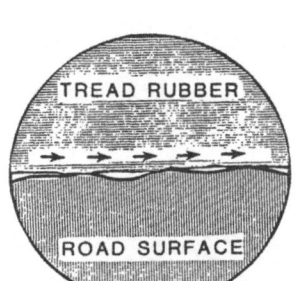

Fig. 3.   Deformation of a tire during sliding.

19

The energy dissipated per revolution is perhaps an indeterminate function of loss compliance, loss modulus, and loss tangent. For our studies, tan δ determined at 40 Hz and at temperature of 60°C (a temperature fairly far above the Tg of rubbers of interest) proved to be the significant material property that relates to energy dissipated per cycle in a rolling tire, i.e., the tire's rolling resistance.

In contrast, the sliding friction, i.e., traction on a wet pavement during braking, involves a high frequency deformation of the tire tread. The deformation, as illustrated in Figure 3, is caused by the fine texture of asperities in the road surface. It is estimated to be of the order of 10^5-10^6 Hz.[2] The viscoelastic properties of rubbers at such high frequencies are difficult to measure. For this purpose, tan δ measured at lower frequencies than 10^5-10^6 Hz, and temperatures in the proximity of the glass transition temperature of rubbers, has been found to be a useful material parameter characteristic of a rubber compound.[7]

A typical loss spectrum versus frequency for a rubber is shown in Figure 4. While the general shape of this curve is characteristic for an elastomeric network, the relative magnitude of the energy loss per deformation cycle at any given temperature can change with changes in the rubber structure as well as the type, level, and distribution of compounding ingredients. The tan δ determined in the vicinity of 100 Hz and at 60°C is an appropriate material property that relates to rolling resistance. The glass transition temperature, Tg, of the rubber can be considered an appropriate material property related to wet traction.

STATE OF OUR KNOWLEDGE ON RELATIONS BETWEEN: (1) TAN δ AND Tg OF CONVENTIONAL TIRE RUBBERS; AND (2) MOLECULAR STRUCTURE AND FUNDAMENTAL MATERIAL PROPERTIES

In Figure 5, tan δ measured at 60°C and 40 Hz on vulcanized unfilled, conventional tire rubbers is plotted versus Tg of the corresponding

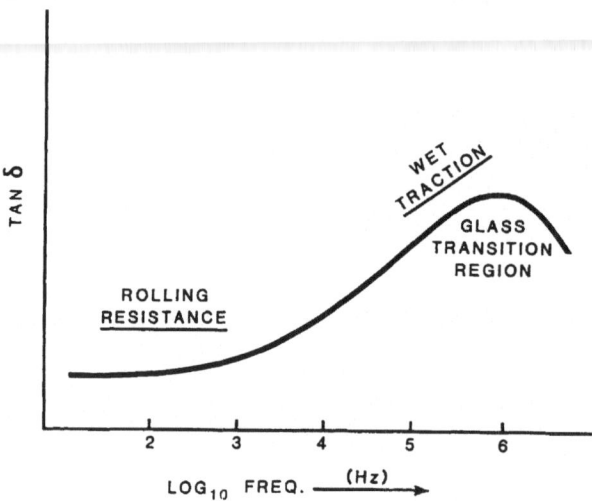

Fig. 4. Relationship between rolling resistance and wet traction to the viscoelastic loss spectrum at 60°C.

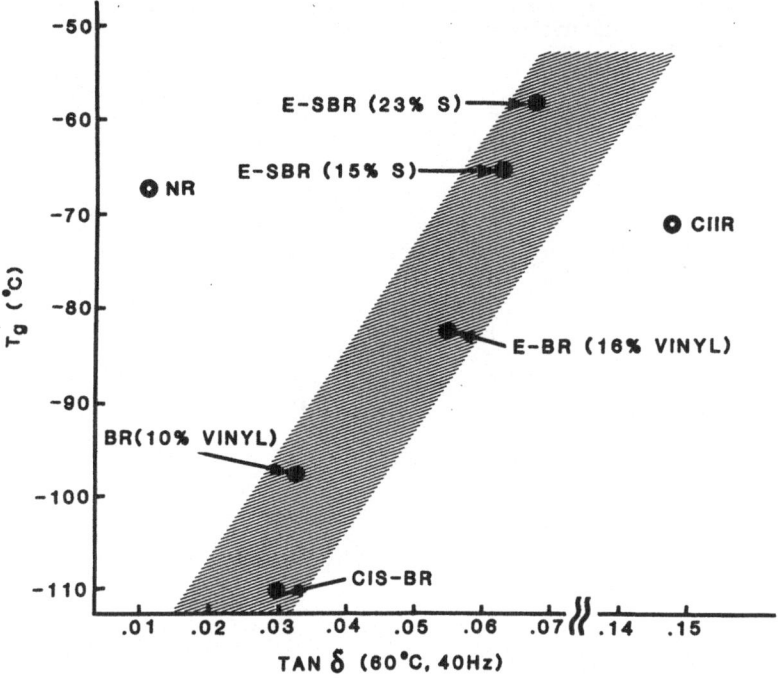

Fig. 5. Tg and tan δ for conventional tread rubber gum
vulcanizates.

unvulcanized rubbers. The butadiene-based rubbers fall within a narrow
band with the characteristic feature that higher Tg is associated with
high tan δ values. Thus, for this class of rubbers, higher traction is
accompanied by higher rolling resistance. Exceptions to this trend are
NR, which in spite of its relatively high Tg compared with polybutadienes
has excellent rolling efficiency but inadequate wet traction, and chloro-
butyl rubber, which at nearly the same Tg as NR shows outstanding traction
but excessive energy loss per cycle under the conditions of a rolling tire.
What the rubber technologist desires is to have vulcanizates that give
responses which fall in the upper left-hand corner of this plot.

A summary of the influence of molecular structural features of styrene/
butadiene rubbers on Tg and loss tangent at low frequencies in the rubbery
plateau region of the viscoelastic spectrum is shown in Table 1. Polymer
composition and microstructure of the butadiene units strongly influence Tg,
while monomer sequence distribution has a somewhat lesser influence. The
effect of molecular weight distribution, crosslink density, and molecular
architecture on Tg is comparatively minor. In contrast, molecular weight,
molecular weight distribution, crosslink density, and chain architecture
strongly influence loss tangent in the rubbery plateau region.

Little information could be located in the literature, as indicated by
the question mark entries in this Table, of the effect of composition, butadiene microstructure, and monomer sequence distribution on tan δ of rubber
networks in the appropriate frequency range. This prompted us to prepare
rubbers of controlled structure that incorporate these features, and to
study their fundamental and functional properties.

Table 1. Molecular Features of Rubbers Derived From
Butadiene and Styrene Related to T_g and Tan δ

MOLECULAR STRUCTURAL FEATURES	T_g	LOSS TANGENT TEMP. >> T_g FREQUENCY-LOW
COMPOSITION	✔	?
CHAIN MICROSTRUCTURE % VINYL, CIS AND TRANS	✔	?
MONOMER SEQUENCE DISTRIBUTION	✔	?
MOLECULAR WEIGHT AND MOLECULAR WEIGHT DISTRIBUTION	✔	✔
CROSSLINK DENSITY	✔	✔
MOLECULAR CHAIN ARCHITECTURE (LINEAR OR STAR)	✔	✔

Table 2. Structural Control in Styrene/Butadiene Solution
Rubbers Using Anionic Catalyst Systems

- COMPOSITION

- MONOMER SEQUENCE DISTRIBUTION (RANDOM, TAPERED, BLOCK)

- MICROSTRUCTURE VARIATION IN POLYBUTADIENE MICROSTRUCTURE (VINYL, CIS AND TRANS)

- MOLECULAR WEIGHT AND MOLECULAR WEIGHT DISTRIBUTION

- MOLECULAR CHAIN ARCHITECTURE (LINEAR OR STAR)

PREPARATION OF STYRENE/BUTADIENE RUBBERS OF CONTROLLED STRUCTURE

The solution polymerization system with anionic catalysts is best
suited for the preparation of rubbers of controlled structure.[8] The
well-known structural control that is attainable in styrene/butadiene
rubbers in such systems is given in Table 2. Of particular interest is

Table 3. Control of Structural Features in Solution
Rubbers by Anionic Polymerization

CHARACTERISTIC OF RUBBER	POLYMERIZATION VARIABLES FOR CONTROL
COPOLYMER COMPOSITION AND COMONOMER SEQUENCE DISTRIBUTION FOR STYRENE/BUTADIENE	● TMEDA/n-BuLi; LOW TO MODERATE TEMP. ● Na t-AMYLATE/n-BuLi; LOW TO MODERATE TEMP. ● RATE OF COMONOMER ADDITION AT HIGH TEMPERATURE
BUTADIENE MICROSTRUCTURE:	
VINYL/(TRANS + CIS)	TMEDA/n-BuLi; LOW TO MODERATE TEMP.
TRANS/CIS AT CONSTANT VINYL	$Ba(t\text{-BUTOXIDE})_2/Mg(BUTYL)_2/Al(ETHYL)_3$
MOLECULAR WEIGHT	$[\text{MONOMER WEIGHT}] / [\text{n-BuLi}]$; EXTENT OF CONVERSION
MOLECULAR WEIGHT DISTRIBUTION AND CHAIN ARCHITECTURE (LINEAR OR STAR)	COUPLING OF ACTIVE POLYMER CHAINS WITH COMPOUNDS, SUCH AS TIN TETRACHLORIDE $(SnCl_4)$

the structural control that can be obtained with respect to composition, monomer sequence distribution, and microstructure of the butadiene units.

EXPERIMENTAL

Materials

Butadiene (El Paso Products Company, 99% purity) was distilled from dimer and inhibitor and condensed into a container under pressure of dry nitrogen.

Styrene (Gulf Oil Company, 99% purity) was purged with dry nitrogen for at least 30 minutes before use.

n-Hexane (Phillips Chemical Company, 85% minimum purity) was dried using Linde 5A molecular sieves.

n-Butyllithium (Foote Mineral Company, 15% by weight in n-hexane) was further diluted, as needed, using purified n-hexane.

Barium di-tert-butoxide-(butyl)(ethyl)magnesium-triethylaluminum complex catalyst was prepared according to the procedure given previously.[9]

Tetramethylethylenediamine (TMEDA) (Eastman Laboratory Chemicals, 99% purity) was used as received.

Polymer Synthesis

We controlled the structure of SBR's in our work with the use of a number of polymerization variables, described in Table 3. Polybutadienes with vinyl contents ranging from 9 to 81% were synthesized in n-hexane with n-butyllithium initiator and TMEDA modifier. The desired vinyl content was obtained by controlling the mole ratio of TMEDA to n-BuLi and polymerization temperature.[10] Solution SBR's having variable styrene content at constant low vinyl (less than 10%) were prepared in n-hexane using n-butyllithium and reacting at 110°C while continuously adding a mixture of monomers to the reactor. This procedure produces an SBR with the required low vinyl content and randomly distributed styrene units, as determined by NMR. At a constant low vinyl content (2-4%), the ratio of trans-1,4/cis-1,4 was adjusted using organometallic compounds of magnesium and aluminum in the presence of a barium salt, such as barium di-t-butoxide. Like n-butyllithium, the barium-based anionic polymerization catalysts, discovered in our laboratory, give "living polymerization" for butadiene and styrene.[9] Consequently, the molecular weight of polymers produced by these catalysts is readily controlled.

The molecular weight distribution and chain architecture for polymers prepared in these living polymerizations can be controlled by post-polymerization reactions, such as by the reaction of the "living ends" with compounds such as tin tetrachloride.

Polymer Characterization

We devoted special attention to characterize all the polymers that were prepared for this study. The characterization techniques listed in Table 4 were used for determining: (1) comonomer composition, butadiene microstructure, and sequence distribution of the monomer units; (2) molecular weight, molecular weight distribution, and chain architecture; (3) glass transition; and (4) relative crosslink density. We made every effort to use state-of-the-art equipment, and to extend the development of techniques by interfacing the equipment with appropriate computer data acquisition systems to obtain reliable and accurate characterization data.

NMR analyses were carried out using a Varian FT-80 ^{13}C NMR spectrometer.[11] NMR spectrometers having higher field strengths were also used to determine more detailed aspects of the rubber structure. Chemical shifts of the rubbers in deuterochloroform were measured in ppm with respect to the reference standard tetramethylsilane (TMS). Styrene sequence distribution information for SBR's was obtained from an analysis of the NMR peak areas for the aromatic ring proton resonances, using a computer program specially developed by us to deconvolute the spectral overlaps. The approach that we used is a refinement of that originally described by Mochel and Johnson.[12]

The ^{13}C NMR spectrum given in Figure 6 below the chemical structure of the rubber represents the saturated carbon region on the right and the unsaturated carbon region on the left. Assignments of specific NMR spectral peaks to the various C/H structure arrangements are shown by the letter coding in this Figure. Many of the structural assignments shown were made in our laboratory by studying model compounds and polymers prepared for this purpose. Using such NMR techniques, we can give a detailed picture of the molecular structure of styrene/butadiene rubbers.

Glass transition temperatures were measured using a Differential Scanning Calorimeter (Perkin-Elmer, DSC-2), using a heating rate of 20°C/minute.

Rubber Vulcanizate Preparation and Characterization

Black filled tread rubber compositions contained: 100 phr polymer; 45 phr N339 carbon black; 10 phr naphthenic oil; 3 phr zinc oxide; 2.5 phr stearic acid; 1.52 phr N-tert-butyl-2-benzothiazole sulfenamide (TBBS); 2.18 phr sulfur. Unfilled rubbers were cured with 1.52 phr TBBS and 2.18

TABLE 4

Table 4. Characterization Methods Used on Tread Rubbers

CHARACTERISTIC OF RUBBER	CHARACTERIZATION METHOD
COMONOMER COMPOSITION, BUTADIENE MICROSTRUCTURE AND COMONOMER SEQUENCE DISTRIBUTION	HIGH RESOLUTION NMR (^{1}H AND ^{13}C)
MOLECULAR WEIGHT MOLECULAR WEIGHT DISTRIBUTION CHAIN ARCHITECTURE	MEMBRANE OSMOMETRY GPC AND LIGHT SCATTERING
GLASS TRANSITION TEMPERATURE	DSC
VOLUME FRACTION SWOLLEN RUBBER	EQUILIBRIUM SWELLING IN n–DECANE

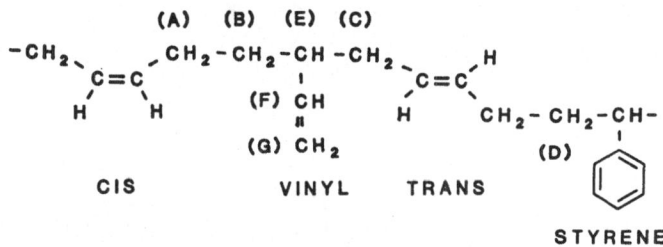

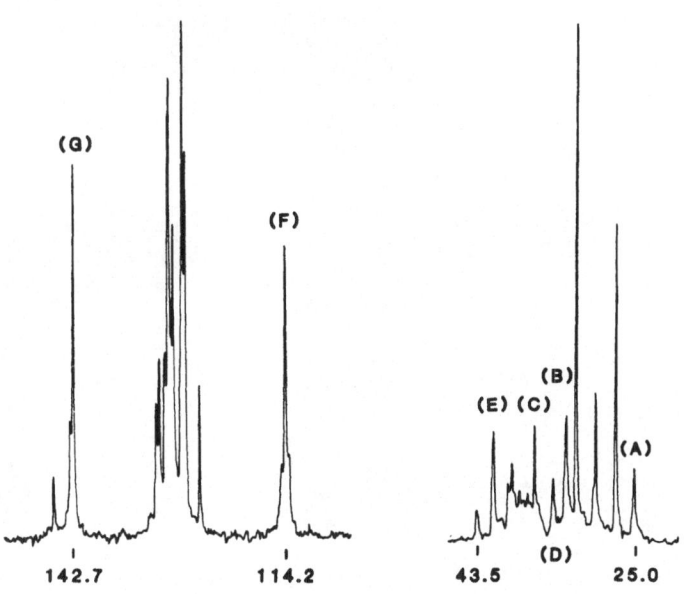

Fig. 6. ^{13}C NMR spectrum of a solution SBR in CDCl$_3$ solvent.

phr sulfur. The compounding ingredients were mixed using a two-roll mill preheated at **82°C** to 93°C. The **stocks** were cured at **160°C.** Cure **times** were based on 100% optimum **as** measured using a Monsanto Rheometer. With **this** curative system, 200% modulus was held at 7.3 ± 0.3 **MPa** for the carbon black filled vulcanizates.

Swelling measurements were carried out by **immersing** the samples in n-decane at **25°C.** The volume fraction (V_r) of rubber present in the **net**work at **swelling equilibrium** was taken as an approximate measure of **cross**link density. Corrections for **polymer/filler** interactions were not applied.

Techniques for Determining the **Dynamic** Mechanical Properties and Some of the Functional Properties of Rubber **Vulcanizates**

The commercially available **MTS** Laboratory Automation System was the **major** tool used in mechanical **characterization.**[13] The MTS system, as shown in Figure 7, **is** equipped with a computer-controlled servohydraulic capability. It **is** a **very** versatile mechanical tester which can be used to test **specimens** of a variety of geometries **under** a broad range of **conditions** of temperature, frequency, and **strain.** The double lap shear specimen used to obtain dynamic mechanical properties data **is** shown in Figure 8.

For rapid determination of the tan δ at low frequencies and room temperature of the large number of rubber vulcanizates tested, we adapted the Yerzley Oscillograph.[14] **This** equipment was one of the **earliest** instruments used for determining the dynamic mechanical properties of rubber vulcanizates.

Fig. 7. **MTS** laboratory automation system.

Fig. 8.   Double lap shear
test **sample**.

The principle on which this instrument **is** based **is** depicted
schematically in Figure 9.   A compression button **is** deformed by a loaded
beam which **is** released to execute a **"see-saw"** motion.   The instrument
measures the damped vibration at a frequency of **about** 2-3 Hz at room
temperature.   The original instrument as **commercially** available was not
precise enough for the tan δ measurements for our studies.   We modified the
basic equipment by instrumenting **it** with appropriate transducers and **inter-**
facing it with a **computer** data **acquisition** System.   A photograph of the
Instrumented Yerzley Oscillograph in our laboratory **is** shown in Figure 10.
It allowed us to screen a large number of experimental **vulcanizates** and
samples taken from different parts of tires for measurements of tan δ.

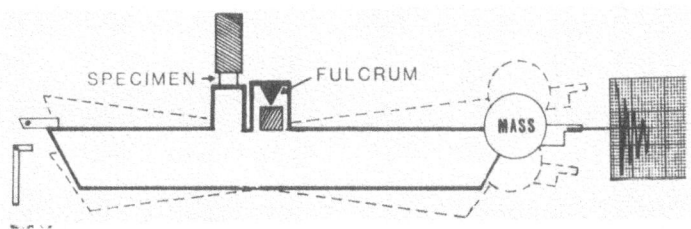

Fig. 9.   Schematic of Yerzley Oscillograph.

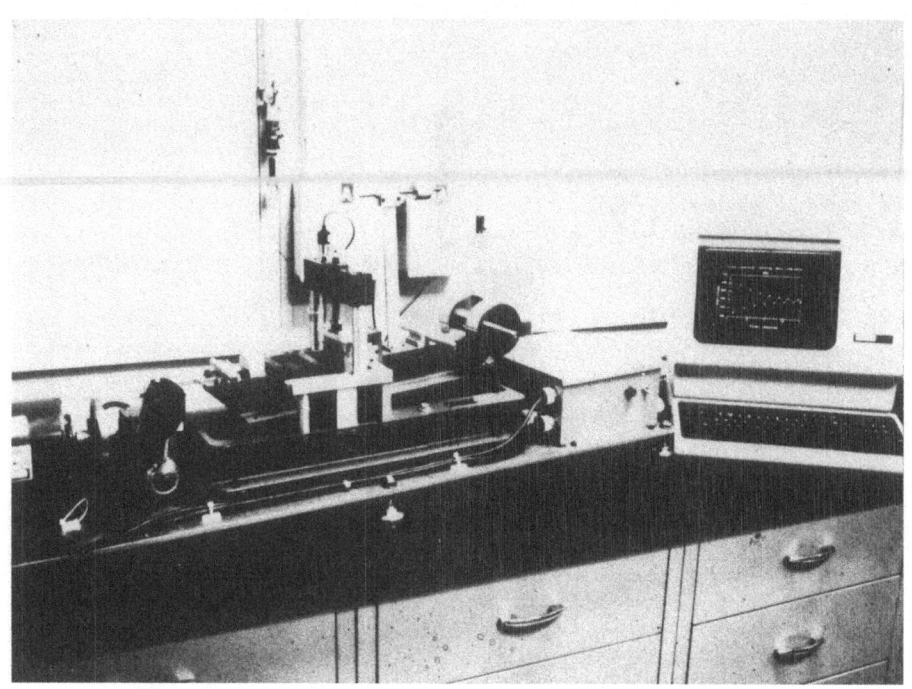

Fig. 10. Instrumented Yerzley Oscillograph (IYO).

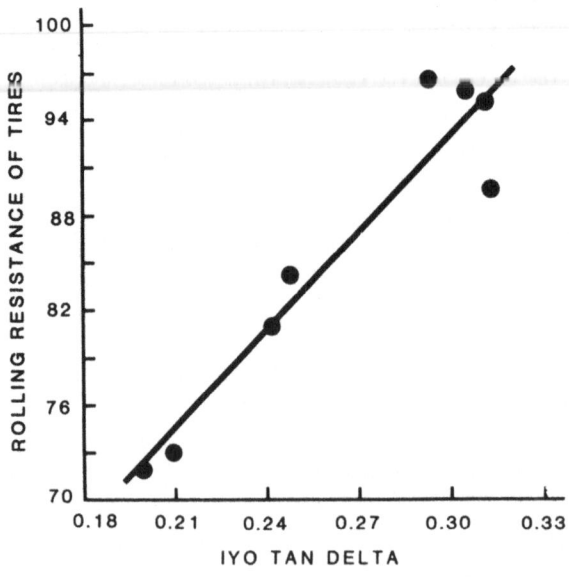

Fig. 11. Yerzley tan delta vs. rolling resistance (twin roll).

The tan δ measured by our Instrumented Yerzley Oscillograph (abbreviated as IYO) has adequate precision, and was found to correlate reasonably well with tan δ measured on rubber samples by the MTS equipment at 40 Hz and 60°C. It also correlates remarkably well with rolling resistance of tires[15] (Figure 11) as determined by the "Twin Roll Method" and used for evaluation and testing of tires.

The Pendulum Skid Tester,[16] interfaced with a data acquisition system,[17] was used for determining wet coefficient of friction. The Instrumented Pendulum Skid Tester (abbreviated as IPST) is shown schematically in Figure 12. The basic principle of this instrument is that the energy loss by the pendulum, as a rubber sample slides over the surface during the test cycle, is equal to the work done in overcoming the friction between the sliding rubber sample and the surface. The coefficient of friction calculated from measurements by this instrument is referred to as IPST number. The IPST number determined by this instrument shows excellent correlation with wet skid coefficient of friction of actual tire tests as shown in Figure 13.

RESULTS AND DISCUSSION

Soon after vinyl BR's were first introduced commercially,[18] it was recognized that the vinyl structure imparted non-typical dynamic properties. In particular, their vulcanizates showed improved resistance to heat build-up when compared to emulsion SBR.[19,20] Before relating the special contribution of the vinyl structure to properties, it is informative to first determine the influence of styrene content (at constant vinyl) on the two fundamental properties of interest: Tg and tan δ, as shown in Figure 14. As expected, both Tg and tan δ increase almost linearly with styrene content. Thus like a conventional tire tread rubber, the rolling resistance of solution SBR's prepared using lower amounts of styrene is accompanied by a decrease in the wet traction of the rubber.

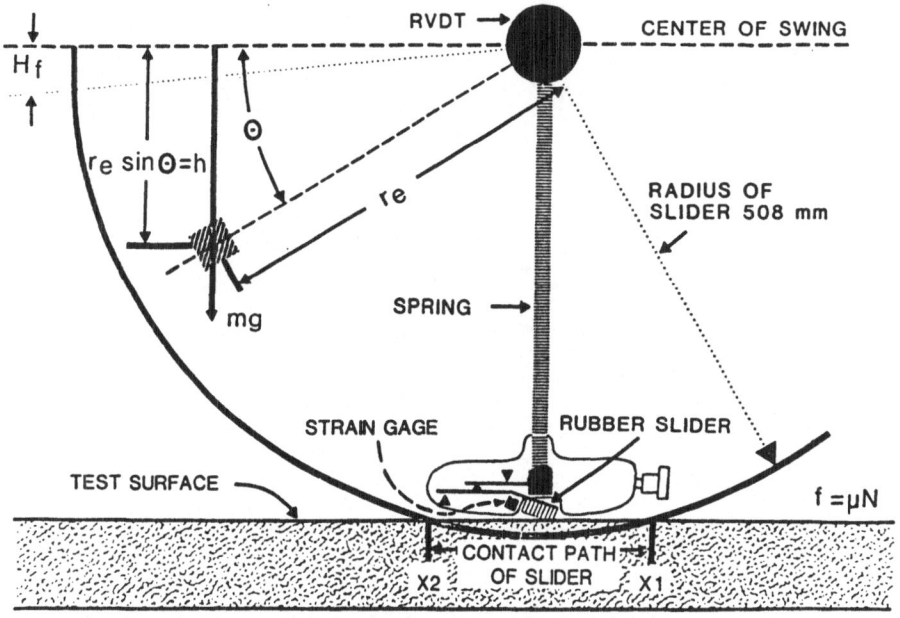

Fig. 12.   Instrumented Pendulum Skid Tester (IPST).

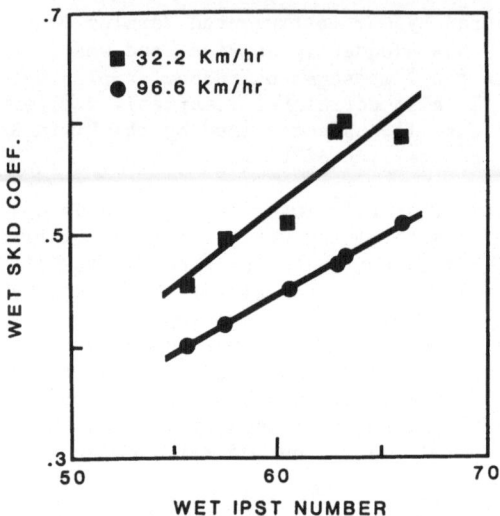

Fig. 13.   Comparison of wet skid coefficient of
friction of tires with wet IPST number
from laboratory measurements.

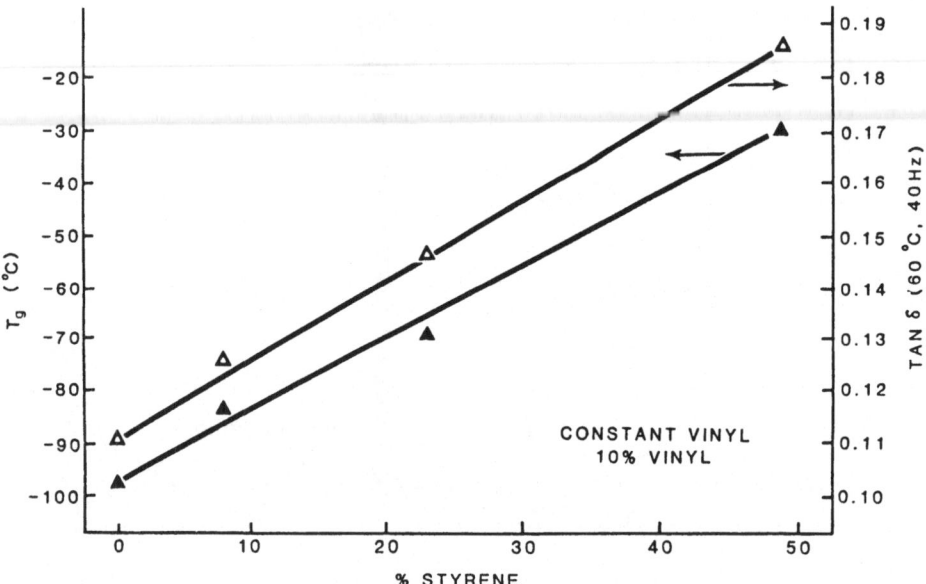

Fig. 14.   Variation of Tg and tan delta with styrene content of solution
SBR's at constant (10%) vinyl structure.

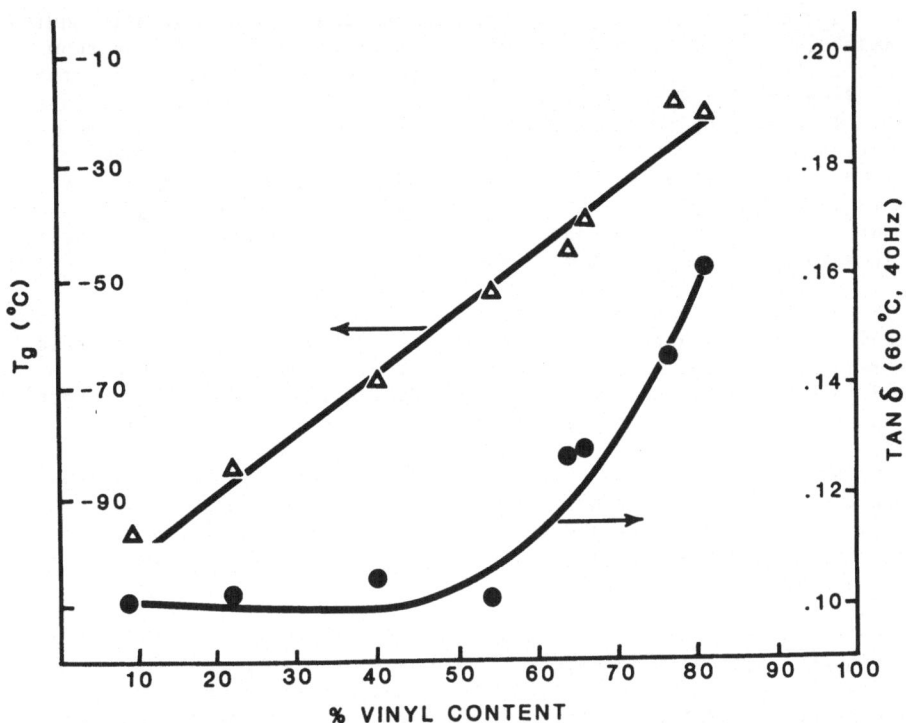

Fig. 15. Variation of Tg and tan delta with vinyl content of
vinyl BR's.

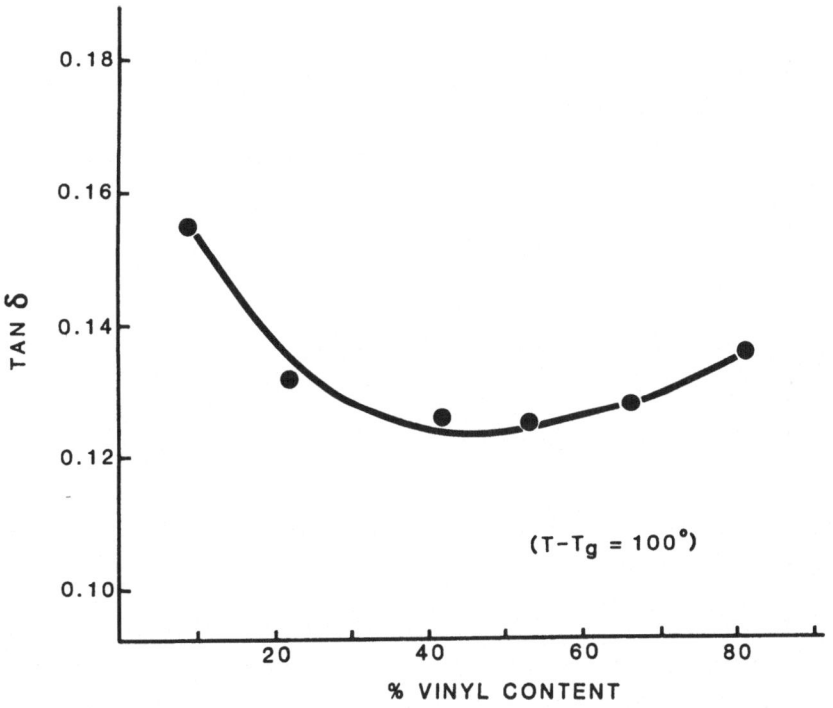

Fig. 16. Dependence of tan delta on vinyl content of vinyl BR's at
temperatures where T-Tg = 100°. T is the test temperature.

The effect of increasing vinyl content on tan δ at essentially equivalent molecular weight is, however, quite surprising and unique, as shown in Figure 15. The glass transition temperature, Tg, of the rubbers increases nearly linearly with increasing vinyl content. The tan δ remains constant with vinyl content up to about 50%. At higher vinyl content, the tan δ increases sharply. The effect of increasing vinyl content on tan δ is further emphasized if we plot tan δ at a constant value of temperature equal to T-Tg = 100°C to eliminate the effect of proximity to Tg, as shown in Figure 16. For vinyl content up to 60%, tan δ decreases substantially with increasing vinyl content of these rubbers.

The conclusion based on the above data is the following: In the low and medium vinyl polybutadiene range, the tan δ related to rolling resistance remains constant as vinyl content increases, while Tg of these rubbers related to skid resistance increases. Thus there is a region where traction (Tg) can be increased without changing rolling resistance (tan δ).

It has been established by a number of studies[21] that in crosslinked rubbers the major contribution to hysteresis in the plateau region comes from the effect of dangling chain ends. Thus it is generally accepted that hysteresis, i.e. tan δ, should decrease with increasing crosslink density.

Therefore, in order to identify a possible effect of crosslink density on loss tangent in these vulcanizates, we determined the swelling ratios of these rubbers. As shown in Figure 17, the volume fraction of the rubber vulcanizates swollen in n-decane, in fact, decreases almost linearly with vinyl content. Thus the crosslink density with increasing vinyl content in these rubbers decreases. The decrease in tan δ with increasing vinyl content in the range up to 60% (Figure 16) cannot, therefore, be attributed to an increase in crosslink density. Thus the effect of the vinyl groups

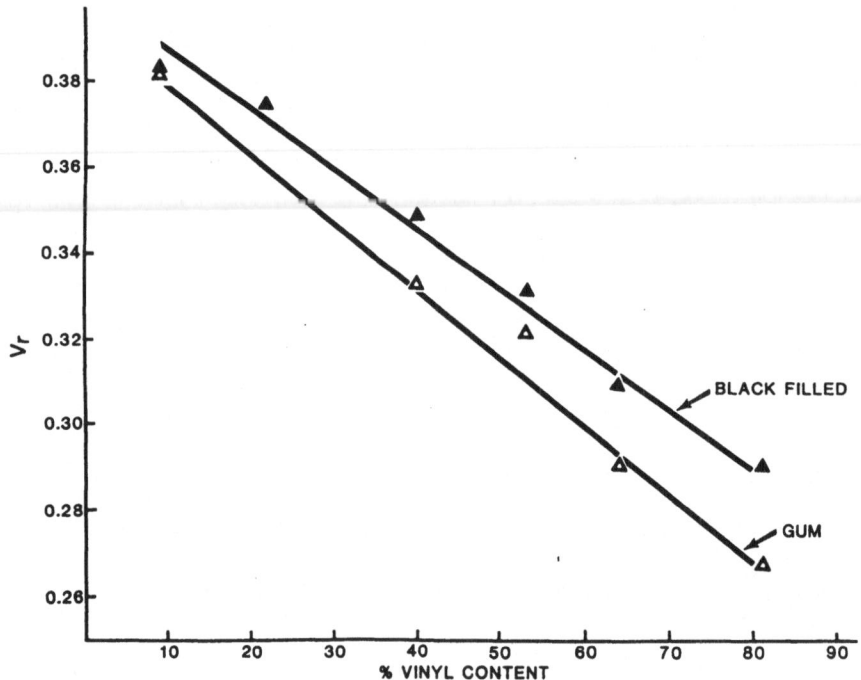

Fig. 17. Variation of volume fraction of swollen rubber with vinyl content of polybutadienes prepared by anionic polymerization.

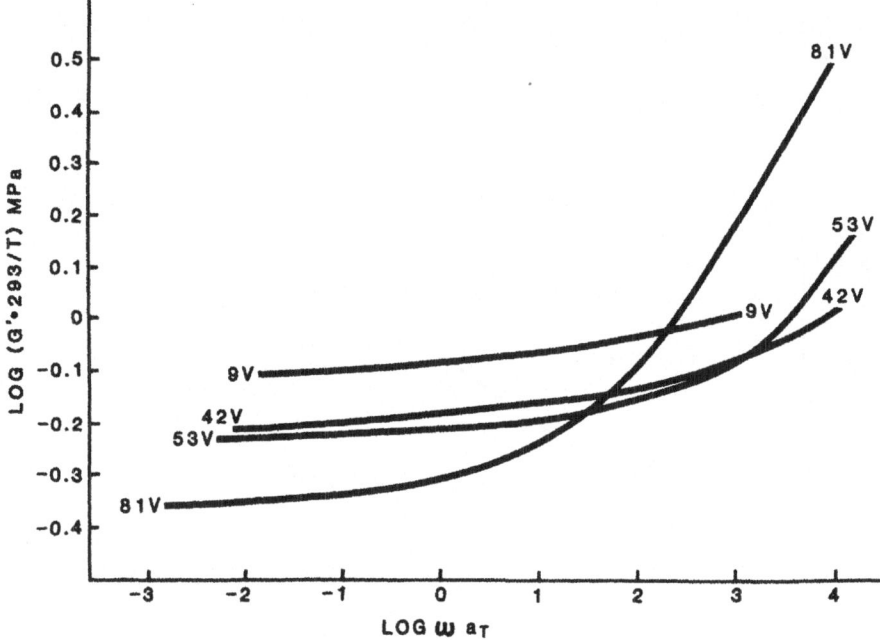

Fig. 18.   Reduced storage modulus versus frequency for vinyl BR gum
           vulcanizates.

in decreasing tan δ dominates the effect of lower crosslink density with
increasing vinyl content, at least in this vinyl range.

To confirm further this unexpected effect, dynamic mechanical properties were run on this same set of rubbers over a range of frequencies and
temperatures. Master curves of log G' vs. log frequency were constructed
for each rubber by shifting to a common reference temperature of 20°C. In
Figure 18, we show these plots for rubbers of increasing vinyl content (from
9% vinyl to 81% vinyl). The observation again is that the plateau modulus
decreases as the vinyl content increases, which is in agreement with the
studies of Carella, Graessley, and Fetters.[22] This confirms that the
crosslink density in these rubbers is decreasing as vinyl content increases.

DEVELOPMENT OF TIRE TREAD RUBBER COMPOSITION WITH COMBINATION OF LOW
ROLLING RESISTANCE AND HIGH WET TRACTION

Guided by the above studies, we have developed a number of experimental
rubbers, and blends of rubbers, for application in tire treads. The experimental rubbers consist of vinyl BR's, SBR's, and high trans SBR's prepared
by the anionic polymerization techniques, previously described in this
paper. Tan δ and Tg results for some of these rubber compositions are
plotted in Figure 19. Also shown are the corresponding results for rubber
compositions based on conventional rubbers used in the past. These data
show that the experimental compositions have significantly lower IYO tan δ,
and have simultaneously higher glass transition temperatures, when compared
to compositions based on conventional rubbers.

We have made tire treads with some of the rubber compositions that
were used for the study of tan δ and Tg, shown in Figure 19. The test
results on coefficient of wet skid resistance of tread rubbers and rolling

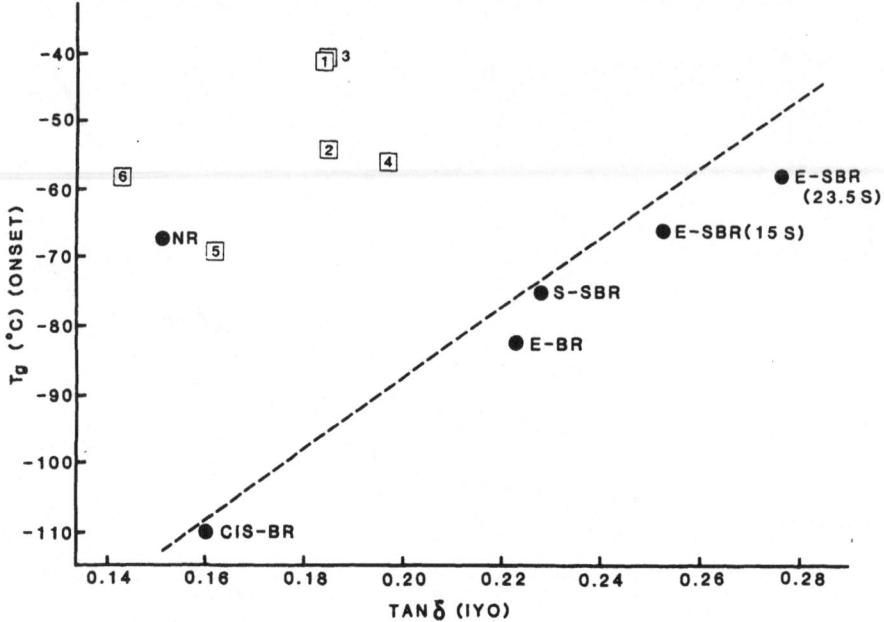

Fig. 19. Plot of Tg versus tan delta for conventional (●) and
experimental (□) tread vulcanizate compositions.

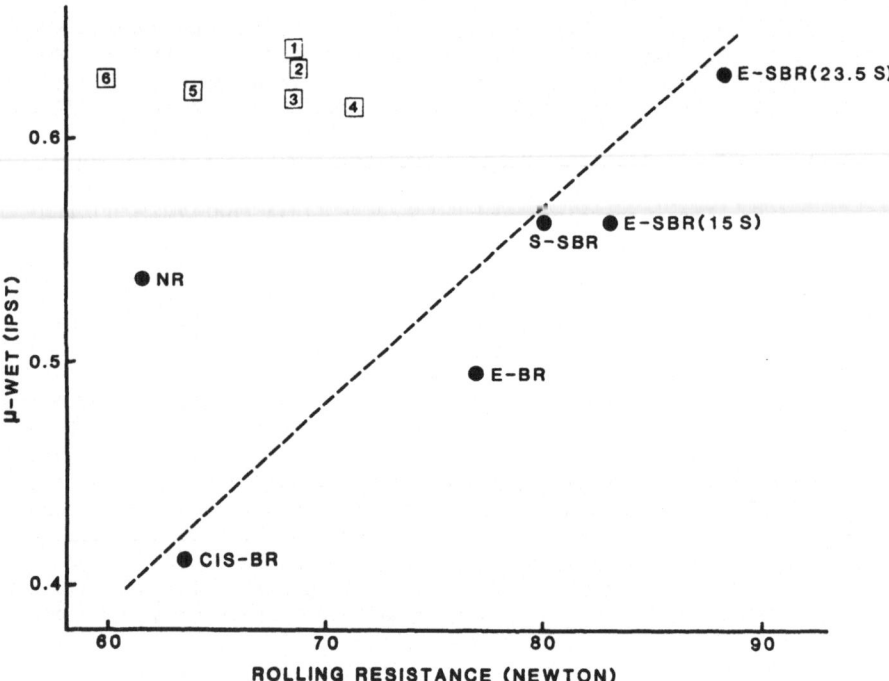

Fig. 20. Plot of coefficient of wet skid resistance versus rolling
resistance for the rubbers in Fig. 19.

resistance are plotted in Figure 20. They show that the new rubber compositions, which evolved from our work, have an identifiably superior combination of low rolling resistance and high wet traction, as compared to those made from conventional rubbers.

We believe that the approach that we have discussed should serve as a guide for the development of new tire rubber compositions of superior performance in the future. The systematic and fundamental approach in polymerization research outlined in the beginning of this paper should be emphasized.

ACKNOWLEDGEMENT

The work reported herein required the combined effort and contributions of several groups with diverse skills and expertise in our laboratories. The authors gratefully acknowledge the contribution of our colleagues from several areas of specialization and polymer research disciplines that were vital to this effort's success. In particular, we are indebted to Mr. John A. Wilson, who contributed significantly to the polymer synthesis studies.

REFERENCES

1. K. H. Nordsiek, Kautsch. Gummi Kunstst., $\underline{38}$, 178 (1983).
2. R. Bond, Proc. Royal Soc., London, $\underline{A399}$, $\overline{1}$ (1985).
3. N. Yoshimura, M. Okuyama, and K. Yamagishi, The Present Status of Research on Rolling Resistance in Japan, p. 51, in Symposium: Tire Rolling Resistance, D. J. Schuring, ed., Rubber Division Meeting, Am. Chem. Soc., 1983.
4. A. Schallamach, Int. Polym. Sci. Techni., $\underline{6}$, T44, T63 (1975).
5. E. Southern and A. G. Thomas, Rubber Chem. Technol., $\underline{52}$, 1008 (1979).
6. A. N. Gent and C. T. R. Pulford, J. Appl. Polym. Sci., $\underline{28}$, 943 (1983).
7. J. D. Ferry, Viscoelastic Properties of Polymers, 2nd ed., Wiley, New York, 1970.
8. H. L. Hsieh, R. C. Farrar, and K. Udipi, in Anionic Polymerization, J. E. McGrath, ed., Symposium Series No. 166, Am. Chem. Soc., Washington, D.C., 1981.
9. I. G. Hargis, R. A. Livigni, and S. L. Aggarwal, in Elastomers and Rubber Elasticity, J. E. Mark and J. Lal, eds., Symposium Series No. 193, Am. Chem. Soc., Washington, D. C., 1982.
10. T. Antkowiak, A. E. Oberster, A. F. Halasa, and D. P. Tate, J. Polym. Sci., Part A-10, $\underline{10}$, 1319 (1972).
11. D. D. Werstler, Rubber Chem. Technol., $\underline{53(5)}$, 1114 (1980).
12. V. D. Mochel and B. L. Johnson, Rubber Chem. Technol., $\underline{43(5)}$, 1138 (1970).
13. MTS 830 Elastomer Test System by MTS System Corp., Box 24012, Minneapolis, MN 55424.
14. F. L. Yerzley, Rubber Age, $\underline{104}$, 27 (1972).
15. D. J. Schuring, Rubber Chem. Technol., $\underline{53}$, 600 (1980).
16. British Standards Institution, British Standard No. 903; 1950, part 22, Methods of Testing Vulcanized Rubber, London (1950).
17. J. M. Giustino and R. J. Emerson, Instrumentation of the British Portable Skid Tester, Paper No. 76, 123rd Meeting Rubber Division, Inc., Am. Chem. Soc., May 10-12 (1983).
18. E. W. Duck and J. M. Locke, J. Inst. Rubber Ind., $\underline{2(5)}$, 223 (1968).
19. K. H. Nordsiek and K. M. Kiepert, Kautsch. Gummi Kunstst., $\underline{33}$, 251 (1980).

20. R. E. Railsback and N. A. Stumpe, Jr., Rubber Age, 107, 27 (1975).

21. R. G. Mancke and J. D. Ferry, Trans. Soc. Rheol., 12, 335 (1968).

22. J. M. Carella, W. W. Graessley, and L. J. Fetters, Macromolecules, 17, 2775 (1984).

POLYMER AND CHAIN END STRUCTURE IN

ANIONIC DIENE POLYMERIZATION

S. Bywater and D.J. Worsfold

National Research Council of Canada
Ottawa, Ontario
Canada K1A 0R9

ABSTRACT

The microstructure of the polybutadiene and polyisoprene produced by anionic polymerization is correlated to the structure of the allylic anion of the active chain end. The charge distribution over this anion is affected greatly by changing counterions, the use of solvating solvents, and the presence of cation chelating additives.

The polymer cis 1,4/trans 1,4 ratio is determined by several factors. Of importance is the cis or trans structure of the ion formed at the moment of reaction, and the rate at which it will isomerize to its equilibrium structure compared to the rate of addition of the next monomer unit. The relative rates of reaction at the two isomeric chain ends can also be important.

Variations in the ratio of 1,4 structures to vinyl structures in the polymers are more difficult to interpret in terms of active chain end structure. Although in general an increase in charge at the γ position in the allyl ion leads to the larger vinyl contents, very wide and apparently irregular variations occur with change in counterion. Some of the largest increases of vinyl structure appear in polymerizations in the presence of chelating diamines, and the complex series of solvates that form are analyzed by means of data from U.V., N.M.R. and kinetic studies.

INTRODUCTION

The polymerization of the dienes butadiene and isoprene by anionic initiators can lead to a wide range of polymer structures according to the conditions under which the polymerization is performed. Various proportions of cis 1,4 structures, trans 1,4 structures and vinyl structures are found in the polymers, Table 1 gives selected values to illustrate this. These structures must be the products of the state of the active chain end at the moment of reaction. The active chain end can be thought of as a substituted allylic anion and can have two forms, the cis and trans forms.

Table 1. Polymer Microstructure Variation with Conditions

Monomer	Solvent	Counterion	Temp.	% Microstructure cis 1,4	trans 1,4	vinyl	Ref
Isoprene[a]	Hexane	Li+	30°	95	0	5	1,2
Isoprene[b]	THF[d]	Li+	20°	—	12	88	2
Isoprene[c]	THF	Li+	20°	—	24	76	4
Isoprene[b]	THF	Na+	0°	6	5	89	5
Isoprene	Dioxane	K+	15°	4	32	64	3
Butadiene[a]	Hexane	Li+	20°	56	37	7	6
Butadiene[b]	THF	Li+	0°	6	6	88	9
Butadiene	Hexane+DIPIP[e]	Li+	5°	—		99+	7
Butadiene	THF	K+	0°	5	28	67	9
Butadiene	DEE[f]	K+	0°	11	34	55	9

a Near 10^{-5} M.Li+ and high monomer concentrations.

b In presence of Boron Tetraphenyl salts to ensure ion pair reaction only

c Free ion reaction.

d THF is tetrahydrofuran.

e DIPIP is 1,2-dipiperidinoethane, D⁻PIP/Li ratio >2, Li $\sim 10^{-3}$ M.

f DEE is Diethyl ether.

trans   cis

The charge on the anion is spread over the end three carbons, particularly the γ and α positions. The relative proportions of the charge on the α and γ positions, as measured by NMR, is sensitive to the counterion and the solvent[10] (Table 2). In THF with a Cs^+ counterion the charge appears fairly evenly distributed between the two positions, and the cis form is heavily preferred. But with lithium counterions in hydrocarbon solvents, where much closer approach of the counterion is possible, there is considerably less charge on the γ position. In this case to accommodate what would be greater charge at the α position, multicentre bonds have been suggested to give aggregates containing more than one active chain end. These, however, at equilibrium favour the trans form of the chain end, and it seems likely that these aggregates must dissociate to propagate. Cis/trans isomerization of the active chain ends may also occur. This is relatively rapid at room temperature with the lithium counterion in both hydrocarbon and polar solvents and comparable to the rate of propagation. The rate of isomerization decreases with the larger counterions.

When an active chain reacts with another monomer molecule, depending on whether the reaction is at the α or γ position of the active chain end, this unit is incorporated as the penultimate unit either as a 1,4 or vinyl structure, respectively. Also if it reacts in a 1,4 sense, the original stereo structure of the active chain end will be incorporated into the polymer chain, thus a trans active chain end would give a trans 1,4 penultimate in-chain unit.

Satisfactory explanations have been given for the variation of cis and trans structures of the polymers in terms of three factors.

a. The relative rates of addition of monomer to the two stereoisomeric chain ends.

b. The preferred mode of addition (i.e., cis or trans) of the monomer to form the corresponding next isomeric active chain end.

c. The position of equilibrium and the equilibration rate of the initially formed active chain end compared to its rate of monomer addition.

This is exemplified well in the case of the polymerization of isoprene in hydrocarbon solvents with lithium-based initiators.[2] The newly formed active chain end has a cis structure, and at low concentration of chain end where propagation is rapid compared to isomerization a very high cis 1,4 polymer is formed. However at high concentrations of chain end or low monomer concentrations the isomerization to the preferred trans configuration is more rapid than propagation and appreciable trans 1,4 structures are incorporated.

Table 2. Variation of Charge on γ Carbon with Conditions in Polymerization of Butadiene

Solvent	Counterion	γ Charge[a]	% Trans[b]
Benzene[c]	Li$^+$	0.22	77
DEE[d]	Li$^+$	0.35	75
THF[d]	Li$^+$	0.40	23
Hexane+DIPIP[c]	Li$^+$	0.51	93
THF[d]	Na$^+$	0.49	23
THF[d]	K$^+$	0.53	–
THF[d]	Cs$^+$	0.52	–

[a] Charges tabulated as fraction of 1ε calculated from ^{13}C NMR spectra.
[b] With K$^+$ and Cs$^+$ equilibrium is probably not easily attained.
[c] 20°.
[d] -20°.

Also in the polymerization of butadiene in THF with lithium counterions,[11] at room temperature a polymerizing system shows a UV spectrum consistent with a large percentage of cis active chain ends as expected for an equilibrium system. But at low temperatures the spectrum appears as largely trans whilst the polymerization is in progress, suggesting that newly formed centers are largely trans and cannot equilibrate at that temperature.

However very often the major change in polymer structure on changing polymerization conditions is in the vinyl content of the polymer (Table 1), and these changes have only been explained in the most general terms. In general it is thought that as the charge moves to the γ position of the active chain end so the vinyl content of the polymer will increase. But in fact systems that show the largest upfield shift of the γ carbon in the ^{13}C NMR spectra, indicating the highest charge, do not by any means always give the highest vinyl content polymers (e.g., K in THF).

Conditions that have been found to give very high polymer vinyl content are the presence of chelating amines such as tetramethylethylene-diamine (TMEDA)[12,13] or dipiperidinoethane (DIPIP)[7,8] with lithium based initiators in hydrocarbon solvents. The latter in the case of butadiene gives virtually 100% vinyl polymer. Complexes between the lithium and the complexing amines have been suggested as the active agents,[12] and evidence for complexes is found in NMR spectra of the active chain ends.[13] Halasa has reasonably suggested that the differences in behaviour of TMEDA from DIPIP is caused by specific steric factors.[8] In the detailed investigation reported here it has been found that an explanation in terms of the formation of a single complex was not viable, and ultimately it has been necessary to propose that a series of complexes form. More than one may be active in the polymerization, and it is the complex kinetic interactions between them which determine the structure under any given set of conditions.

The equilibria proposed are as follows. Starting with an initial tetrameric unsolvated active chain end aggregate (T)

Scheme 1. T + d ⇌ T.d   Solvated tetramer
           T.d + d ⇌ 2D.d   Solvated dimer
           D.d + d ⇌ 2M.d   Solvated monomer

first this forms a solvate (T.d) which with more diamine breaks down to a solvated monomeric form (M.d), via a solvated dimer (D.d). In each case only one solvating molecule appears necessary.

RESULTS AND DISCUSSION

The evidence for the above complexes comes from three principal sources, UV spectra, NMR spectra, and kinetics. Although in fact the evidence from the three sources are inevitably very interdependent, they will be treated separately.

U.V. Spectra

If a solution of the active chain end from the polymerization of butadiene by butyllithium in cyclohexane is taken and a series of additions of DIPIP made, very large changes in the UV spectrum appear (Figure 1). The initial peak associated with butadienyllithium in hydrocarbon solvents at 276 nm disappears and is replaced with a peak at 328 nm. There is, however, no real isosbestic point which would indicate that only two species are involved. There appears to be at first a general broadening of the initial peak, with an increase in absorption at 250 nm, and then a sharp rise in the peak at 328 nm and a fall at 250 nm. There is evidence of the presence of at least two other products as well as the original material.

A solution of the active polymerization mixture with an approximately 1:1 ratio of DIPIP to Li has an intermediate spectrum. If then the overall concentration of the solution is varied keeping this ratio constant, it is found that the shape of the spectrum is unchanged over the range of concentration 10^{-2} to 10^{-4} M. Evidently the ratio of the various components in the mixture is unchanged, and the equilibria connecting them are concentration independent. This is consistent with the latter two equilibria in scheme 1 where the number of molecules is the same on both sides of the equilibria.

Similar behaviour is found for isoprenyllithium except that in this case, at these complexing agent ratios, the final complex does not form completely (Figure 2). The two sets of spectra are at a hundredfold different concentration, and again similar spectra are found at similar ratios of DIPIP to lithium, again indicating their independence of the global concentration.

TMEDA in less complete studies also shows the appearance of more than one other species as it is added to solution of active chain ends.[5]

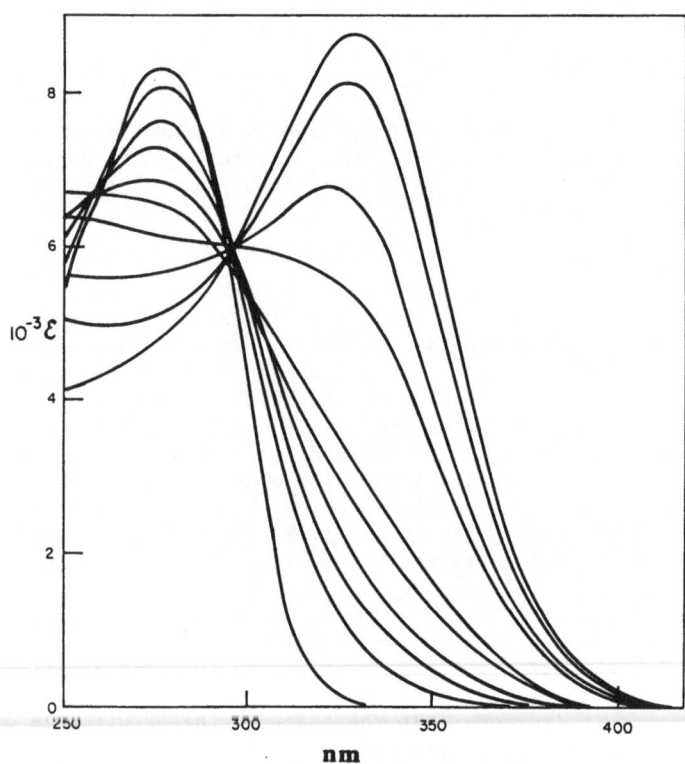

FIGURE 1. U.V. Spectra of Butadienyllithium in Cyclohexane solution at
5°, 1.2×10^{-3} M, in presence of DIPIP. DIPIP/Li ratio from
bottom to top at 320 nm, 0, 0.10, 0.18, 0.39, 0.49, 1.16,
3.43, 4.50, 5.60.

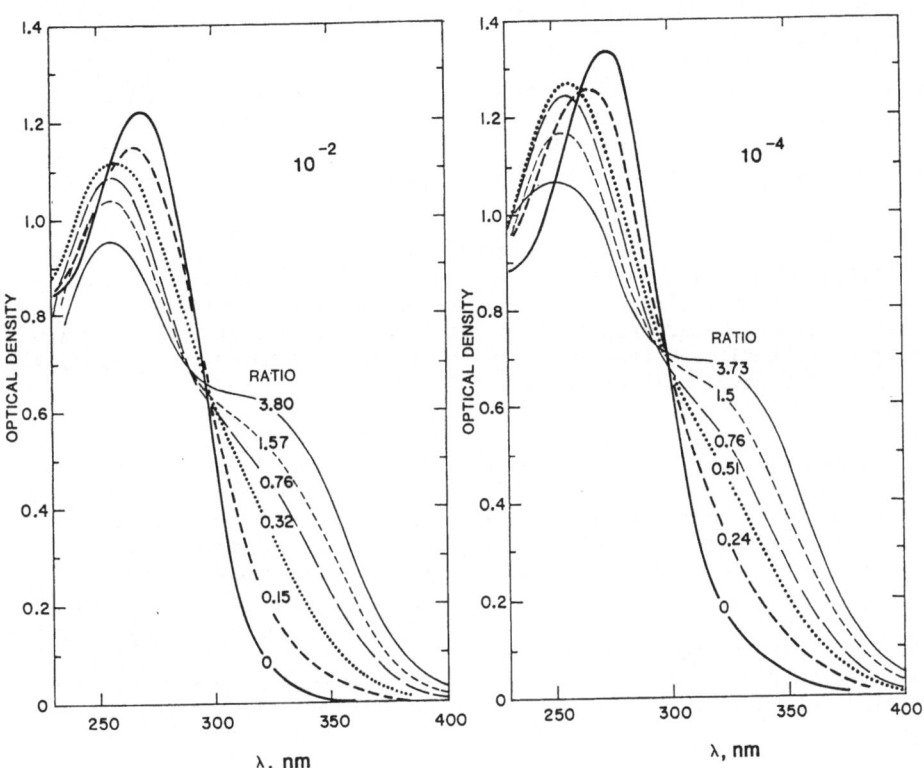

FIGURE 2. U.V. Spectra of Isoprenyllithium in Cyclopentane solution at
21°, 10^{-4} and 10^{-2} M, in presence of various DIPIP/lithium
ratios.

NMR spectra because of their higher resolution can readily differentiate between the cis and trans forms of the active chain end which is more difficult with UV spectra. But because of the much slower time scale of NMR spectra, averaging of signals between the various associates often occurs in these rapidly equilibrating systems. This happens in the NMR spectra of butadienyllithium at various DIPIP to lithium ratios at room temperature where only average cis and trans peaks are seen. Table 3 shows how the position of these peaks in the ^{13}C NMR spectra change with added DIPIP. There is a general movement downfield of the averaged shifts of the α peak and upfield of the γ peak as the concentration of DIPIP increases showing delocalization of the charge away from the end carbon to a more balanced distribution between the α and γ positions. The proportion of the cis and trans forms also changes to give largely the trans form. This butadiene case is unusual in that more often it is the cis form of the active chain end whose proportion is increased when polar co-ordinating agents are present. Figure 3 illustrates this for TMEDA and DIPIP, THF has also been found to give about 77% cis chain ends with butadiene and effectively 100% cis with isoprene in the presence of Li^+ counterions.

TMEDA with both butadienyllithium and isoprenyllithium causes the NMR shift of the γ peak to move upfield to near the 70 ppm region as does DIPIP with butadienyllithium below. DIPIP with isoprenyllithium behaves differently. Figure 4a illustrates the position of the cis and trans γ peaks with increasing DIPIP ratio, and it is seen that although the trans (minor) peak behaves normally and finally moves to ~70 ppm, the cis peak only moves part of the way. The cis peak is the major peak, and if the temperature of the measurements is dropped the rate of exchange between the solvates is slowed down sufficiently so that at −20° this peak splits into three separate peaks. These change in proportion as the DIPIP/Li ratio increases (Figure 4b). The initial absorption (a) in the absence of DIPIP decreases very rapidly and a peak at 97.5 ppm, (b) little shifted from the original, grows rapidly to be replaced by the final peak, (c) which at higher ratios declines in proportion as the trans peak increases. The first intermediate peak to appear (b) has its maximum concentration when the DIPIP/Li ratio is 0.25, and it is supposed that it is the initial tetramer solvated with one molecule of DIPIP. This is a fairly strongly formed complex and is nearly completely formed at this ratio as is seen from the rapid decline of peak a. There is of

Table 3. Variation of ^{13}C shift of γ-carbon for butadienyllithium as a function of DIPIP:Li ratio

Ratio DIPIP/Li	% trans active centres	γ-carbon shift	
		trans	cis
0	77	101.6	102.0
0.15	77	98	98
0.61	76	82.4	89.7
1.09	86	74.0	85.4
2.10	93	70.9	82.2
2(TMEDA)	61	72.2	72.2
BULK THF	35	79.6	81.9

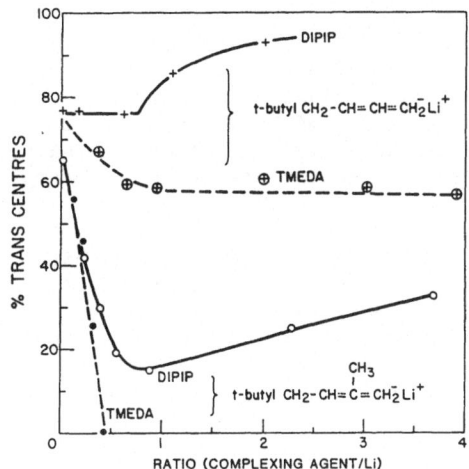

FIGURE 3. Percentage trans isomer in active chain end models for
butadienyllithium and isoprenyllithium. DIPIP is
1,2-dipiperidinoethane and TMEDA is tetramethylethylene-
diamine.

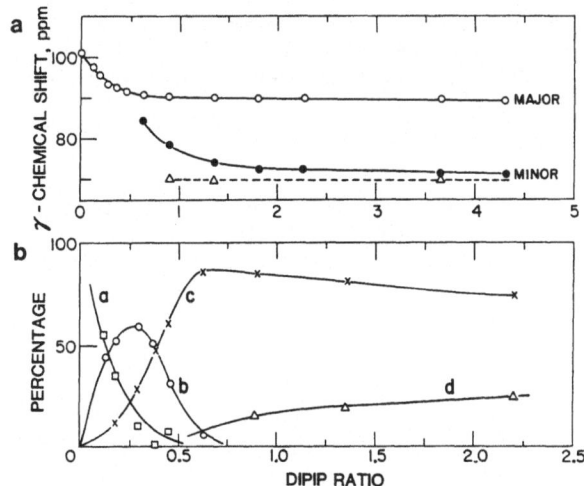

FIGURE 4. a) Influence of DIPIP/Li ratio on ^{13}C chemical shifts of the γ
carbon in isoprenyllithium in cyclopentane at 20°. Major
peak-cis isomer, minor peak-trans isomer.

b) Influence of DIPIP/Li ratio on the relative proportions of
above peaks when split by cooling to -20°. a,b,c components
of major cis peak, d peak due to trans isomer.

course a possibility that the complex is in fact two dimers of isoprenyl-lithium bridged by the bidentate DIPIP molecule, but it would still act effectively as a tetramer.

The next peak (c), at 90.2 ppm, has its maximum concentration at about a DIPIP/Li ratio of 0.5 before it starts to decline slowly. By the same reasoning a singly solvated dimer may be assumed and apparently this cis dimer at these ratios of DIPIP/Li does not break down to give a solvated cis monomer active chain end. This species is the major species present and must be reponsible for the major UV absorption shown in Figure 2 at 276 nm to account for the high optical density found, assuming a normal extinction coefficient. The relatively small change in the UV absorption maximum and the minor upfield shift of the γ peak in the NMR spectrum, are both considered characteristic of a dimer. The trans form however does appear to favor a solvated monomeric active chain end with its UV and NMR absorptions comparable to those of the TMEDA and THF solvates. The slow increase in the proportion of the trans form on increasing the DIPIP ratio further is then caused by the mass action effect in the equation.

$$\text{cis D.d.} + \text{d} \rightleftharpoons 2 \text{ trans M.d}$$

The rate of cis/trans isomerization with a lithium counterion, although slow on the NMR timescale, does allow equilibrium to be established fairly rapidly. It is interesting that mixed cis/trans dimers do not appear at the higher ratios of DIPIP/Li.

Kinetics

The most complete study has been made on the system DIPIP/polybutadienyllithium. The addition of quite small amounts of DIPIP to the polymerization of butadiene in hydrocarbon solvents causes substantial increases in propagation rate and in the vinyl content of the polymers, Table 4. The same is true over a range of initial lithium concentrations (Figure 5).

Table 4. Variation of rate of propagation and microstructure of polybutadiene with DIPIP/Li ratio

DIPIP/Li Ratio	$k_p \times 10^6$ [a]	% vinyl
0	2.59	11
0.095	5.33	62
0.190	8.11	77
0.284	12.0	82
0.497	17.9	89
0.663	24.7	93
0.948	40.8	97
1.42	55.2	98
1.90	59.2	99

[a] At an active chain end concentration of 1.2×10^{-3}M, 5°C.

Because of the low kinetic order of the propagation rate in the chain end concentration for the unsolvated chain ends, as these chain ends are depleted by solvation this rate will not fall markedly at low

DIPIP/Li ratios. Thus at low DIPIP/Li ratios, <1, making the assumption that all the reaction from the portion of the reaction rate in excess of the above unsolvated rate is producing completely vinyl polymer, it is possible to calculate the vinyl content of the polymer and a good correlation is found with experiment. Also if the excess rate is plotted against DIPIP concentration, for a wide range of lithium concentration a linear correlation is found, indicating that the DIPIP is at these low concentrations rate determining. At DIPIP/Li ratios in the range 3-6, at 10^{-3}M in Li, the rate no longer depends on this ratio, and evidently solvation is complete. At this point the rate is first order in chain ends. This rate constant may be used to calculate the concentration of the chain end which is vinyl-active at intermediate DIPIP/Li ratios assuming only one such active species forms. However, if this value is inserted into equations for a simple one-stage dissociation of the initial active chain end complex, no consistent value of the dissociation content over the whole range of concentrations covered experimentally can be found for any likely scheme. To fit the experimental kinetic results it is necessary to postulate more than one active complexed chain end. It is in fact possible to computer fit the results to scheme 2.

$$\text{Scheme 2} \qquad T + 2d \xrightleftharpoons{K_1} 2\ D.d \xrightarrow{k_1} \text{polymer}$$
$$Dd + d \xrightleftharpoons{K_2} 2\ M.d \xrightarrow{k_2} \text{polymer}$$

Because this involves two disposable association constants as well two rate constants, the flexibility this allows permits only to say this scheme is possible. The more complex scheme 1 is of course also possible as that has even more disposable constants. It is even possible to allow the intermediate complex to give polymer with high but less than 100% vinyl content. In view of the extra evidence from the UV and NMR spectra it is taken that scheme 1 is the effective mechanism.

Some kinetic measurements have also been made on isoprene systems (13,5), and the behaviour is somewhat different in that there is not the immediate increase in rate of propagation and change in microstructure with very small ratios of complexing agents to Li. This difference in behaviour can be attributed to the higher rate of polymerization of isoprene in the absence of DIPIP or TMEDA, because in the presence of an excess of these amines the rates are not changed greatly. Whereas in the case of butadiene the rates in the presence of the complexing agents are comparable to those of isoprene, but because the uncomplexed rate is very much lower a large increase in rate is observed. This explains the difference in shape of the curves for butadiene and isoprene in Figure 3.

Hence in general all three types of evidence concur with scheme 1. This requires that the principal active agent is a mono-solvated monomeric active chain end which gives a high vinyl polymer, nearly 100% in the case of DIPIP with butadiene. But also that an intermediate solvated dimer can also propagate, although with not necessarily as high a stereospecificity. It is unusual that an aggregated species should be called upon to propagate, and it is suggested that the bidentate chelating agent is a special case. Not only is it possible to postulate a solvated dimer, but also a form in which two separate monomeric active chain ends could complex with either end of the diamine. This would give a species that could react as if it was a chain end solvated with a single amine, but kinetically would behave as a solvated dimer reacting. The structure of the polymer formed, however, would be expected to be similar to a chain end solvated with a normal tertiary amine, with a less highly vinyl structure. But at the higher amine/Li ratios the highly vinyl directed solvated monomeric chain end would take over.

EXPERIMENTAL

All reactions were performed under vacuum. The kinetic and UV measurements were made in vessels prewashed with butyllithium under vacuum.[14] The reagents were added via breakseals or fragile bulbs. The reaction rates were followed by UV spectroscopy and the polymer structures were determined by NMR. The hydrocarbon solvents were washed with concentrated sulphuric acid, passed over activated silica gel, and stored on the vacuum line on calcium hydride. Before use they were distilled in the vacuum apparatus from butyllithium. The monomers were pretreated with and distilled from butyllithium on the vacuum line before use. The amines were distilled from calcium hydride.

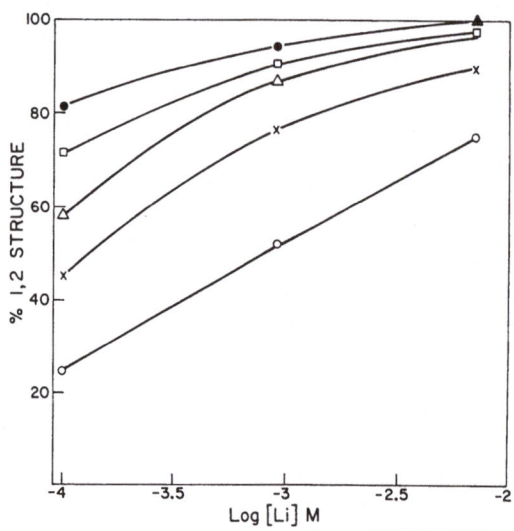

FIGURE 5. Change in vinyl content of polybutadiene with varying Li concentration and different DIPIP/Li ratios. DIPIP/Li molar ratios from bottom to top 0.1, 0.31, 0.50, 0.73, 1.1.

REFERENCES

1. F.W. Stavely and co-workers, Ind. Eng. Chem., 48, 778 (1956).
2. D.J. Worsfold and S. Bywater, Macromolecules, 11, 582 (1978).
3. R. Salle and Q-T. Pham, J. Polym. Sci. Polym. Chem. Ed., 15, 1799 (1977).
4. D.J. Worsfold and S. Bywater, Can. J. Chem., 42, 2884 (1964).
5. S. Bywater, to be published.
6. M. Morton and J.R. Rupert, Initiation of Polymerization, F.E. Bailey, Jr., ed., ACS Symp. Ser. Am. Chem. Soc., Washington, D.C., 212, 283 (1983).
7. A.F. Halasa, D.N. Schulz, D.P. Tate, and V.D. Mochel, Adv. Organomet. Chem., 18, 55 (1980).
8. A.F. Halasa, D.F. Lohr and J. Hall, J. Polym. Sci. Polym. Chem. Ed., 19, 1357 (1981).

9. S. Bywater, Y. Firat, and P.E. Black, J. Polym. Sci. Polym. Chem. Ed., $\underline{22}$, 669 (1984).
10. S. Bywater and D.J. Worsfold, J. Organometal. Chem., $\underline{159}$, 229 (1978).
11. A. Garton and S. Bywater, Macromolecules, $\underline{8}$, 694 (1975).
12. A.W. Langer, Polym. Prepr., Div. of Polymer Chemistry, Am. Chem. Soc., $\underline{7}$(1), 132 (1966).
13. V. Collet-Marti, S. Dumas, J. Sledz, and F. Schué, Macromolecules, $\underline{15}$, 251 (1982).
14. S. Bywater, D.H. MacKerron, and D.J. Worsfold, J. Polym. Sci. Polym. Chem. Ed., $\underline{23}$, 1997 (1985).

POLYURETHANE ELASTOMERS WITH MONODISPERSE SEGMENTS AND THEIR MODEL PRECURSORS: SYNTHESIS AND PROPERTIES

Claus D. Eisenbach, Martin Baumgartner, and Claudia Günter

Polymer-Institute, University of Karlsruhe
P.O.Box 6980, D-7500 Karlsruhe 1, FRG

ABSTRACT

Uniform model compounds for the soft and hard segment of polyurethane (PU) elastomers and the corresponding tailor-made elastomers with monodisperse segment length distribution were synthesized and characterized with the objective of getting a better understanding of the structure and morphology as well as the structure-property relationships of multiphase segmented PU elastomers from the study of such clearly defined model systems.

Monodisperse α-hydro-ω-hydroxypoly(oxytetramethylene) and hydroxy-terminated oligomers of 1,4-butanediol and 4,4'-diphenylmethanediisocyanate (soft and hard segment precursors) as well as their bis-(diphenylmethaneurethanes) (soft and hard segment model compounds) were obtained in a stepwise synthesis and in combination with chromatographic fractionation techniques, and in some cases (oligourethanes) by employing the tetrahydropyranyl protecting group. The hydroxy-terminated segment precursors were reacted to the PU elastomers with strictly monodisperse hard segments and narrow soft segment length distribution by a modified prepolymer process which largely suppresses the pre-extension reaction; it was established that the prepolymer formation in the melt doesn't obey Flory statistics and that the deviations also vary with the molecular weight of the starting polyether.

The soft segment model compounds with terminal urethane groups exist in two crystalline modifications with different hydrogen bond strength between neighbouring urethane groups. Polymorphism is also observed in the case of the hydroxy-terminated hard segment precursors and is connected with deviations from the planar zig-zag conformation of the butanediol urethane part. There is no evidence for chain folding, and cocrystallization of oligourethanes of varying length only occurs between oligomers of successive length and not until a critical chain length of three repeat units is exceeded. Furthermore, the urethane group has been found to undergo a rapid transurethanization (urethane interchange) reaction already in the oligourethane melting temperature range and even without any catalyst, rendering monodisperse into polydisperse systems. Consequently, the hard segment length distribution and thus the hard domain morphology of segmented PU elastomers are affected by the thermal history in that the usual processing temperatures result in an equilibration of the segment length distribution and thus a reorganization of the hard segments.

PU elastomers with a narrow (monodisperse) hard segment length distribution show a better phase separation and higher degree of crystallinity in the hard domains as compared to elastomers with broad segment length distribution, even though the structure and morphology of the semicrystalline hard domaines is still complex and not yet fully understood. Concerning the differences in the mechanical properties between elastomers with narrow and broad segment length distributions it can be stated that monodisperse hard segments generally result in improved properties, e.g., a higher and flatter plateau modulus over a wider temperature range, and a higher softening temperature as compared to elastomers with polydisperse hard segment length distribution are characteristic for the dynamic mechanical behaviour; the soft segment length distribution has a comparatively minor effect and only the average soft segment length is of some importance in connection with the possibility of soft segment crystallization.

INTRODUCTION

Polyurethane (PU) elastomers are multiblock-copolymers $\{AB\}_n$ with an alternating sequence of so-called soft and hard segments. The distinct differences in the chemical structure of the segments A (polyether or polyester) and B (polyurethane) contribute to segmental incompatibility, which results in forming a multiphase polymer system upon cooling from the melt or solvent evaporation, as sketched schematically in Figure 1. The morphology can be portrayed in simplified terms by assuming only two phases[1,2]: segregated polyurethane segments form domains in a more or less continuous soft segment phase, and these so-called hard domains act as multifunctional, thermoreversible crosslinks, and as fillers. The elastic properties of these materials are based on the differences in the glass transition temperature (T_g) and melting temperature (T_m) of the two phases; the working temperature range is limited towards low temperatures by the soft phase T_g, and - for crystallizable hard segments - the upper temperature limit is given by the hard domain T_m.

It is quite obvious that this class of thermoplastic elastomers is not only interesting from processing considerations, but also because of the potential possibilities to adjust the material properties to different demands simply by varying the primary structure and average molecular weight of the segments, both of them influencing the superstructure of the system. On a microscopic view and on a molecular scale the basic controlling factors certainly are the packing order of the hard segments (two possibilities are contrasted in Figure 1), the dimensions of the domains and interface region as well as the pollution of either phase with the respective thermodynamically incompatible segment, the extent of which being related to the relative difference in the interaction parameters of the two types of segments and their segment length distribution.

However, a detailed knowledge is lacking regarding the structure/property relationships in these physically cross-linked elastomers, especially as far as the size and shape of hard domains, the interphase between the hard and soft phase,

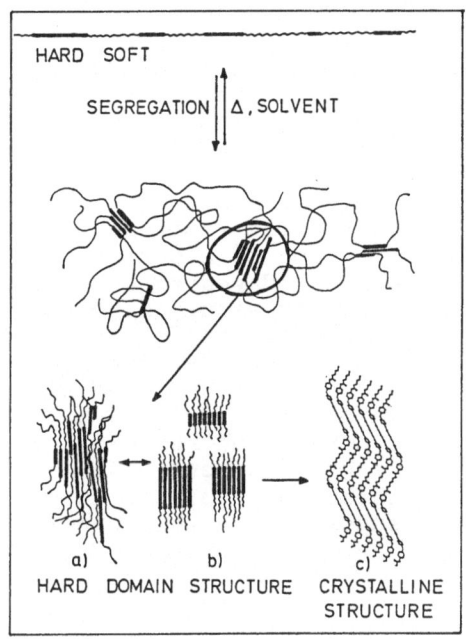

Fig. 1. Schematic representation of the segregation and domain structure of segmented PU elastomers; a: random hard segment segregation; b: fractionation after hard segment length; c: crystalline structure of hard segment phase as proposed by Blackwell[5] and Hespe.[6]

the conformations of the entire polymer chains and particularly of the hard segments, the role of hydrogen bond formation, and last but not least the segment size and polydispersity are concerned. In this context it has to be emphasized that this gap has to be ascribed essentially to the complex and only roughly known primary structure of these multiblock copolymers which originates in the statistics of the polyaddition reaction of three components (hydroxy-terminated polyether or polyester, diisocyanate and diol). It is obvious from the reaction scheme in Figure 2 representing the two-step method of segmented PU elastomer synthesis that, e.g., the extent of the so-called pre-chain extension reaction given by the numeral y not only changes the average soft segment length (polyether) but also effects the average hard segment length through the corresponding fraction of unreacted diisocyanate, not to mention the resulting sequence length distribution which is obscure as well. There are various approaches to solve these still open questions and to relate these parameters to polymer properties,[2-17] but we are convinced that only the synthesis and study of well characterized, ideally monodisperse model systems and analogous segmented polymers will lead to a better understanding of this matter.[18] This will also allow us to establish correlations between the segment sizes or segment length distribution and the polymer superstructure and properties instead of having to relate this to the total diiso-cyanate (hard segment) content, and finally facilitate a con-trolled synthesis of PU elastomers with tailored properties.

A laborious method for the synthesis and isolation of high-ly pure oligourethanes was first reported by Kern et al.[19] The synthesis and characterization of a series of PU-elastomers with strictly monodisperse hard segments were first reported by Harrell[20] (cf.[21]). Although these urethane hard segments based on piperazine and butanediolbischloroformate differ from the con-ventional diisocyanate/diol based systems by the lack of hydro-gen bond formation, they can be considered as model systems for multiblock copolymers in general. Following this view, we have further developed the synthesis of these non-hydrogen bond-forming oligourethanes and elastomers;[22-23] the results of this work[23,24] will not be included here, but it will be evident from the subsequent discussions in this paper that these sy-

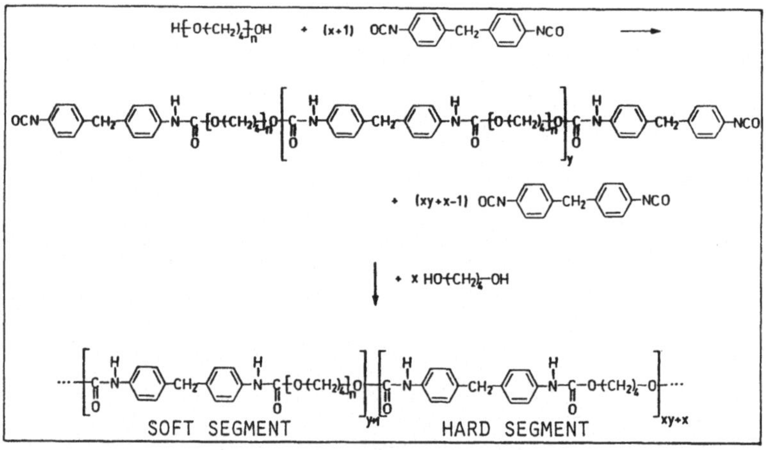

Fig. 2. Synthesis of segmented PU elastomers via the pre-polymer (two-step) procedure.

stems may be advantageous in some respects for particular model experiments.

The problem of synthesizing hydrogen bond-forming urethane hard segment model compounds on the basis of 4,4'-diphenylmethanediisocyanate (MDI) and 1,4-butanediol (BDO) as well as the corresponding polyether-polyurethane block copolymers with narrow hard segment length distribution has been tackled in several laboratories,[25-29] and more or less simultaneously with our own work in this area.[18,23,30-32] However, some of these so-called model systems have not been properly characterized regarding purity and segment length dispersity. Recently, the synthesis of monodisperse oligomers of BDO and 2,4-toluenediisocyanate and their use for elastomer synthesis have also been reported.[33-34]

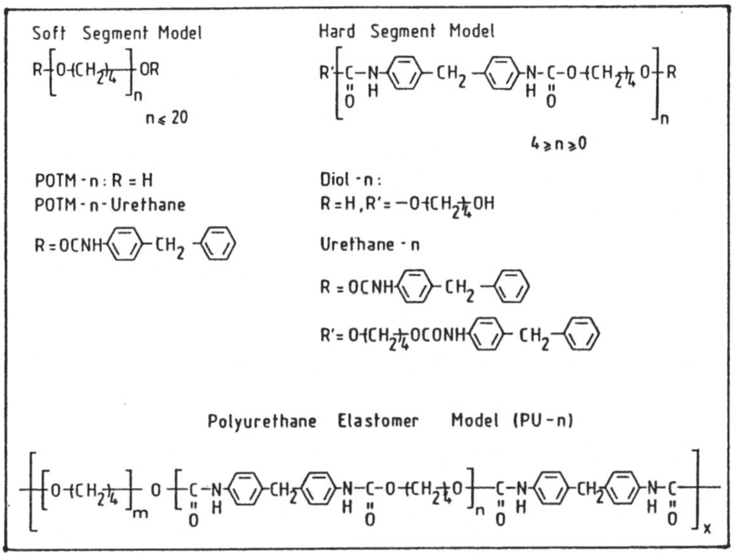

Fig. 3. Chemical structures of the oligoether soft segment and oligourethane hard segment model compounds, and of the segmented PU elastomers made from these segments.

In this paper, we will shortly discuss our approach to prepare hydroxy-terminated MDI/BDO-based urethane oligomers and their 4-diphenylmethane carbamates,[18] and describe in more detail the synthesis of monodisperse hydroxy-terminated oligo(oxytetramethylenes), their use for the investigation of the prepolymer formation, and the synthesis of linear PU-elastomers with narrow (monodisperse) soft and/or hard segment length distribution by using the hydroxy-terminated segment precursors. The chemical structures of the model systems studied in this work are represented in Figure 3. Some of the characteristics of these urethane and ether oligomers and of the corresponding segmented PU elastomers will be discussed and their implications for the morphology and properties of this class of multiblock copolymers will be shown.

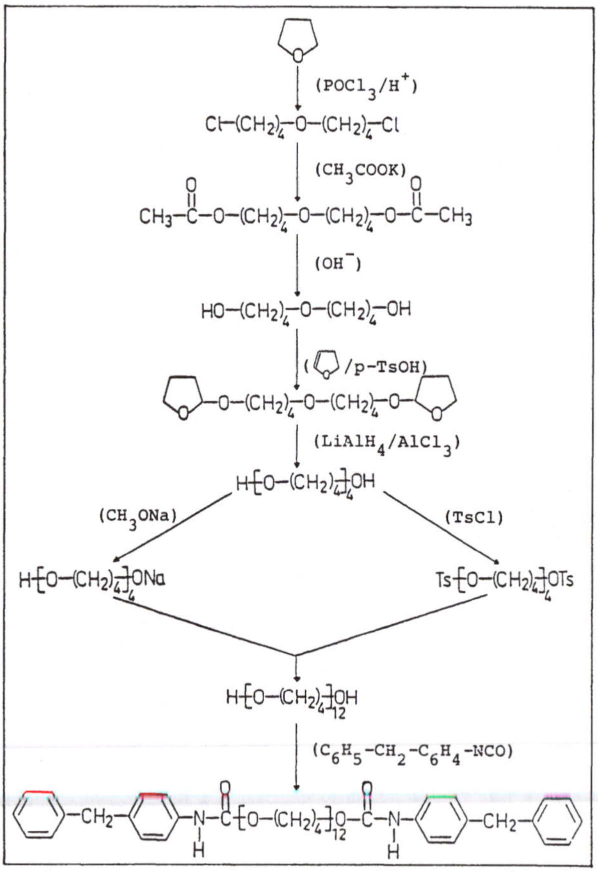

Fig. 4. Synthesis of monodisperse α-hydro-ω-hydroxypoly-(oxytetramethylene) (POTM-n) and their diphenyl-methane-4-carbamate derivatives (POTM-n-Urethane).

EXPERIMENTAL

Materials

α-Hydro-ω-hydroxypoly(oxytetramethylene) soft segment precursors (POTM-n) and the corresponding diphenylmethane-4-isocyanate-endcapped POTM-n-Urethane soft segment model compounds were synthesized in a multistep procedure as shown in Figure 4 for the example of the dodecamer POTM-12. α-Hydro-ω-hydroxytetra(oxytetramethylene), the actual starting material, was obtained by an improved synthesis[35] from tetrahydrofuran (THF) via 4,4'-dichlorodibutylether, 4,4'-diacetoxydibutylether, 4,4'-dihydroxydibutylether and reductive cleavage of the bisacetal 4,4'-bis(2-tetrahydrofuryloxy)dibutylether following the route described by Bill et al.[36] The tetramer (POTM-4) was converted to the bistosylate by reaction with p-toluenesulfonyl chloride. This pure bistosylate (0.01 mole) was reacted in THF at 40°C with two parts of the monosodium alcoholate of POTM-4 and two parts of POTM-4 diol for 52 hrs, yielding 10.6 g (88% conversion) of an oligomer mixture containing 35 per cent of the dodecamer. After saponification of traces of unreacted sulfonic acid ester and removal of sodium sulfonate, the α,ω-bishydroxyoligoether mixture (Figure 5, curve 2) was fractionated in 2-3 g portions over a Waters Prep-Pak-500/C_{18} column first with absolute methanol and then with methanol/water (80/20 v/v) as eluent.

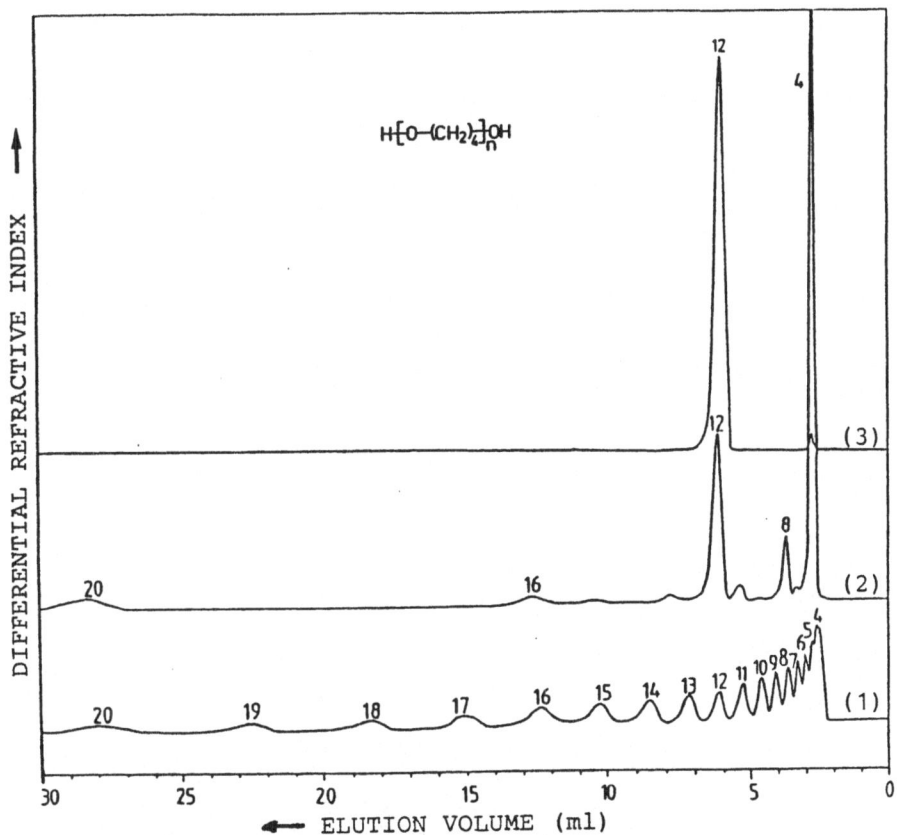

Fig. 5. Reversed Phase (RP)-HPLC analysis of commercial POTM-14 (1, number average molecular weight 1000), α,ω-bishydroxy-polyether reaction products POTM-n (2), and pure α-hydro-ω-hydroxydodeca(oxytetramethylene), POTM-12 (3).

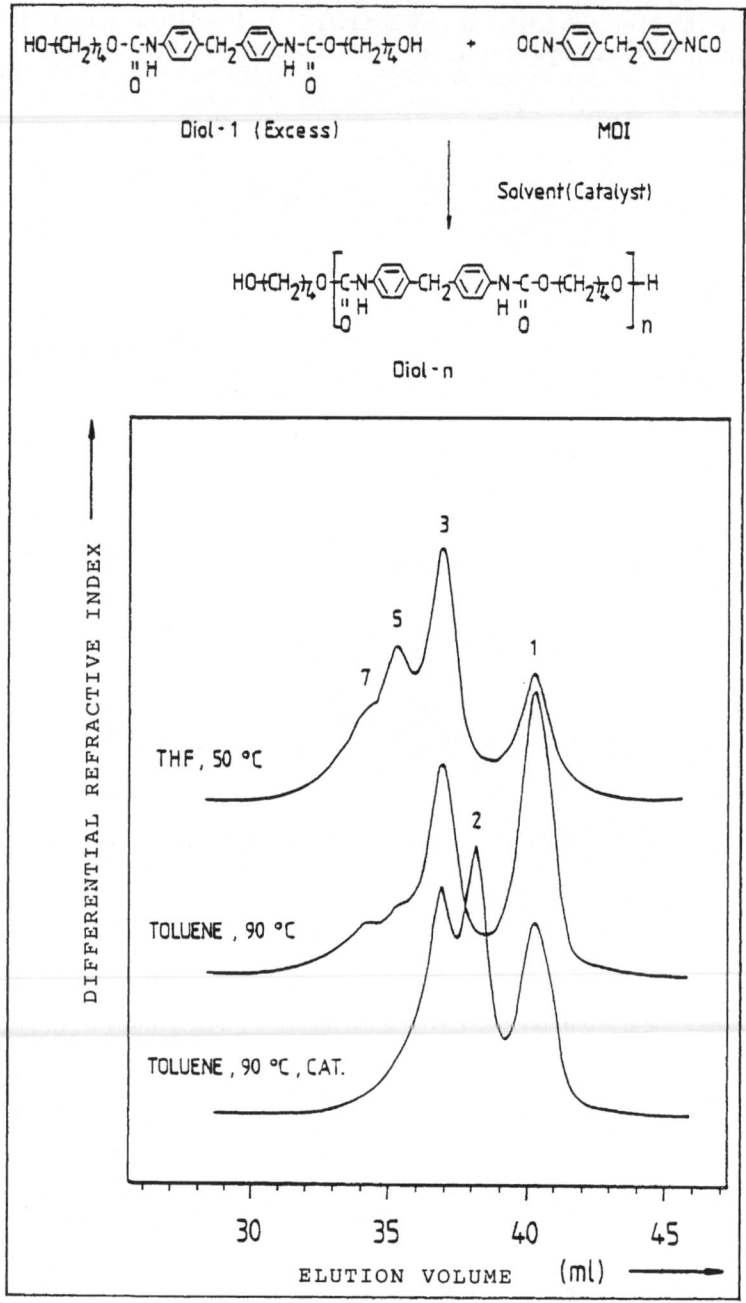

Fig. 6. Synthesis and HPLC analysis of hydroxy-terminated oligomers (Diol-n)
of 1,4-butanediol (BDO) and 4,4'-di-
phenylmethanediisocyanate (MDI);
starting materials: Diol-1 and MDI;
reaction variables indicated; the
peak marking means the numeral n of
the corresponding Diol-n.

The yield of the ultrapure dodecamer (Figure 5, curve 3; $T_m=43^\circ$C) isolated after the second preparative reversed phase (RP)-HPLC run was about 30 % of the charged crude oligomer mixture. The other pure oligomers, namely the octamer ($T_m=308$ K), the hexadecamer ($T_m=319$ K), and the eicosamer ($T_m=320$ K) could also be isolated by the same fractionation process, and most of the unconverted tetramer starting material POTM-4 could be recovered.

The α,ω-bis(4-diphenylmethaneurethanes) of the poly(oxytetramethylene)diols, POTM-n-Urethane, were obtained by reacting the corresponding POTM-n with 20 mole % excess of diphenylmethane-4-isocyanate in THF solution and recrystallization from ethanol.

4,4'-Diphenylmethanediisocyanate (MDI)-1,4-butanediol (BDO) hydroxy-terminated hard segment precursors (Diol-n) and the corresponding diphenylmethane-4-isocyanate-endcapped derivatives (Urethane-n) were synthesiszed in a step-by-step procedure using tetrahydropyranyl ether as protecting group for one hydroxyl group of BDO, Diol-1 or Diol-2 in the reaction with MDI, or by using an excess of one of the reaction components.[18,30]

Diol-2 and Diol-3 were conveniently synthesized by reacting MDI (0.9 mmole) with excess Diol-1 (3.6 mmole) in 180 ml toluene in the presence of dibutyltin dilaurate catalyst (0.001-0.1 mmole) at 90°C for 2 h (Figure 6). After evaporation of the solvent, unreacted Diol-1 was removed from the residue by repeated extraction with cold methanol (ca. 0.6 g pure Diol-1 recoverable); the remaining residue was boiled with methanol,

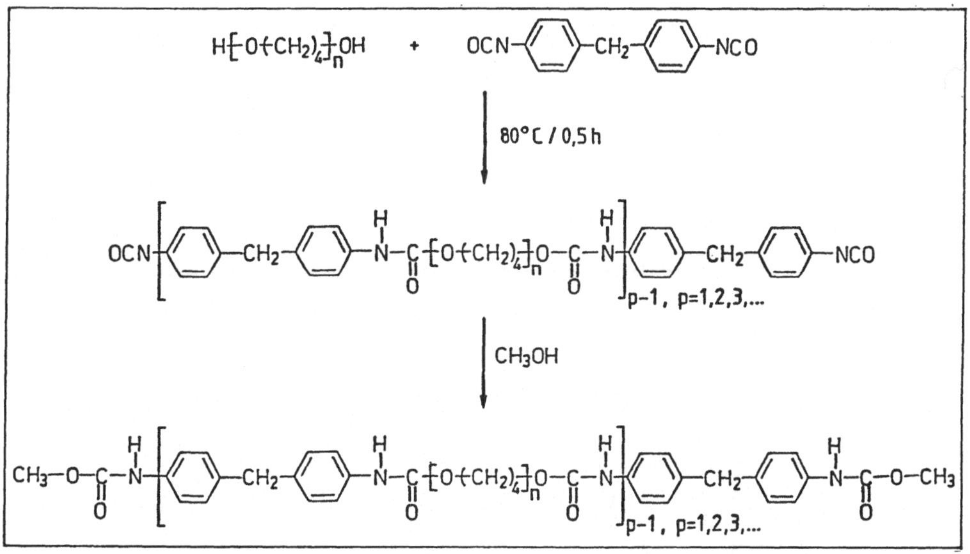

Fig. 7. Reaction scheme for the prepolymer formation and conversion of the macrodiisocyanate into non-reactive methylurethane.

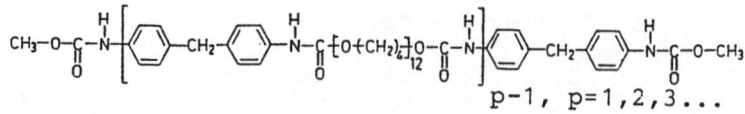

p-1, p=1,2,3...

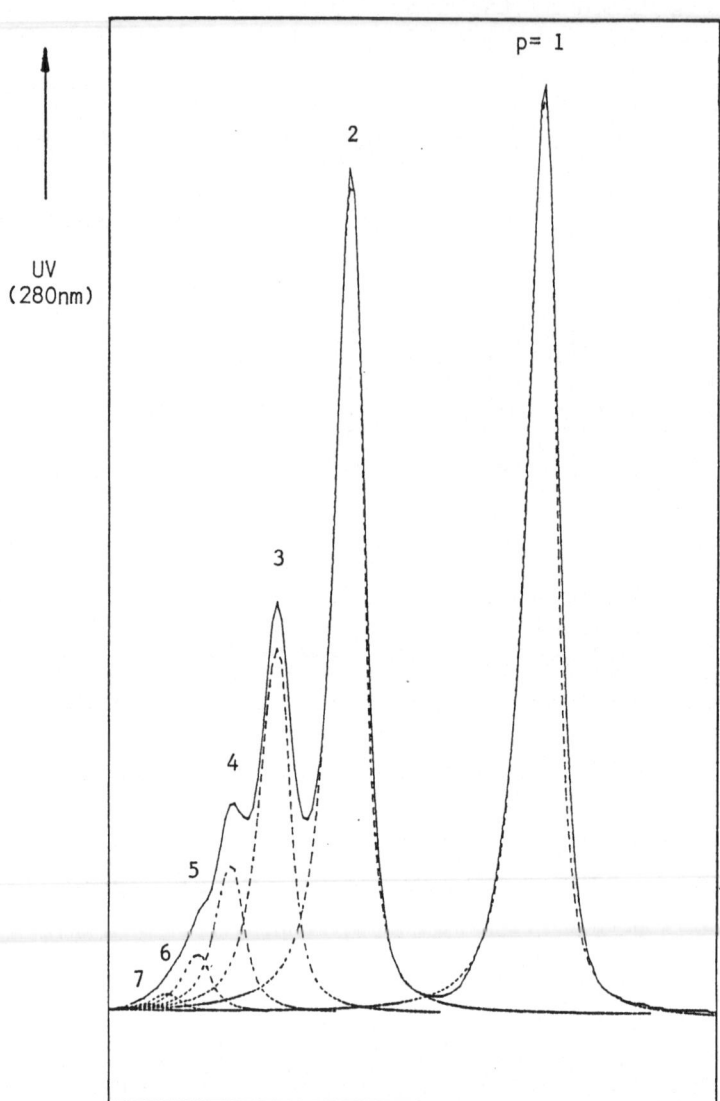

Fig. 8. HPLC analysis of the prepolymer of α-hy-
dro-ω-hydroxydodeca(oxytetramethylene)
(POTM-12) and MDI (molar ratio 1:3) after
conversion into methylurethane (cf. scheme
Fig. 7); Peak 1: p=1 (dimethyl-4,4'-di-
phenylmethanedicarbamate); peak 2-7:
POTM-12-MDI polyaddition products (p=2,
3,...6).

and 0.4 g Diol-2 (equilibrium melting temperature, T_m^o=468 K) precipitated from the filtrate upon cooling. The methanol-insoluble residue was dissolved in N,N-dimethylacetamide (DMA) and poured into boiling methanol; the yield of pure Diol-3 (T_m^o=486 K) was 0.5 g (51.2% with regard to employed MDI).

The bis(4-diphenylmethaneurethanes), Urethane-n, were either synthesized by a separate method starting from 4-hydroxybutyldiphenylmethane-4-carbamate[18,30] or similar to POTM-n-Urethane synthesis by reacting the Diol-n with 4-diphenylmethaneisocyanate.

The prepolymer formation (Scheme Figure 7) was investigated with the α-hydro-ω-hydroxyoligo(oxytetramethylene) tetramer and dodecamer polyols (0.001-0.01 mole) and MDI in the melt at 80°C under argon; the molar ratio MDI/polyol was varied from 2 to 10. The tetramer and dodecamer oligomers were dried in the reaction vessel first in the melt at 80° for 48 hrs in high vacuum, and then freeze dried twice from benzene solution; MDI was vacuum distilled twice and always distilled again under argon prior to use. All samples were reacted with absolute methanol prior to HPLC-analysis (Figure 8) in order to convert unreacted isocyanate groups to methylcarbamates. The mole fractions of unreacted MDI (p=1) and of the polyaddition products were determined from the areas of the resolved HPLC traces.

Polyether-polyurethane block copolymers with polydisperse segment length distribution (PU-n̄, see Reaction Scheme in Figure 2): α-hydro-ω-hydroxypoly(oxytetramethylene), POTM-$\overline{28}$ (2.5 mmole, $\bar{M}_n$=2000, obtained from Bayer AG) was dried in the reaction vessel as described above for the monodisperse oligomer and reacted with MDI in molar ratios of 1:n (n=2-7; n=x+1 in Figure 2) in bulk at 80°C for 30 min. under Ar. The prepolymer was dissolved in DMA (dried over CaH$_2$ and vacuum distilled) and extended with the appropriate amount of BDO (distilled over Na). The course of the chain extension reaction was determined by GPC analyses of samples taken from the reaction mixture and reacted with methanol (Figure 7).

Model PU elastomers with monodisperse hard segment length distribution (PU-n) were prepared in a modified two-step method (Figure 9) by employing a macrodiisocyanate prepolymer completely free of MDI and the above mentioned monodisperse Diol-n oligomers. The reactants were purified as described above, and the preformed hard segment units Diol-n used in the chain extension reaction were dried in a vacuum oven at 40°C for 12 hrs. POTM-$\overline{28}$ (~3 mmoles) was reacted for 30 min at 80°C in the melt with a ten fold excess of MDI. The unreacted MDI was removed from the macrodiisocyanate by repeated extraction with 30 ml dry cyclohexane; the extent of MDI-extraction was checked by GPC analysis of samples reacted with methanol (Figure 10a). After evaporation of cyclohexane, the MDI-free prepolymer was dissolved in 10 ml DMA and the extender (BDO, Diol-n; 0.2 molar in DMA) was added gradually in order to balance the isocyanate groups of the prepolymer. The progress of the chain extension was again followed by GPC analysis of samples taken from the reaction mixture (Figure 10b); the elastomers were obtained in 2-4 g quantities per experiment.

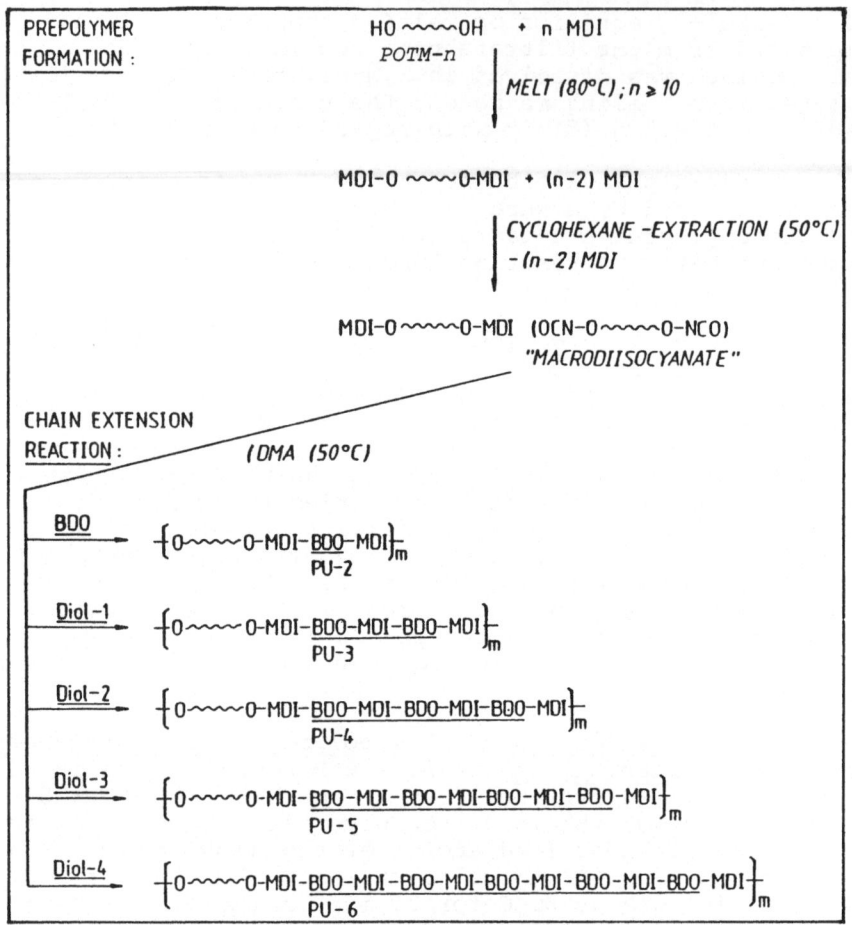

Fig. 9. Reaction scheme for the synthesis of PU elastomers with strictly monodisperse hard segments
and narrow soft segment length distribution
made with hydroxy-terminated polyurethanes
Diol-n and polyethers POTM-n (see Figure 3).

The elastomers were purified by precipitation of DMA solutions with excess methanol. Polymer films were cast from DMA
solution and dried in vacuum at 50°C for 8 days and stored in
a dessicator at room temperature.

Characterization

Chromatographic analysis and fractionation of the oligomeric and polymeric samples were carried out by applying High
Pressure Liquid Chromatography (HPLC) and Reversed Phase (RP)-
HPLC. The model compounds and the elastomers in DMA solution
were analysed at 74°C with DMA as eluent (flow rate: 1 ml/min)
using a Knaur High Temperature Gelchromatograph FR 30, equipped
with 6 ultrastyragel columns ($50, 10^2, 500, 10^3, 10^4, 10^5$Å, Waters
Associates) and a Knaur refractive index (RI-) or UV-detector
(variable wave length: $\lambda=280$ nm used for the prepolymer analysis. The polyether model compounds were further analysed using
a similar instrument equipped with a Merck-Lichrosorb RP-18

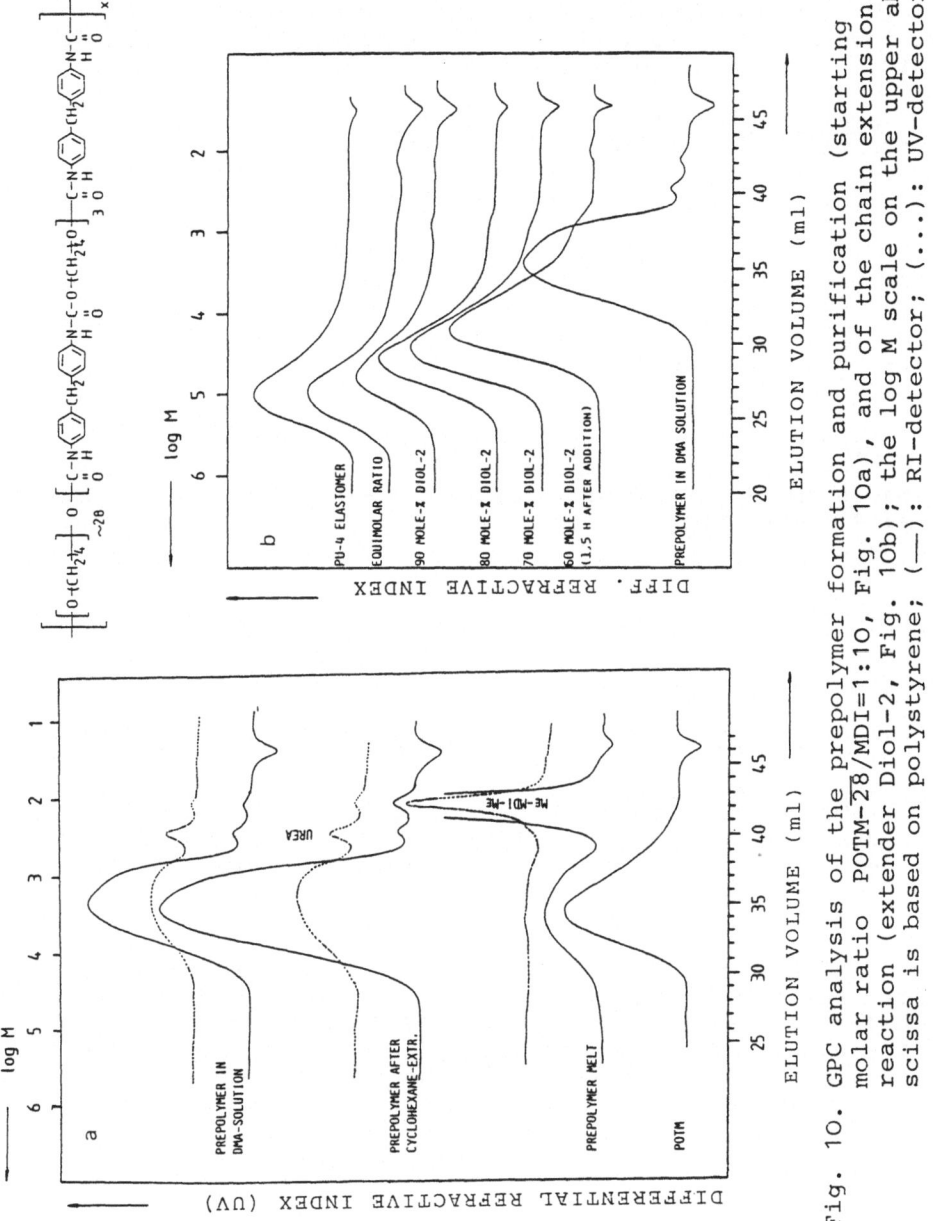

Fig. 10. GPC analysis of the prepolymer formation and purification (starting molar ratio POTM-$\overline{28}$/MDI=1:10, Fig. 10a), and of the chain extension reaction (extender Diol-2, Fig. 10b); the log M scale on the upper abscissa is based on polystyrene; (——): RI-detector; (...): UV-detector.

column (methanol/water 90/10 v/v as solvent, flow rate 1 ml/min, room temperature). The preparative scale fractionation of the POTM-n oligomer mixtures were performed with a Waters Model Prep LC/System 500 equipped with a Waters Prep Pak-500/C18 column (solvent: methanol and methanol/water; flow rate 150-250 ml/min; room temperature).

Infrared (IR) spectra were recorded with a Bruker IFS 113 V FTIR spectrometer. The spectra of the model oligomers were obtained using KBr-pellets or bulk samples (melt). Thin films cast from DMA solution were used for elastomer analysis.

Small angle X-ray scattering (SAXS) data of the model oligomers were obtained from microcrystalline samples with a Kiessig-camera (sample-film distance 170 mm) by using Ni-filtered CuK_α radiation (40 KV, 35 mA, λ=1,5418 $\mathring{A}$).

Differential Scanning Colorimetry (DSC) was performed with a Perkin-Elmer DSC II; the heating rate varied from 2.5-20 K/min, and the sample weight was 1-10 mg. Equilibrium melting points, T_m^o, were obtained by extrapolation to zero heating rate. Gallium, azobenzene, acetanilide and Indium were used as calibration standards.

Dynamic mechanical measurements were carried out with a Lonza Torsion Pendulum by using solution (DMA)-cast films of ca. 35x5x0.5 mm size; (temperature range: -130 to +200°C; heating rate: 1 K/min; frequency: 1 Hz).

RESULTS AND DISCUSSION

Preparative Aspects

Monodisperse poly(oxytetramethylene) soft segment precursors (POTM-n, Figure 3) were prepared by Williamson's ether synthesis method by using α-hydro-ω-hydroxytetra(oxytetramethylene)(POTM-4) as the starting material[31,35] which can be easily synthesized in large quantities.[36] The principal difference between this method and the one described by Bill et al.[36] is that we start from the tetramer, i.e., the monosodium alcoholate and the bistosylate of POTM-4 (Figure 4), and convert them in a controlled polycondensation reaction into a homologous series of POTM-n which differ by four oxytetramethylene repeat units. The main reaction product depends on the molar ratio of POTM-4/POTM-4 monosodium alcoholat/POTM-4 bistosylate (see Experimental Section and Figure 5). It has to be emphasized that in this way the omnipresent question in oligomer fractionation of how to separate oligomers differing only by one repeat unit in length (curve 1, Figure 5) is reduced to the problem of having to fractionate oligomers differing by four repeat units.[37]

The most efficient fractionation technique is the reversed phase chromatography which allows the separation of highly pure oligomers in gram-quantities in only two fractionation runs; this is demonstrated with the RP-HPLC-traces in Figure 5, representing the analytical chromatograms of the polycondensate (curve 2) and of the pure dodecamer POTM-12 (curve 3). It is also evident that the RP-chromatography easily allows the isolation of the higher oligomers since the differences in the elution volume increase with increasing chain length (curve 1, Figure 5) , in contrast to conventional gel permeation chromatography (GPC). The only practical limitation is the increasing amount of eluent required.

Monodisperse polyurethane hard segment precursors (Diol-n), i.e., hydroxy-terminated oligomers of 1,4-butanediol (BDO) and 4,4'-diphenylmethanediisocyanate (MDI) were prepared in a step-by-step synthesis using the tetrahydropyranyl (THP) group to protect hydroxy functions, or by using one of the reaction compounds (diol, or diisocyanate) in excess[38] in order to suppress the polyaddition reaction.[18,30] The usefulness of a trityl-protecting group has also been demonstrated in the synthesis of the very same model urethane compounds, although the experimental approach to the hydroxy-terminated oligomers was quite different.[26,28]

The employment of a protecting group is not necessarily required, and in this particular case the reaction of non-stoichiometric amounts of bifunctional components, Diol-1 and MDI, under suitable reaction conditions can even be advantageous: The use of a catalyst (dibutyltin dilaurate) and the choice of a reaction medium (toluene) in which the higher Diol-n oligomers are completely insoluble only leads to the two next homologues Diol-2 and Diol-3 as shown in the GPC trace Figure 6 (bottom curve).[30] Without any catalyst or by using a solvent in which the higher homologues are also soluble, an appreciable fraction of the reaction product consists of higher oligomers which cannot be separated from Diol-3 by conventional fractionation techniques. It is obvious that the reaction of excess Diol-1 and MDI should only lead to the Diol-n oligomers with odd numbers, and the formation of Diol-2 in the catalyzed reaction system has to be attributed to a trans-urethanization between Diol-3 and Diol-1. This type of side reaction will be further examined in connection with the thermal properties of the polyurethane model compounds and elastomers.

Monodisperse Urethane-n oligomers, the pure models for the polyurethane hard segments, can either be synthesized by reacting Diol-n with 4-diphenylmethaneisocyanate. A second route is to first convert 4-(4-diphenylmethaneurethane)-1-butanol with an excess of MDI to the monoisocyanate-terminated analogue of Urethane-1, and then to react this unsymmetric intermediate (after having extracted the unreacted MDI) with BDO or any of the Diol-n compounds in a stoichiometric ratio, which results in the formation of Urethane-3 or the corresponding higher Urethane-n models, respectively.[18,30] The advantage of the latter method is that the bishydroxy-terminated starting material Diol-n can be converted quantitatively in a single step into a hard segment model with - technically speaking - four more MDI-units.

Segmented polyurethane (PU) elastomers can be prepared either by the one-shot[39] or by the two-step (prepolymer) method,[1,40] where it has been shown that the two-step polyaddition theoretically results in a narrower distribution of hard blocks than the single-step process of the same stoichiometry.[41] It is obvious that the one-shot process never will lead to elastomers with controlled segment length distribution and that the only way to obtain multiblock copolymers of well-defined block-length (ideally monodisperse) is to utilize a sort of two-step process.

A widely held opinion is that the segment length distribution in the two-step process follows the most probable form no matter what the reaction conditions of the polyaddition are. However, this is only reasonable to be assumed if the reaction is carried out in solution,[41,42] and there are indications in the literature[43] that more or less pronounced deviations have to be considered for the reaction in bulk. Since the detailed knowledge of the actual segment length distribution under particular reaction conditions is an essential prerequisite for the tailor-making of multiblock copolymers, we have first investigated the statistics of the prepolymer formation in bulk and based thereon developed a synthetic approach to segmented PU elastomers with controlled (monodisperse) segment length distribution. The monodisperse oxytetramethylene polyols are excellent models for the investigation of the statistics of the prepolymer formation in the two-stage elastomer synthesis, since the polydispersity of the prepolymer can be exactly determined from the GPC or RP-HPLC traces of the prepolymer.

The principles of the prepolymer formation and of the analysis of the composition of the prepolymer have already been described in the Experimental Part. The prepolymer formation in the melt was investigated with monodisperse polyols of different chain lengths and for molar ratios of MDI/polyol varying between 2 and 35. In order to facilitate the chromatogrhic analysis and to avoid side reactions of the macrodiisocyanate samples during the analytic procedure the isocyanate groups were converted into inactive methylurethanes (Figure 7) prior to the HPCL analysis. A typical HPLC trace of the prepolymer resulting from the system POTM-12 and MDI (molar ratio 1:3) is depicted in Figure 8. It is evident that the composition of the prepolymer with regard to the mole fraction of residual MDI (p=1) and of the macrodiisocyanate without (p=2) and with pre-extension (p≥3) can be determined with high accuracy from the chromatogram resolved into the individual sequences containing two terminal diisocyanate residues ("ideal macrodiisocyanate", p=2) and the pre-extended oligomers with one (p=3) or more (p≥4) internal diisocyanate units.

The unambiguous results of these model experiments are that the macrodiisocyanate prepolymer obtained in the bulk does not show a most probable molecular weight distribution, as exhibited from the histograms in Figure 11. The findings are that firstly the fraction of unreacted MDI in the prepolymer is lower (up to 10%) than that calculated by assuming Flory statistics[44] for this polyreaction (cf.41,42,45), secondly that the concentration of simple extended polyols (p=2) is proportionally much higher in relation to the multiple extended ones, and thirdly that the extent to which these deviations occur varies for each degree of oligomerization, n, of the polyol POTM-n which further complicates the situation. The lower extent of the pre-extension during the conversion of POTM-n to the macrodiisocyanate in bulk also means that the average hard segment length is shorter than that calculated by using the residual MDI in the prepolymer for Flory statistics. The above findings mean that several assumptions and findings in the literature have to be revised concerning the segment length distribution of both the hard and soft segments (an internal MDI-unit is not considered

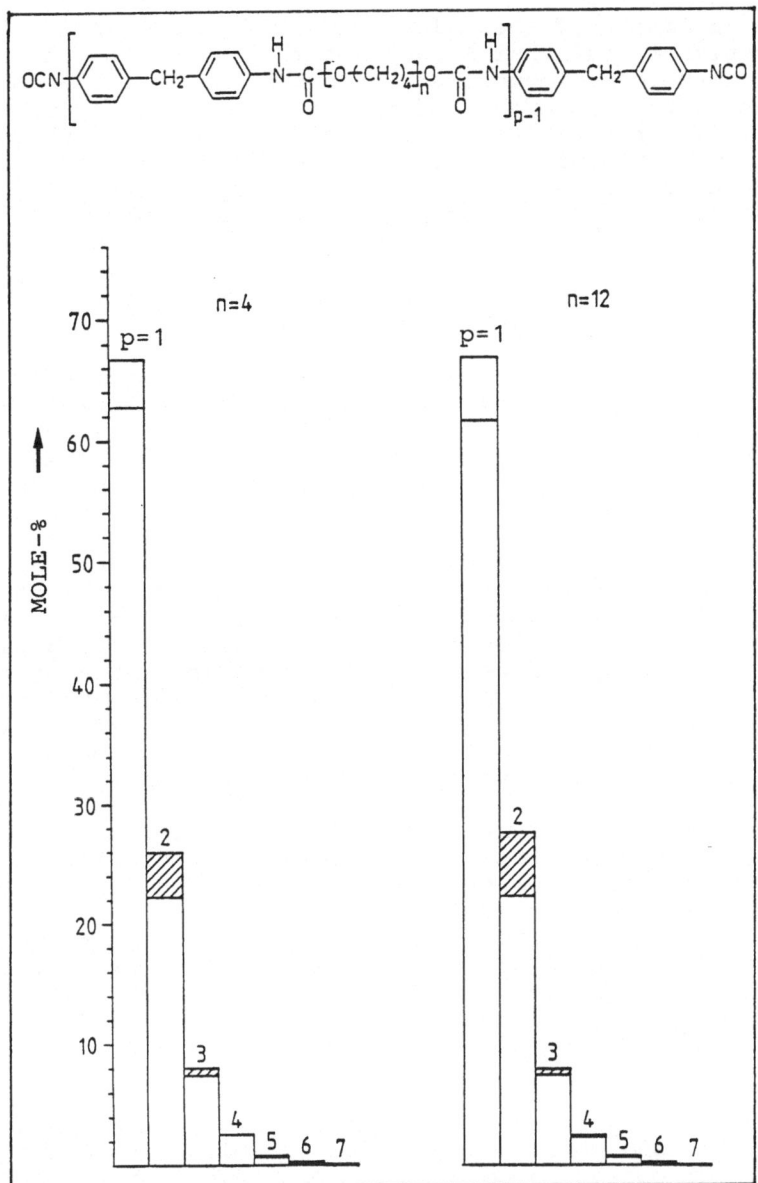

Figure 11. Histograms representing the prepolymer
composition (mole %) obtained with
POTM-4 or POTM-12 (see Fig. 8) and MDI
(molar ratio 1:3). The open cut-off
areas of the columns represent the theoretically expected additional portion,
the hatched areas the experimentally
found additional portion of (macro)di-
isocyanates in the prepolymer by assuming or comparing to most probable statistics,[41,42,45] respectively.

as a hard segment) in PU elastomers[41-43,45] and conclusions derived therefrom.

The variation of the molar ratio POTM-n/MDI in the prepolymer formation in bulk has shown that the pre-extension detectable within the experimental error becomes neglectable only at more than fifty-fold excess of MDI; however, a ten-fold excess of MDI already reduces the extension of POTM-n to such an extent that over 90% of the prepolymer consists of "ideal" macrodiisocyanate molecules, i.e., POTM-n endcapped with MDI, and only less than 10% of the prepolymer molecules are pre-extended. This is considered to be an acceptable compromise between the objective to prepare model PU elastomers with a soft segment distribution as narrow as possible and a reasonable experimental effort.

A completely MDI-free prepolymer (macrodiisocyanate) with less than 10% pre-extended macrodiisocyanate molecules can be isolated from the initially obtained bulk prepolymer by extraction of the residual MDI with cyclohexane. Since the macrodiisocyanate, in contrast to MDI, is appreciably insoluble in cyclohexane due to the urethane groups obtained as connecting links between the polyether and MDI, a two-phase system is formed upon addition of cyclohexane to the crude prepolymer: a lower macrodiisocyanate phase polluted with MDI and cyclohexane, and an upper cyclohexane phase containing mostly MDI and polluted with predominantly the longer macrodiisocyanates (originated from the high molecular weight POTM-n fraction). Six or at most ten extractions of the prepolymer give the desired MDI-free macrodiisocyanate with an even narrower molecular weight distribution than the starting POTM-n, in case a polydisperse polyol had been used, as is shown by the HPLC traces in Figure 10a.

The segmented PU elastomers with a soft segment length distribution nearly identical to the one of the starting POTM-n and a strictly monodisperse hard segment length distribution have been obtained by reacting this macrodiisocyanate in DMA solution in a quantitative reaction with stoichiometric amounts of either BDO or the hydroxy-terminated Diol-n hard segment precursors (see the reaction scheme in Figure 9). Multiblock copolymers up to a molecular weight $\overline{M}_W$ of 80,000-100,000 as determined by light scattering were prepared by this method. The conversion of the polyaddition reaction can easily be followed via HPLC analysis (Figure 10b), and this allows the synthesis of a family of segmented PU elastomers (PU-n) with monodisperse hard segments and corresponding overall molecular weight, a requirement for obtaining comparable data, e.g., in the analysis of mechanical properties.

Structure and properties

Soft segment precursors POTM-n as well as the soft segment models POTM-n-Urethane both show very narrow melt endotherms in the DSC as one should expect for monodisperse compounds. Whereas the melting temperatures of the POTM-n bishydroxyoligomers increase with increasing degree of oligomerization (extrapolation to infinite chain length after Flory-Vrij[46] yield $T_m^0 = 327 \pm 3K$ (cf.[36,47])), the opposite is observed for the corresponding endcapped POTM-n-Urethane oligomers as illustrated in Figure 12a (extrapolation to infinite degree of oligomerization yields $T_m^0 = 325 \pm 3K$ (cf.[36])). This is due to the decreasing influence of

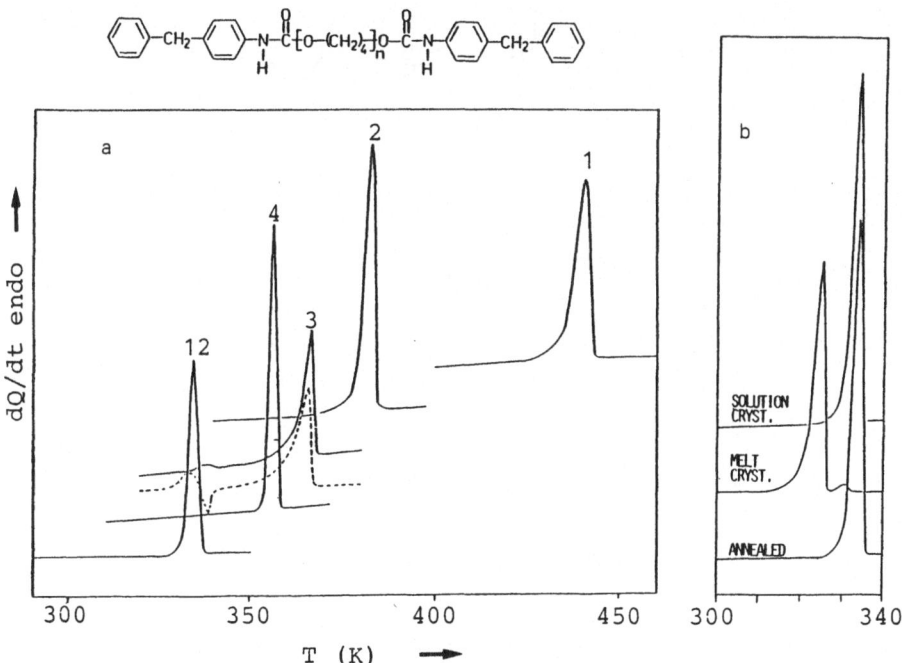

Fig. 12. DSC curves of diphenylmethane-4-carbamates of
α-hydro-ω-hydroxy-poly(oxytetramethylene) (POTM-n-
Urethane) as crystallized from ethanol (Fig. 12a;
dashed curve: melt crystallized sample) and of
POTM-12-Urethane crystallized under different conditions (Fig. 12b); heating rate 20°C/min; the
numerals 1,2,3,4 and 12 in Fig. 12a represent the
number n of $-O(CH_2)_4$-units in POTM-n-Urethane.

the urethane endgroup (melting temperature enhancing effect because of hydrogen bond formation between urethanes) on the melting behaviour with increasing length of the polyether chain.

One of the most interesting findings in the characterization of the soft segment model compounds with urethane end
groups is that there exist two different crystalline modifications depending on the crystallization conditions[24,35]: crystallization of the dodecamer from the melt leads to a metastable modification, characterized by a lower melting temperature
(T_m^o=327 K) as compared to the POTM-12-Urethane crystallized
from ethanol solution (T_m^o=334 K), which can be converted to the
higher melting modification by annealing at 332 K (Figure 12b);
besides, the d_{001}-spacing as determined from X-ray scattering
in the low angle range is found to be larger for the melt crystallized sample (7.52 nm) than for the solution crystallized
one (7.14 nm).

The IR spectra of the two modifications of POTM-12-Ure-
thane in the N-H and carbonyl region given in Figure 13 (curves
a and d) clearly show from the shift of the absorption bands of
the melt crystallized sample to higher wave numbers that the
N-H... O=C hydrogen bond is weakened as compared to the solution crystallized sample (shift of 10 and 8 cm^{-1}, respectively),

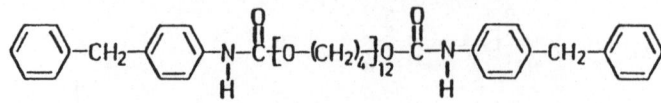

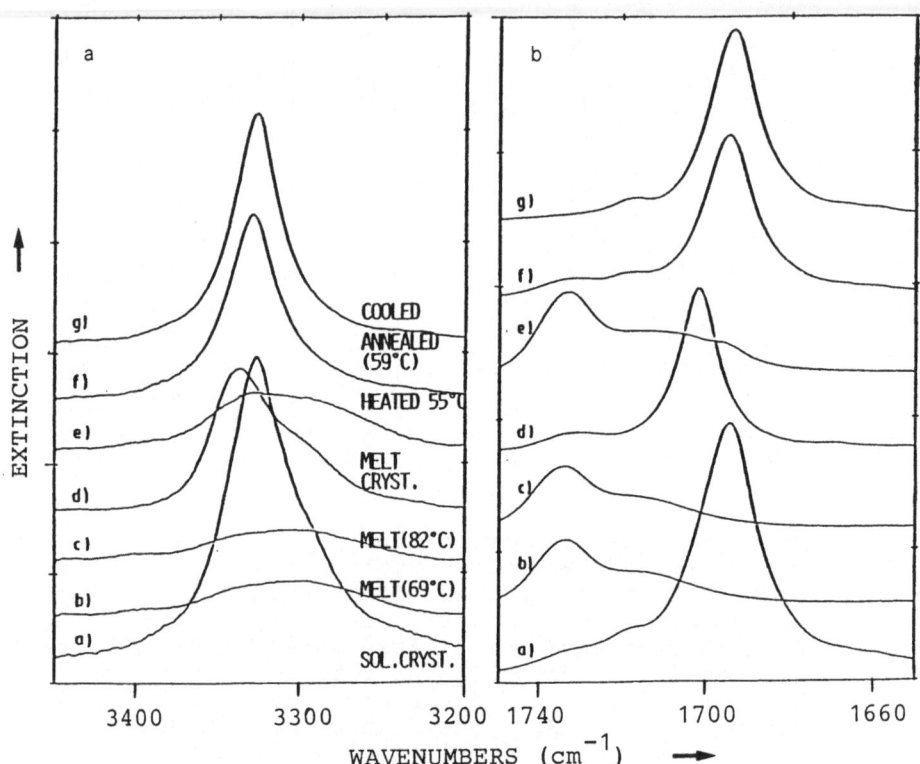

Fig. 13. IR spectra of the diphenylmethane-4-carbamate of α-
hydro-ω-hydroxydodeca(oxytetramethylene), POTM-12-
Urethane, in the NH (Fig. 13a) and carbonyl region
(Fig. 13b) after different thermal treatment as indicated (temperatures of spectra recording - if not
room temperature - given in parentheses).

that all hydrogen bonds are destroyed in the melt (curves b,c,
and e: non hydrogen-bonded N-H and C=O), and that the thermodynamically more stable modification is obtained again through
annealing (curves f and g).

From these findings of the existence of a metastable modification of the urethane-endcapped polyol which is characterized by a lower melting point, larger d-spacings and a weaker
hydrogen bonding as compared to the higher melting modification,
it can be inferred that metastable crystalline areas of non-
ideally (in view of a straight N-H...O=C hydrogen bonding)
packed hard segment portions adjacent to the soft segment may
occur in the interface between the hard and soft phase.

Hard segment elements are very desirable models for studying the structure and properties of the hard phase particularly with regard to the segregation behaviour of the hard segments of the multiblock copolymer, and the nature and degree of order within the hard domains which are the controlling factors for the material properties of PU elastomers.

The melting behaviour of a series of hydroxy-terminated polyurethane oligomers Diol-n and the corresponding urethane endcapped Urethane-n hard segment models is illustrated by the DCS traces in Figure 14. The melting point, T_m, of the precipitated and dried materials increases with increasing degree of oligomerization and extrapolates[46] to $T_m^0=523.5$ K for the infinite long chain (hard segment polymer). It has been pointed out earlier[18] that this finding is in contradiction to the data reported in literature for similar models[25], but our results have been also confirmed by others.[27,28] The investigation of the melting of the model compounds based on MDI and BDO is complicated by the fact that polymorphism exists, i.e., besides the planar zig-zag conformation of the tetramethylene sequence of the BDO subunit established by single crystal X-ray analysis of Urethane-0[6,16,49] a gauche-trans-gauche conformation is also discussed;[12,13,17] the latter hasn't yet been proven by single crystal X-ray analysis. The crystalline structure of the higher melting modification of Diol-1 (see dashed trace, Figure 14)

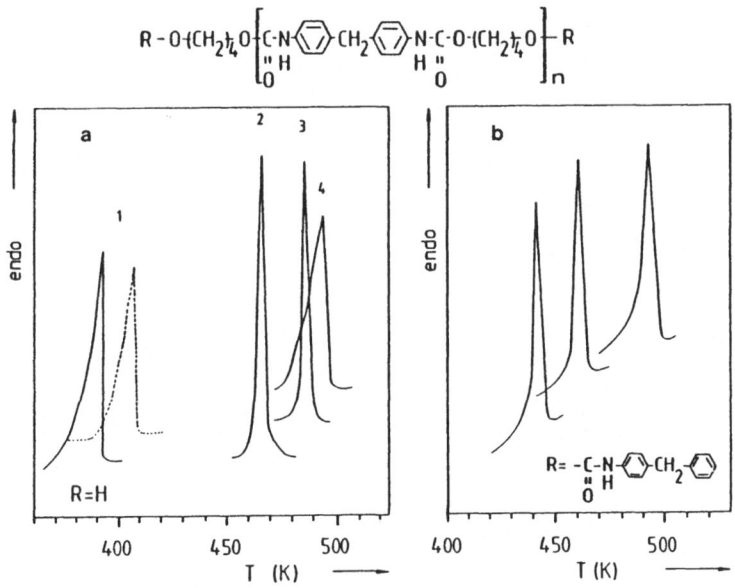

Fig. 14. DSC curves of the polyurethane hard segment model compounds Diol-n (Fig. 14a) and Urethane-n (Fig. 14b); heating rate 20°C/min; dashed curve is the crystalline modification of Diol-1 with the contracted conformation.[48]

differs from the extended conformation in that the butanediol
unit is oriented gauche to the adjacent urethane group.[48] The
contracted conformation postulated for the MDI-BDO based urethanes actually has been proven for other polyurethane model
systems based on piperazine and 1,4-butanediolbischloroformate
by single crystal analysis.[24]

The variation of the d_{001}-spacings obtained from powder
diffraction patterns of microcrystalline samples of Diol-n (n=
1-4) and Urethane-n (n=1-3) with the degree of oligomerization
is depicted in Figure 15. This clearly shows that the oligomers
crystallize without chain folding. From the slope of the plot
one obtains for the length of the repeat unit 1.75 nm (Diol-n)
and 1.85 nm (Urethane-n), respectively. By assuming an extended
conformation of the tetramethylene sequence of the BDO subunit[49]
an angle of inclination of the molecular axis relative to the
a/b-axis of about 29° and 36.5°, respectively, is calculated.

Besides the important question if chain folding of hard
segments does occur (cf.[11]) or not, it is just as important for
the understanding of the hard domain structure to know the mix-

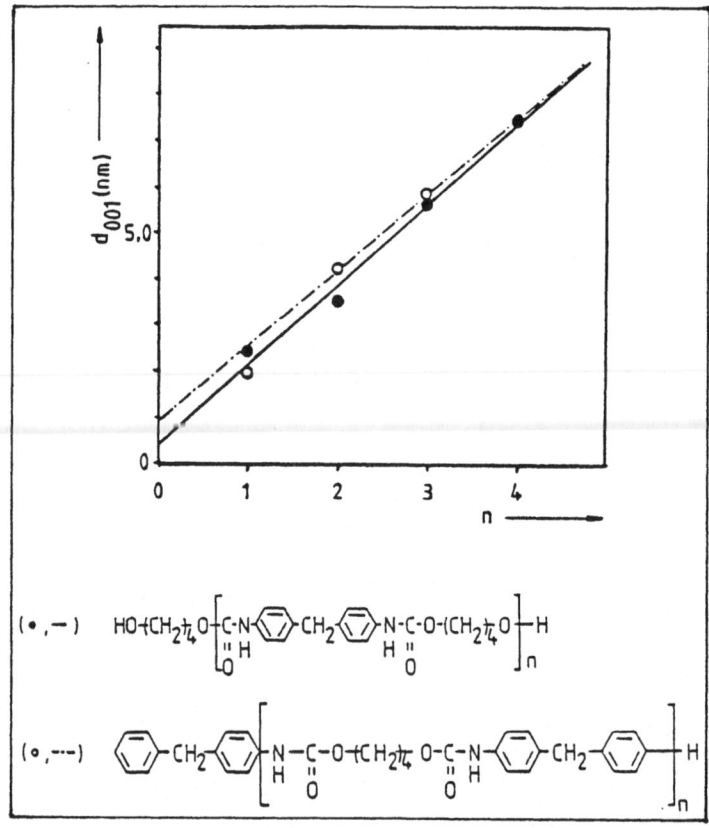

Fig. 15. Variation of the d_{001}-spacings with
the degree of polymerization, n, for
the polyurethane hard segment model
compounds Diol-n (●,——) and Urethane-n (o, -·-).

ing behaviour of hard segments of different length. The direct access to solve this problem is to make the phase diagrams of mixtures of different oligomers by DSC analysis. The melting point/ composition diagrams for the binary systems Urethane-0/ Urethane-1,[18] Diol-1/Diol-2, Diol-2/Diol-3 and Diol-3/Diol-4 are given in Figures 16,17 and 18.

Irrespective of the type of the urethane oligomer, the first numbers up to a degree of oligomerization of 3 or 4, do not form mixed crystals, but eutectic type phase diagrams are obtained (Figure 16 and 17). Contrary to the Urethane-n system (Figure 16) the phase diagrams of the Diol-n systems are somehow more complicated and it reflects the coexistence of the eutectic of the two crystalline modifications (fully extended tetramethylene sequence and gauche conformations) of the Diol-n oligomers. Similar phase diagrams have been obtained with the lower oligomers of piperazine/1,4-butanediolbischloroformate, and from their data[24] and results obtained with the MDI/BDO-based systems not presented here it could be proven that binary mixtures of the higher oligomers ($n>3$, $\Delta n=2$) do not form mixed crystals as well but again a eutectic.

The higher oligomers starting with $n=4$ (or perhaps already to a certain extent with $n=3$, see Figure 18) are able to cocrystallize, but only as long as the difference in chain length does not exceed one repeat unit; if the oligomers in a binary systems differ by more than 1 repeat unit (e.g., trimer/pentamer), again a pure eutectic is observed as already mentioned above. This is the unambiguous conclusion from our studies of

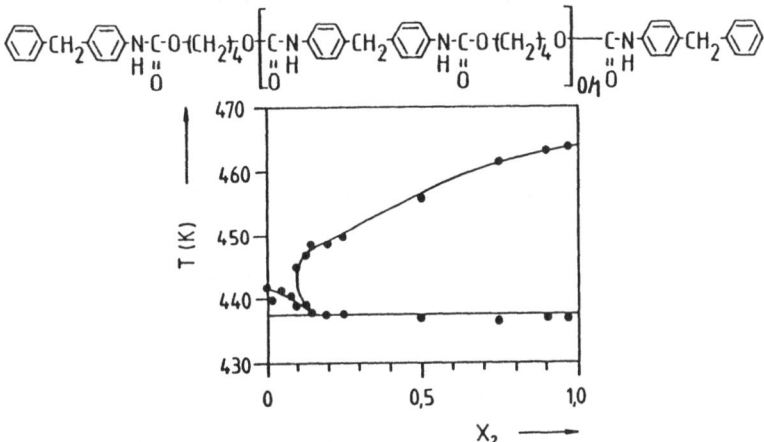

Fig. 16. Melting point-composition diagram (DSC analysis, heating rate 20°C/min) of the two-component system Urethane-0/Urethane-1 (0/1 in formula); X_2 = mole fraction Urethane-1.

73

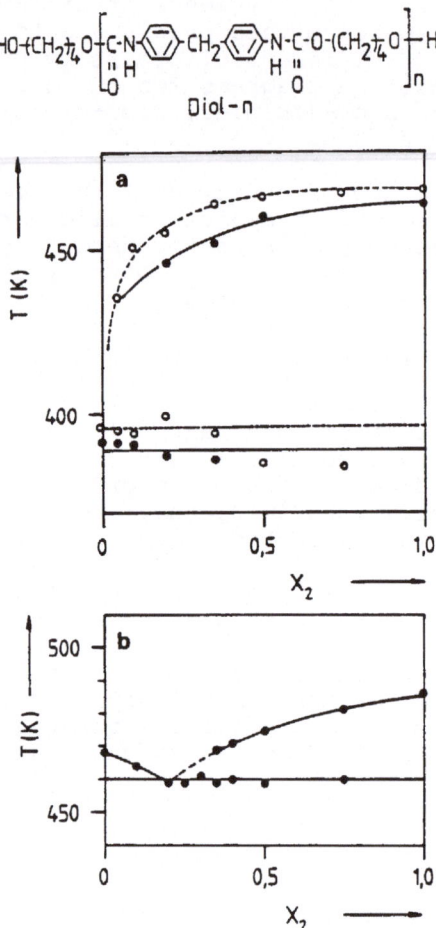

Fig. 17. Melting point-composition diagram
(DSC analysis, heating rate 20°C/
min) of the Diol-1/Diol-2 binary
system (Fig. 17a; (o): contracted
conformation; (●): extended conformation) and of the Diol-2/Diol-3 binary system (Fig. 17b); X_2 = mole
fraction Diol-2 (a) and Diol-3 (b).

the model oligourethanes based on piperazine and butanediolbis-chloroformate.[24,50] In the case of the MDI/BDO-based systems discussed here this certainly holds true too, but cannot be proven just as clearly because of two facts: firstly, the MDI/BDO-based oligomers are chemically unstable when heated above $\sim 180°C$, i.e., a relatively fast transurethanization reaction takes place which will be further exemplified below; secondly polymorphism (two crystalline modifications or even mesomorphic structures) seems to be always present, and pure modifi-

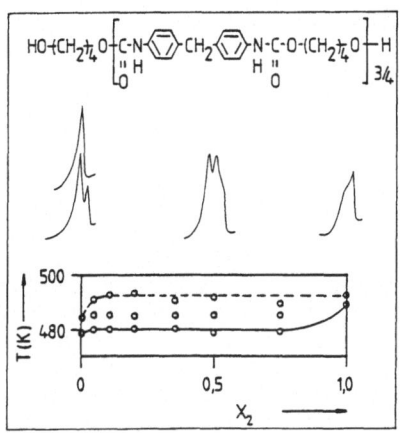

Fig. 18. Melting point-composition diagram and selected DSC traces (heating rate 20°C/min) of the Diol-3/Diol-4 binary system; X_2= mole fraction Diol-4; data points represent temperatures at which a peak or shoulder appeared in the DSC trace; dash line and solid line connect the highest and lowest transition temperature of each sample (peak or shoulder in the DSC trace).

cations required for establishing transparent phase diagrams are not attainable through annealing since the transurethanization beginning in this temperature range means decomposition of the monodisperse oligomer.

These difficulties are evident from the DSC traces obtained with the binary system Diol-3/Diol-4 and the melting point-composition diagram derived therefrom (Figure 18): a single sharp endotherm is only obtained for the pure trimer, whereas the tetramer could not be obtained purely in any modification and therefore shows a broader melting endotherm with a distinct shoulder. The polymorphic character of this system is reflected from all the DSC traces of the oligomer mixtures as shown exemplarily in Figure 18 for the 50 mole% mixture. It cannot be decided from this type of DSC trace if the system is consisting of mixed crystals and/or of a eutectic of different structure, or at which ratio both (mixed crystals and eutectic) coexist. The highest and lowest transition temperatures deduced from the endotherms (peak or shoulder) were tentatively connected by a curve simply to elucidate the differences between the "phase diagram" of this binary system involving the tetramer (Figure 18) and those only involving the shorter oligomers (Figure 16 and 17). From the comparison it can be reasonably assumed that the tetramer represents the lower critical chain length to allow cocrystallization[24] even when taking into consideration that the DSC traces not only might reflect transitions but also chemical conversion.

An important property of the hydrogen bond forming polyurethane, and among those for the MDI/BDO-based systems in particular, is their chemical instability which is apparent in a rapid transurethanization reaction. To the best of our knowledge this phenomenon has been first investigated by us[30] (cf.[51]), and simultaneously to our first studies Camberlin et al.[25] reported on a limited thermal stability of oligomers of similar structure. The effect of thermal treatment is best illustrated by the GPC trace in Figure 19 which was obtained from a formerly monodisperse Urethane-1 model compound (highly pure, no catalyst) only after 20 min annealing in the melt at 190°C[30] (cf.[24,31]). The chromatogram can be resolved completely by a superposition of the chromatograms of a homologous series of Urethane-n (n=0-7), and the evaluation of the relative amounts of the individual oligomers in the reaction product leads to the conclusion that the monodisperse system has been changed to a polydisperse system with nearly most probable distribution:

The decomposition is so fast that Urethane-n oligomers are already formed in considerable amounts when the sample is only heated to complete melting (heating rate 20 K/min) and then quenched (maximum cooling rate). In various model experiments it has been found that this reaction already takes place in the solid state, i.e., already at about 5-7 K below the crystalline melting point of the sample, where no melting could be observed under the microscope.

It is known that the urethane group is thermally unstable

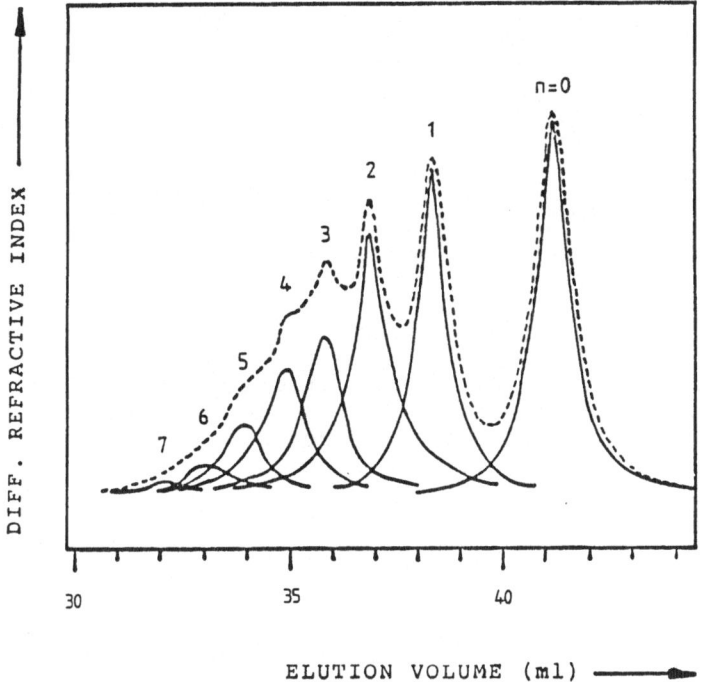

Fig. 19. GPC analysis of the polyurethane hard
segment model compound Urethane-1
annealed 20 min at 463 K; (---): gel
permeation chromatogram; (—): curve
resolution by a homologous series of
Urethane-n oligomers.

at elevated temperatures,[52,53] and that the free isocyanate and
hydroxyl group can be formed in a retro-urethane reaction, a
process used in blocking-deblocking of isocyanates, e.g., with
phenols. However, for the system investigated here, i.e., the
Urethane-1 in particular which was used in the model experiments, the formation of free isocyanate groups could not be
clearly detected by FTIR spectroscopy up to temperature of
195°C, although the decomposition to a polydisperse system proceeded within minutes. Contrary to this, it was found in comparative model experiments with binary mixtures of different
4-alkylphenyl urethanes which are known to form free isocyanates, that only less than 5% of the crossreaction products
were formed even after 60 min in the melt at 190°C.[54] This and
the fact that the decomposition reaction already occurs in the
solid state leads to the conclusion that the decomposition of
these particular polyurethanes based on MDI and BDO does not go
through a dissociation to the isocyanate and alcohol but through
a concerted, four-center type reaction with a more or less simultaneous breaking and reforming of urethane bonds as sketched
schematically below. This reaction could even be favoured by the
fact that a gliding of molecules (in the solid state: lattice
planes) along each other by only half the distance of the repeat unit brings all the urethane groups in a favorable position for this transurethanization.

$$\sim\underset{H}{N}-\overset{O}{\overset{\|}{C}}-O\sim \qquad \xrightarrow{\Delta/Cat} \qquad \sim\underset{H}{N}|\overset{O}{\overset{\|}{C}}-O\sim$$
$$\sim O-\underset{\underset{O}{\|}}{C}-\overline{\underset{H}{N}}\sim \qquad\qquad \sim O-\underset{\underset{O}{\|}}{C}|\underset{H}{N}\sim$$

   Of course, the formation of free isocyanate and alcohol groups cannot be excluded but is considered to play only a minor part in this equilibration reaction. At temperatures much above 195°C the dissociation of the urethane group might well be the dominant process as stated by Macosko et al.[55]

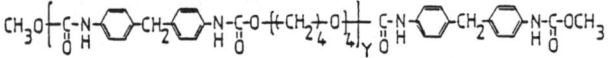

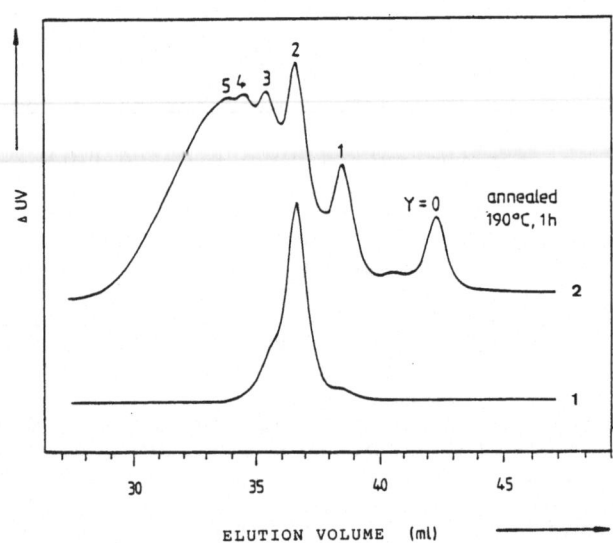

Fig. 20. GPC traces of a fraction of a prepolymer made from α-hydro-ω-hydroxytetra(oxytetramethylene) (POTM-4) and MDI, end-capped with methanol (fraction soluble in hot methanol) before (1) and after annealing for 1 h at 190°C (2).

This type of reaction has to be assumed to occur also in catalyst-free multiblock copolymers or even during the formation of the polyurethane segments in conventional or reaction-injection-moulding (RIM) elastomer synthesis. This has been demonstrated in an annealing experiment with a narrow fraction of a prepolymer obtained from the reaction of POTM-4 with MDI. It is clearly seen from the comparison of the two chromatograms in Figure 20 that, irrespective of the nature of the starting material (see Figure 19), a transurethanization reaction finally will lead to a kind of random distribution of the segment length. This has always to be kept in mind in sample preparation or when repeating an experiment with the same sample, since a single event of exceeding the critical temperature range (170-190°C) destroys the monodispersity.

It has to be emphasized that all the above findings were obtained with systems completely free of any catalyst. The effect of an added catalyst (e.g. dibutyltin dilaurate) is first that the temperature at which the equilibration reaction takes place is much lower, and secondly that this process proceeds much faster than when the catalyst is excluded. This has been already found useful in the synthesis of the Diol-2 and Diol-3

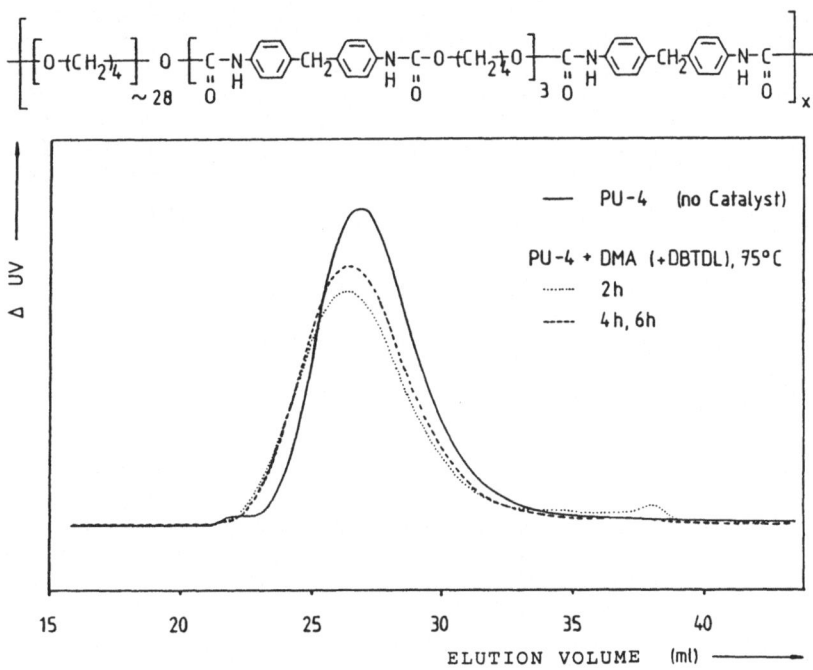

Fig. 21. GPC trace of the segmented PU elastomer PU-4 with monodisperse hard segment length distribution (cf. reaction scheme Fig. 10) before (—) and after heating in DMA solution (1 weight % PU-4) at 75°C in the presence of dibutyltin dilaurate catalyst (0.15 weight %) for 2 hr (···) and 4 or 6 hr ((---), identical chromatograms).

with Diol-1 and MDI as starting materials. Of course, it is not known how the catalyst effects the mechanism proposed above and to what extent a direct exchange reaction between the urethane and the alcoholic end group occurs. If a solution of a segmented PU elastomer is kept at higher temperatures for a longer period of time, no change in the molecular weight distribution was observed. The addition of a catalyst leads to a distinct broadening within only a few hours (Figure 21), and this effect is even more pronounced in the presence of a diol such as BDO. This means that a catalyst always must be excluded if it is the objective of a synthesis to obtain a material with specific properties, e.g., with regard to the segment length distribution by employing preformed hard segment units as in our case; if this precaution is not followed[29] any conclusions on the effect of segment length distribution on polymer properties are questionable.

Segmented polyether-polyurethane elastomers with monodisperse hard and/or soft segments (PU-n; n represents the number of MDI-units in the hard segment) are the only suitable materials to study the effect of segment length and segment length distribution on polymer properties. The differences in the properties between the elastomers with monodisperse and polydisperse hard segment length distributions are primarily due to the formation of more perfect hard domains in the former systems. This is evident from IR spectroscopy, X-ray scattering and the melting behaviour, but will not be discussed here. We will briefly comment only on the melting of PU-n elastomers and the kind of information deducible from the thermograms, and focus on the dynamic mechanical properties to demonstrate the significance of the hard segment length (distribution) on the elastomer properties. Of course, since the overall polymer average molecular weight affects polymer properties,[56] all the elastomers with different primary structure and segment length distribution used in this investigation had comparable molecular weights, as was made possible by the polymer synthesis method and had been checked by GPC (see above).

The melting behaviour of the elastomers with monodisperse hard segment length distribution is of particular interest since in the past a lot of conclusions have been drawn from the thermograms.[3,29,57] It certainly holds true that the level of the melting temperature of the hard segments is related to the hard segment length and the degree of order within the hard domains, as has been confirmed by our studies of a series of PU elastomers with increasingly long hard segments in monodisperse (PU-n) and polydisperse (PU-n̄) systems. The melting temperature generally increases with increasing hard segment length; the monodisperse systems melt higher than the polydisperse ones and are characterized by a narrower endotherm. This is about all one can deduce from the DSC traces and any further interpretations should be made very carefully if made at all.

The restrictions in the interpretation of DSC data outlined above are clarified by the thermograms in Figure 22 with the example of PU-3 giving a melting endotherm with a shoulder (or even a peak) at the low temperature side which was never observed for the other homologues of this PU-n series. The origin of this multiple endotherm in the first heating cycle is not clear and may be connected with changes in the morphology.

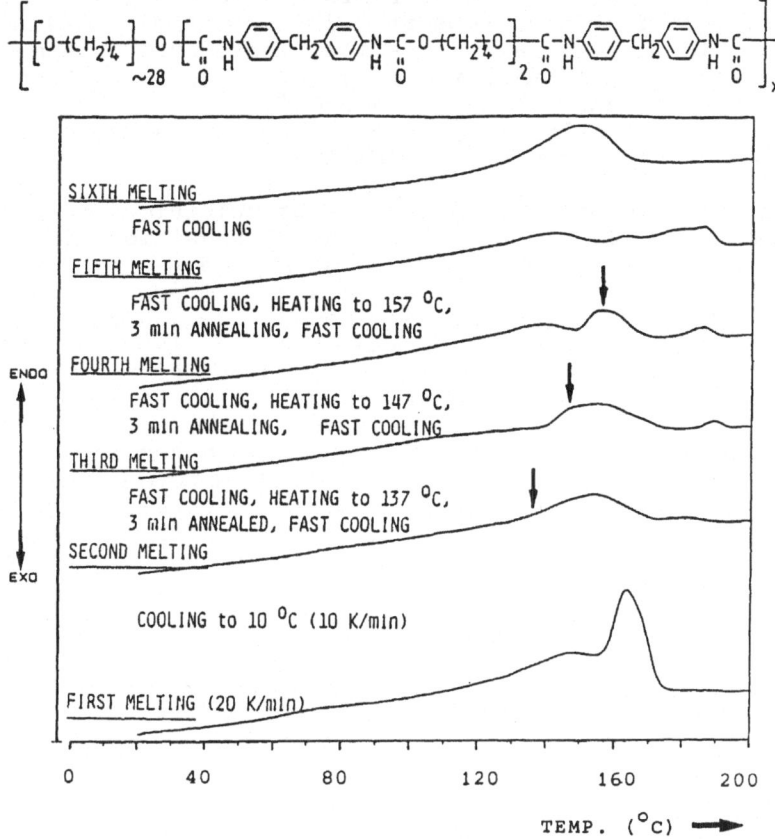

Fig. 22. DSC curves of the segmented PU elastomer PU-3
with monodisperse hard segment length distribution after different annealing histories as indicated; arrow indicates annealing temperature
before next heating cycle.

In the second heating cycle a single, relatively broad melting
endotherm is observed, and the heat of fusion is diminished
by about 30%. Annealing of the sample at various temperatures
changes the thermograms in a complex way. The kind of the thermogram obtained with the pristine elastomer sample cast from
solution cannot be reproduced again, but only the thermogram
obtained after the first heating cycle (see top curve).

This behaviour can only be explained by the complete de-
struction of the formerly monodisperse hard segments to a polydisperse system with different characteristics due to the transurethanization reaction. For a polydisperse system, annealing
or, more generally speaking, the thermal history greatly in-
fluences the superstructure in terms of phase separation and
the hard domain morphology as well as composition (eutectic
and/or mixed crystals of segments of different length). The

shape of the thermogram not only reflects the melting of crystallites of different size together with recrystallization or phase transition phenomena, but is also influenced by the chemical changes during heating, i.e., the transurethanization reaction. It simply means that one shouldn't speculate too much about the origin of multiple peak endotherms of samples with polydisperse segment length distributions.

The dynamic mechanical properties of multiblock copolymer systems are primarily determined by their transition temperatures and the relative weight fractions of the two phases.[4,6] This is seen for both the PU elastomers with monodisperse (PU-n) and polydisperse (PU-n̄) hard segment length distribution. The most striking differences between the two classes of elastomers are seen in the plateau modulus and the softening temperature (melting of hard segment crystallites) as shown in Figure 23. The primary relaxation observed in the dynamic mechanical behaviour at low temperature (maximum in the log decrement, Λ, at ∼-50°C) is related to the glass transition temperature of the soft segments, and the high temperature relaxation process (130-150°C) is linked with the melting of hard segment crystallites. The shoulders seen in the temperature dependence of the storage modulus of the sample with polydisperse hard segments at about -10°C are assigned to recrystallization and sub-

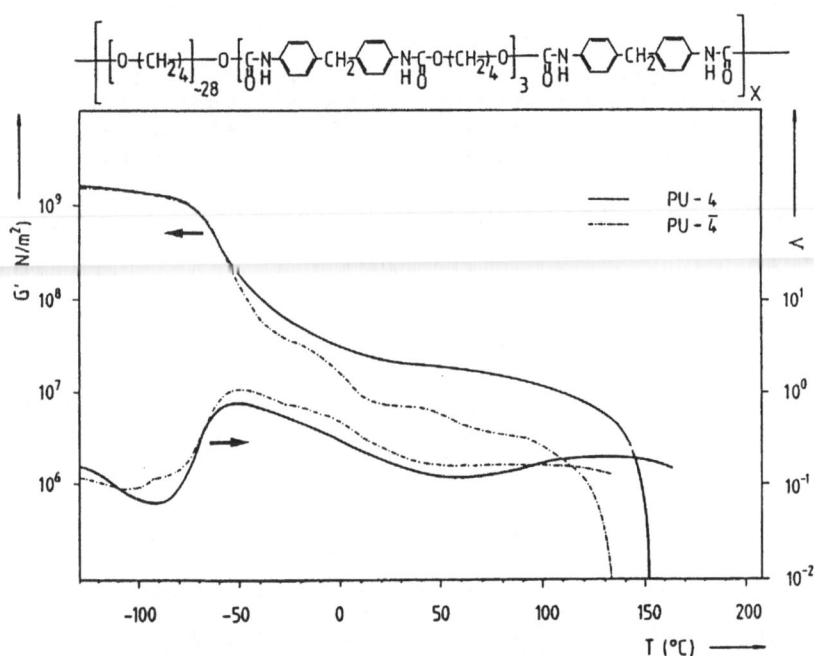

Fig. 23. Dynamic mechanical properties of the segmented PU elastomer PU-4 with monodisperse (n=3, cf. formula in Fig. 3; (——)) and PU-4̄ with polydisperse hard segment length distribution (POTM-28̄/MDI/BDO=1:4:3).

sequent melting of soft segments (to be discussed below) and to phase transition phenomena in the hard phase, respectively.

In general, the modulus of the elastomers with monodisperse hard segments in the plateau region is higher (cf.[21]) by a factor of ~3 and does not change very much over a temperature range of about 150 K; the softening of the hard domains is distinctly shifted to higher temperatures as compared to the elastomer samples containing polydisperse hard segments. These differences are even larger if one considers that the actual average hard segment length in the polydisperse elastomer (molar ratio Polyol/MDI/BDO = 1:4:3) is higher (4.5 MDI units per hard segment) than for the elastomer with monodisperse hard segment (4 MDI units only) due to the pre-extension reaction of the polyol during the prepolymer synthesis, as discussed above.

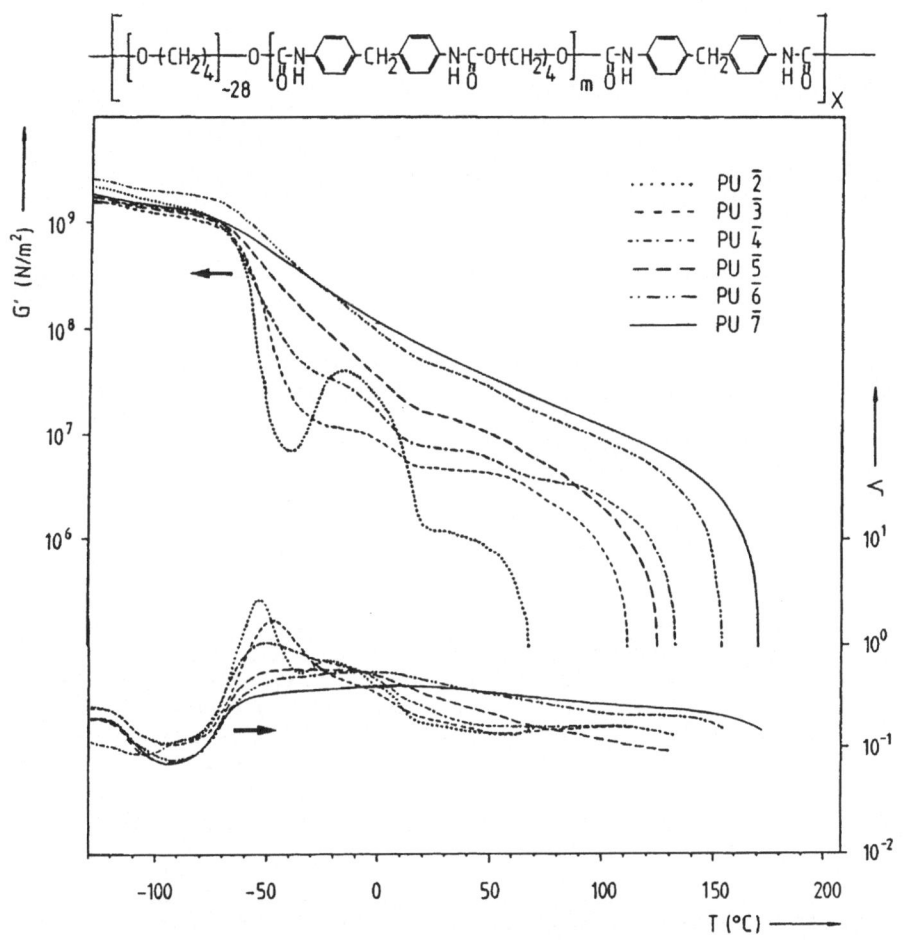

Fig. 24. Dynamic mechanical properties of segmented PU elastomers PU-$\bar{n}$ with polydisperse hard segment length distribution (POTM-$\overline{28}$/MDI/BDO=1:n:(n-1); n varying from 2 (PU-$\bar{2}$) to 7 (PU-$\bar{7}$)).

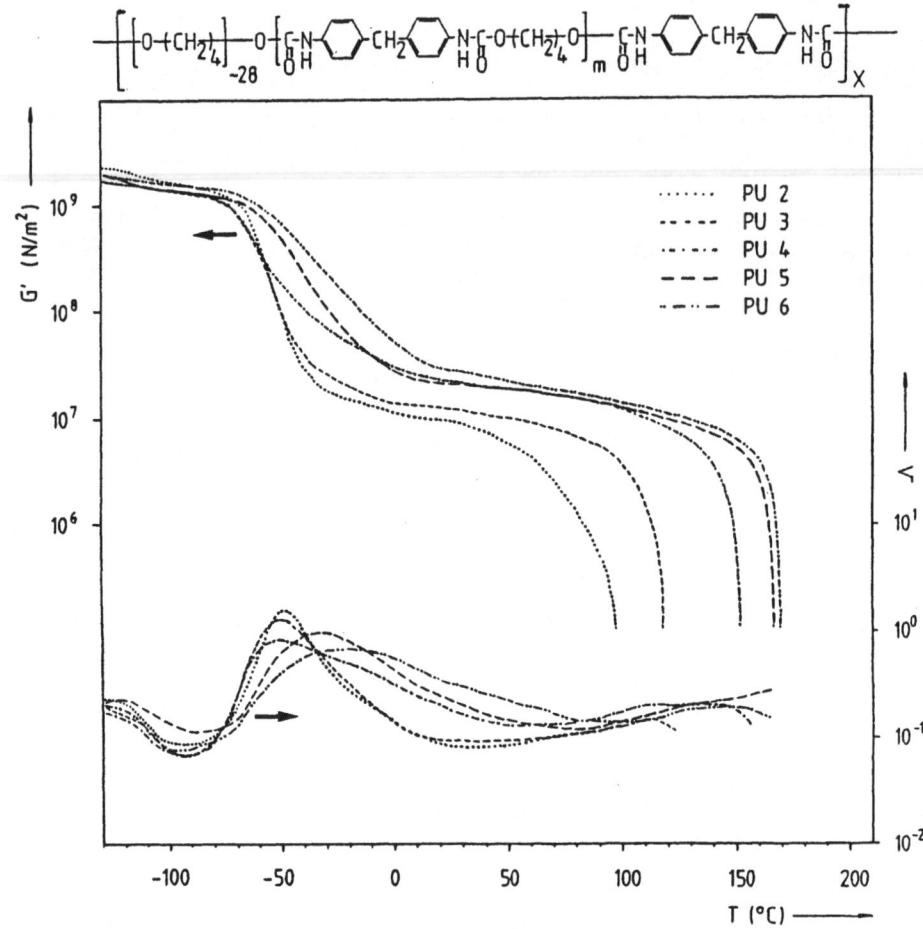

Fig. 25. Dynamic mechanical properties of segmented PU elastomers PU-n with POTM-$\overline{28}$ soft segment and strictly monodisperse hard segment length distribution; n varying from 1 (PU-2) to 5 (PU-6).

The effect of the pre-extension of the polyol POTM-$\overline{28}$ during the prepolymer synthesis is more pronounced for the softer types of PU elastomers with polydisperse hard segment length distribution, in particular for PU-$\overline{2}$ prepared with POTM-$\overline{28}$ and MDI at a molar ratio of 1:2 (see Figure 24). In this case the soft segment is pre-extended in the prepolymer formation to a number average molecular weight of about 3,500 (corresponding to a POTM-$\overline{50}$).[58] It is known that this type of the soft segments in polyurethanes crystallize on exceeding a critical molecular weight of about 3,000,[59] which causes a drastic increase in the modulus between -50 and 0 °C due to recrystallization of soft segments above the soft segment T_g.[60] The reason for not being able to observe this phenomenon in the corresponding elastomers with monodisperse hard segment length distribution is not due to the different hard segment length dispersity but is a result only of the different soft segment length: contrary to the usual two-step PU elastomer synthesis applied for the preparation of the polydisperse systems, the pre-chain extension was completely suppresses in the special synthesis of the PU elastomers with monodisperse hard segments due to the large excess of MDI employed in the prepolymer formation.

The temperature dependence of the dynamic mechanical properties of the family of elastomers with polydisperse and monodisperse segment length distribution is illustrated in Figures 24 and 25. It is evident at first glance that in both cases the so-called plateau modulus increases with increasing hard segment content (length) as does the softening temperature. The main difference is that the elastomer with monodisperse hard segments exhibits a well-defined plateau modulus which doesn't change very much over a broad temperature range, whereas in the case of elastomers with polydisperse hard segment length distribution a distinct plateau modulus does not exist. For the latter systems not only the drop of the storage modulus in the glass transition temperature region of the soft phase is increasingly smeared to higher temperatures with increasing hard segment content, but the storage modulus also falls steeper with increasing temperature as compared to the PU elastomers with monodisperse hard segments, and finally ends up with an interconnecting line between the soft segment glass transition and hard segment melting temperatures for the harder types, i.e., PU-5̄, PU-6̄ and PU-7 (cf.[6,61,62]). This distinct influence of the hard segment length dispersity on the dynamic mechanical properties is explained by the better phase separation and hard segment ordering of the elastomers with monodisperse segments, i.e., these systems behave more like crosslinked rubbers containing filler (especially with regard to the temperature dependency of the plateau modulus); polydispersity of the hard segment length leads to less phase separated systems and the properties of the resulting elastomers (PU-6̄ and PU-7) are best understood by assuming an interpenetrating network with a soft phase highly polluted with hard segments.

ACKNOWLEDGEMENT

Financial support of this work by the Deutsche Forschungsgemeinschaft (Sonderforschungsbereich 60, University Freiburg) and the Fonds der Chemischen Industrie is greatfully acknowledged. We are also grateful to G. Lieser for his assistance in the X-ray scattering analysis.

REFERENCES

1) H. Oertel, Bayer Farben Revue, 11, 1 (1965); Chemiker Ztg., 98, 344 (1974).
2) G.M. Estes, R.W. Seymour, and S.L. Cooper, Macromolecules, 4, 452 (1971).
3) H. Hespe, E. Meisert, U. Eisele, L. Morbitzer, and W. Goyert Colloid Polym. Sci., 250, 797 (1972).
4) R. Bonart, Angew. Makromol. Chem., 58/59, 259 (1977).
5) J. Blackwell and K.H. Gardner, Polymer, 20, 13 (1979).
6) L. Born, H. Hespe, J. Crone, and K.H. Wolf, Colloid Polym. Sci., 260, 819 (1982).
7) Y. Camberlin, J.P. Pascault, M. Letoffe, and P. Claudy, J. Polym. Sci. Polym. Chem. Ed., 20, 1445 (1982).

8) Y. Camberlin and J.P. Pascault, J. Polym. Sci. Polym. Chem. Ed., 21, 415 (1983).

9) R.E. Camargo, C.W. Macosko, M. Tirrell, and S.T. Wellinghoff, Polym. Commun., 24, 314 (1983).

10) C.D. Eisenbach and W. Gronski, Makromol. Chem. Rapid Commun., 4, 707 (1983).

11) J.T. Koberstein and R.S. Stein, J. Polym. Sci. Polym. Phys. Ed., 21, 1439 (1983).

12) J. Blackwell and C.D. Lee, J. Polym. Sci. Polym. Phys. Ed., 21, 2169 (1983).

13) R.M. Briber and E.L. Thomas, J. Macromol. Sci. Phys. Ed., B22, 509 (1983).

14) K.K.S. Hwang, D.J. Hemker, and S.L. Cooper, Macromolecules, 17, 307 (1984).

15) L.M. Leung and J. Koberstein, J. Polym. Sci. Polym. Phys. Ed., 23, 1883 (1985)

16) L. Born and H. Hespe, Colloid. Polym. Sci., 263, 335 (1985)

17) R.M. Briber and E.L. Thomas, J. Polym. Sci. Polym. Phys. Ed., 23, 1915 (1985).

18) C.D. Eisenbach and Cl. Günter, Proc. Div. Polym. Mat. Sci. Eng., 49, 239 (1983).

19) W. Kern, K.J. Rauterkus, and H. Sutter, Makromol. Chem., 44, 78 (1961).

20) L.L. Harrell, Jr., Macromolecules, 2, 607 (1969).

21) N.N. Ng, A.E. Allegrezza, R.W. Seymour, and S.L. Cooper, Polymer, 14, 255 (1973).

22) H. Nefzger, Diploma thesis, University of Freiburg, 1984.

23) C.D. Eisenbach, in "35 Jahre Fonds der Chemischen Industrie", Verband d. Chem. Indust., Ed., Frankfurt 1985, p.77.

24) C.D. Eisenbach, H. Nefzger, M. Baumgartner, and Cl. Günter, Ber. Bunsenges. Phys. Chem., 89, 1190 (1985).

25) Y. Camberlin and J.P. Pascault, J. Polym. Sci. Polym. Chem. Ed., 20, 383 (1982).

26) Z.Y. Qin, C.W. Macosko, and S.T. Wellinghoff, Proc. Div. Polym. Mater. Sci. Eng., 49, 475 (1983).

27) K.K.S. Hwang, G. Wu, S.B. Lin, And S.L. Cooper, J. Polym. Sci. Polym. Chem. Ed., 22, 1677 (1984).

28) Z.Y. Qin, C.W. Macosko, and S.T. Wellinghoff, Macromolecules, 18, 553 (1985).

29) J.A. Miller, S.B. Lin, K.K.S. Hwang, K.S. Wu, P.E. Gibson, and S.L. Cooper, Macromolecules, 18, 32 (1985).

30) Cl. Günter, Diploma thesis, University of Freiburg, 1982.

31) C.D. Eisenbach, M. Baumgartner, and Cl. Günter, Polym. Prepr., American Chemical Society, Div. Polym. Chem., 26(2), 7 (1985).

32) C.D. Eisenbach, Cl. Günter, and U. Struth, Spez. Ber. KFA Jülich, 316, 107 (1985).

33) B. Fu, C. Feger, W.J. MacKnight, and N.S. Schneider, Polymer, 26, 889 (1985).

34) B. Bengtson, C. Feger, W.J. MacKnight, and N.S. Schneider, Polymer, 26, 895 (1985).

35) M. Baumgartner, Diploma thesis, University of Freiburg, 1983.

36) R. Bill, M. Dröscher, and G. Wegner, Makromol. Chem., 182, 1033 (1981).

37) B. Böhmer, Dissertation, University of Mainz, 1968.

38) W. Kern and K.J. Rauterkus, Makromol. Chem., 28, 221 (1958).

39) C.S. Schollenberger, U.S. Patent 2,871,218, filed Dec. 1, 1955, issued Jan. 27, 1959 (to BF Goodrich Company).

40) R.M. Carvey and D.E. Witenhafer, Brit. Patent 1,087,743, filed June 16, 1965, issued Oct. 18, 1967 (to BF Goodrich Company).

41) L.H. Peebles, Jr., Macromolecules, 9, 58 (1976).

42) L.H. Peebles, Jr., Macromolecules, 7, 872 (1974).

43) H. Suzuki, J. Polym. Sci., A-1, 9 (1971).

44) P.J. Flory, J. Amer. Chem. Soc., 58, 1877 (1936).

45) R. Bonart and P. Demmer, Colloid Polym. Sci., 260, 518 (1982).

46) P.J. Flory and A. Vrij, J. Amer. Chem. Soc., 85, 3548 (1963).

47) H. Tadokoro, J. Polym. Sci. Polym. Symp., 15, 1 (1966).

48) P.G. Forcier and J. Blackwell, Acta Cryst. B37, 286 (1981).

49) L. Born, J. Crone, H. Hespe, E.H. Müller, and K.H. Wolf, J. Polym. Sci. Polym. Phys. Ed., 22, 163 (1984).

50) H. Nefzger and C.D. Eisenbach, to be published.

51) C.D. Eisenbach, Cl. Günter, M. Baumgartner, and U. Struth, Makromol. Kolloquium, Freiburg, February 1984.

52) O. Bayer, Das Diisocyanatpolyadditionsverfahren, Hanser, München, 1963, p. 14.

53) J.H. Saunders and K.C. Frisch, Polyurethanes, Chemistry and Technology, High Polymer Series Vol. XVI/1, Interscience, New York, 1962, p. 103.

54) C.D. Eisenbach, Cl. Günter, and H. Hespe, to be published.

55) W.P. Yang, C.W. Macosko, and S.T. Wellinghof, Polym. Prepr. American Chemical Society, Div. Polym. Chem., 26 (2), 321 (1985).

56) C.S. Schollenberger and K. Dinbergs, J. Elastomers Plast., 11, 58 (1979); C.S.S. and K.D., Adv. Urethane Sci. Technol., 7, 1 (1979).

57) R.S. Seymour and S.L. Cooper, Macromolecules, 6, 48 (1973).

58) M. Baumgartner and C.D. Eisenbach, to be published.

59) C.G. Seefried, J.V. Koleske, and F.E. Critchfield, J. Appl. Polym. Sci., 19, 2493 (1975).

60) J.L. Illinger, N.S. Schneider, and F.E. Karasz, Polym. Eng. Sci., 12, 25 (1972).

61) D.S. Huh and S.L. Cooper, Polym. Eng. Sci., 11, 369 (1971).

62) C.G. Seefried, J.V. Koleske, and F.E. Critchfield, J. Appl. Polym. Sci., 19, 2503 (1975).

RELATIONSHIP BETWEEN CHEMICAL COMPOSITION AND HYSTERESIS IN POLYURETHANE

ELASTOMERS

Catherine A. Byrne

U.S. Army Materials Technology Laboratory
Polymer Research Division
Watertown, Massachusetts 02172-0001

ABSTRACT

Polyurethane elastomers have been prepared from 1,4-cyclohexane-
diisocyanate, hydroxy terminated poly(tetramethylene oxide), 1,4-butane-
diol and trimethylolpropane. Selected sample blocks (63.5 x 63.5 x
15.9mm) have been tested for hysteretic heat build-up during compressive
cycling. During 200,000 cycles at 18.5 Hertz under a maximum load of
24.5 kg and a ratio of minimum load to maximum of 0.1, the internal
temperatures for different samples ranged from 50 to 160°C. Thermomechanical analysis indicates final softening values of 250°C for most
samples. Mechanical properties are maintained up to at least 180°C in
dynamic mechanical analysis. The samples have low weight percents of hard
segment, approximately 20 percent, but have high Shore A hardness values
in the 80-90 range. Trouser tear strengths range from 4kN/m for a
heavily crosslinked sample to 30 kN/m for a linear sample. The low
internal temperature generated during compressive fatigue testing, as
well as other exceptional properties of these polyurethanes are attributed to the compact nature of the diisocyanate and its ability to form
very small hard segment crystallites with melting temperatures above
200°C.

INTRODUCTION

Polyurethane elastomers are promising materials for solid tires and
track pads for tanks. There is little systematic study reported of the
relationship between the specific chemical compositions of a series of
chemically similar polyurethanes and their hysteresis and other properties thought to influence elastomer performance on heavy vehicles.[1] In
this work, the characteristics of a group of polyurethanes are reported,
including the thermal properties, mechanical properties at elevated
temperatures, hardness, tear strength and compressive fatigue behavior.
Abrasion resistance is also an important property, but the correlation
between laboratory tests and road tests is frequently poor. Abrasion
resistance will not be discussed here.

The polyurethanes chosen for this study are based on 1,4-cyclohexane-
diisocyanate (CHDI, 99 % trans isomer, Elate 166, Akzo Chemie America),
poly(tetramethylene oxide) soft segment diol of different molecular
weights (PTMO), chain extender 1,4-butanediol (BD) and crosslinker tri-

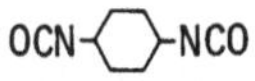

OCN—⬡—NCO

1,4-CYCLOHEXANEDIISOCYANATE (CHDI)

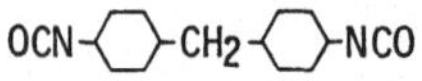

OCN—⬡—CH_2—⬡—NCO

METHYLENE BIS(4-CYCLOHEXYL ISOCYANATE) (H_{12}MDI)

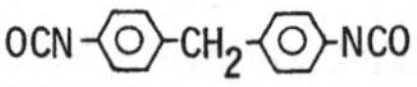

OCN—⬡—CH_2—⬡—NCO

METHYLENE BIS(4-PHENYL ISOCYANATE) (MDI)

HO—$(CH_2CH_2CH_2CH_2O)_n$—H

POLY(TETRAMETHYLENE OXIDE) (PTMO)

$HOCH_2CH_2CH_2CH_2OH$

1,4-BUTANEDIOL (BD)

$CH_3CH_2C(CH_2OH)_3$

TRIMETHYLOLPROPANE (TMP)

Fig. 1. The chemical structures of the starting materials for the polyurethanes.

methylolpropane (TMP). The starting materials discussed in the paper are shown in Fig. 1. Polyurethanes prepared from CHDI have been compared to those obtained from several other diisocyanates in a study by Wong and Frisch.[2] In general, the stress properties of the CHDI samples were poorer at room temperature but better retention of strength occurred at higher temperatures, in the range examined, up to 150°C. A linear CHDI-derived sample with Shore A hardness of 97 showed a stress of approximately 15 MPa at 300 % elongation and 25°C, and 8 MPa at 150°C. A sample prepared from methylene bis(4-phenylisocyanate) (MDI) of similar hardness, but higher hard segment content, showed a stress of 17 MPa at 25°C and about 4 MPa at 150°C.

Polyurethanes from CHDI also[3] exhibit better retention of tear resistance at elevated temperatures. A comparison was made of polyurethanes prepared from PTMO 1000 (approximate number average molecular weight of 1000) and MDI having 33.9 weight percent hard segment and a similar CHDI polymer having 27.2 weight percent hard segment. Split tear resistance (ASTM D 1938) values of 44.0 kN/m (252 pli) at 25°C and 2.8 kN/m (16 pli) at 150°C were observed for the MDI polyurethane and 37.5 kN/m (214 pli) and 25.2 kN/m (144 pli) at the corresponding temperatures for the polymer prepared from CHDI. While the tear resistance of the latter materials is lower at room temperature it retains 67% of that value at 150°C. The polymer prepared from MDI retains only 6% of its tear resistance at 150°C.

A few samples were also prepared from methylene bis (4-cyclohexyliso-cyanate) (H_{12} MDI, Desmodur W, Mobay),[4] in place of CHDI, for comparison purposes, because it was convenient to do so. These samples typically exhibit poor properties at elevated temperatures, but possess hardness values in the range of interest, below 80 Shore A, similar to reinforced natural rubber vulcanizates.

EXPERIMENTAL

The synthesis involved formation of a prepolymer by reacting the PTMO with CHDI at 100°C for one and one half hours. The chain extension step was performed in a polyethylene beaker in which the preheated (100°C) prepolymer, diol, and triol, were mixed thoroughly using a laboratory mechanical stirrer at the highest setting for one minute. The sample was degassed under vacuum for one to two minutes and poured into a mold. The samples were cured in the mold for sixteen hours at 100°C. Two shapes of sample were made, 1.3 mm (0.05 in) thick sheets and 63.5 x 63.5 x 15.9 mm (2-1/2 x 2-1/2 x 5.8 in) blocks for compression testing. Some of the sheets were prepared using 9×10^{-4} weight percent of dibutyl tin dilaurate in 2-butanone (T-12, M&T Chemical). The preparation of 100 g batches of polyurethane and hand casting as blocks was not possible when a catalyst was used, because the larger samples set up too quickly. The CHDI is intermediate in reactivity between MDI and hydrogenated MDI (H_{12} MDI) and so good elastomers could be made without a catalyst.[2] The infrared spectra of thin films of the samples prepared with a catalyst showed a trace of free isocyanate at 2255 cm^{-1}. When the attenuated total reflectance method was used the polyurethane surfaces showed no free isocyanate.

Thermomechanical Analysis (TMA) was performed using a Perkin-Elmer TMS-1 with a Perkin-Elmer Model UU-1 Temperature Program Controller. Samples were heated in a helium atmosphere from -100°C to the softening point of the sample at 20°C/min. The weight used in the TMA was 40 grams in most cases, 20 grams for one or two samples.

A Perkin-Elmer DSC-2 cooled with liquid nitrogen and purged with helium was used for DSC. Sample size was 20-40 mg. Heating rates were 10°C/min. or 20°C/min. A Perkin-Elmer Thermal Analysis Data Station was used to determine the temperatures of the transitions.

A Rheometrics Dynamic Spectrometer, RDS-7700, was used for measurement of dynamic mechanical properties. The samples were cut to 13 by 64 mm and were 1.3 to 1.6 mm thick. A liquid nitrogen controller was used to achieve the desired temperature. Measurements were taken at 10° temperature increments, with an equilibration time of two minutes at each temperature. The range studied was from -100 to 160°C, with a strain setting of 1 percent and a rate of 6.28 radians per second.

During Rheometrics analysis, a torsional motion is imposed on the sample and the torque and normal forces resulting from the motion are measured by a transducer. From these values the storage modulus, G', the loss modulus, G" and the ratio of the latter to the former, the loss tangent (tan δ) can be calculated.

Tear resistance was measured at 22°C using a trouser tear specimen described in ASTM D 470-82. The crosshead speed was 500 mm/min. The data reported are averages of results for six specimens. A hand-held Shore Durometer, hardness type "A-2" was used to determine the hardness of the samples, according to ASTM D 2240-75. Specimens were rectangular, 25 mm in length by 13 mm wide. Four plies of sample were used to achieve a thickness of 6.4 mm. Three readings were taken, 6 mm apart on the surface of the four plies. Hardness values measured on the blocks tended to be as much as five units lower.

Internal heat generation was measured during fatigue loading of a series of the polyurethane elastomers described above. The sample dimensions were chosen because the block had a shape factor similar to that of a tank track pad. This shape factor is calculated by the formula (width x length) divided by [(2 x thickness) x (width + length)] and for these samples is equal to 1.0.[5] Compression-compression fatigue testing was performed on sample blocks at 18.5 Hertz with a maximum load of 20,000 N (4500 pounds) and a ratio of minimum to maximum load of 0.1. The testing continued for 200,000 cycles and a time of 180 minutes.

An Instron Model 1322 servohydraulic test machine was used for all fatigue tests. Throughout the test, load and stroke data were acquired using a Nicolet Digital Oscilloscope. These data were stored on floppy disks for post-test reduction. The internal temperature of each block was measured using a J-type thermocouple inserted through a drilled hole into the center of the block. Temperature was recorded by an Electronic Controls Design Datalogger Model DL 2020 at time intervals which varied depending on the rate of temperature rise, every minute when the temperature changed rapidly and every ten minutes as it levelled off.

The Nicolet Oscilloscope is microprocessor based and data reduction can be performed internally. The data were reduced by displaying and plotting load versus stroke at various times during each fatigue test. Hysteresis was calculated from the area between the loading and unloading curves using a program supplied by Nicolet. Each value represents an average of five to six successive cycles of loading.

RESULTS AND DISCUSSION

Initially, the TMA curves for a number of samples were examined. It

Table 1. Thermomechanical Analysis Results for Polyurethanes

Moles/1 Mole/Moles/Moles CHDI/ PTMO/ BD /TMP	Transition Temperatures(C°)
1.7 - 2000[a] - 0.80 - 0.13	-84, 48, 174
1.9 - 2000 - 0.80 - 0.13	-72, 217; -84, 231
2.0 - 2000 - 0.80 - 0.13	-71, 225; -81, 225
2.1 - 2000 - 0.80 - 0.13	-71, 244
1.7 - 2000 - 0.90 - 0.07	-83, 45, 182
1.9 - 2000 - 0.90 - 0.07	-78, 233
2.1 - 2000 - 0.90 - 0.07	-72, 256
2.0 - 2000 - 0.60 - 0.27	-74, 241
2.1 - 2000 - 1.20 - 0.20	-81, 37, 199
2.4 - 2000 - 1.20 - 0.20	-83, 246
2.5 - 2000 - 1.20 - 0.20	-70, 251; -83, 9, 246
2.6 - 2000 - 1.20 - 0.20	-85, 5, 244
2.8 - 2000 - 1.20 - 0.20	-81, 255
2.1 - 2000 - 1.35 - 0.10	-79, 251
2.4 - 2000 - 1.35 - 0.10	-75, 187, 238
2.5 - 2000 - 1.35 - 0.10	-74, 254
2.6 - 2000 - 1.35 - 0.10	-79, 258
2.8 - 2000 - 1.35 - 0.10	-75, 258
2.6 - 2000 - 0.90 - 0.40	-74, 251
2.5 - 2000 - 1.50 - 0.00	-81, 254
2.6 - 2000 - 1.50 - 0.00	-80, 250
2.6 - 1000 - 1.20 - 0.20 PTMO 1000[a]	-68, 265
2.6 - 2000 - 1.20 - 0.20 Postcure 125°	-74, 247
2.6 - 2000 - 1.20 - 0.20 Postcure 150°	-74, 236

[a]These numbers represent the approximate number average molecular weight of poly(tetramethylene oxide).

is important that the samples have final softening temperatures well above the temperature experienced by the samples in compressive cycling. All the samples tested exhibit at least two transitions shown in Table 1, one due to the T_g of the PTMO soft segment and the other to the final softening, considerably above the temperatures occuring during compressive cycling and also higher than final softening values for many polyurethanes. An increase in NCO/OH equivalent ratio leads to a higher final softening. This could be due to allophanate crosslinking, but the presence of allophanate groups in these samples has not been confirmed. There are difficulties associated with the determination of the amount of allophanate crosslinking in these samples. All of the samples are insoluble in organic solvents and they swell only to a very small extent in hot solvents such as N, N- dimethylformamide.

An assumption was made at the outset, that the best hysteresis properties would be observed for crosslinked samples. All the samples discussed in this work except one are crosslinked by TMP. It was assumed that crosslinking would reduce the compressibility of the samples and the heat generated would be lower. The assumption may not be justified for samples prepared with such a compact diisocyanate, since the 2.5-2000-1.5-0.0 in Table 1 has a final softening temperature as high as similar crosslinked samples.

An increase in weight percent of hard segement, effected either by increasing the amount of CHDI or by using PTMO 1000, causes an increase in final softening. Post cures of six hours at 125 and 150°C in air caused a lowering of the final softening temperatures, to the extent of 5 to 20°C. A post cure had been suggested as a method of improving the mechanical properties of these polymer. The reduced value for the final softening temperature of the sample post cured at 150°C and aged two weeks suggests some degradation. Post cured plymers have not been included in the compression testing to be discussed later. It should be noted that some improvement in mechanical properties has been reported for post cured samples.[6] In the present work, insufficient data has been collected with which to evaluate the effectiveness of a post cure.

DSC curves for sample 1.7-2000-0.8-0.13 are shown in Fig 2. The low temperature scan in Fig. 2. a. shows two transitions due to the soft segment, the T_g at -83°C and an endotherm due to melting of the crystalline soft segment at -5°C. The soft segment is very well phase segregated because its T_g is very close to that for the pure soft segment, -85°C. The presence of crystallinity is also an indication of the purity of the soft segment in the polyurethane. The high temperature scan in Fig. 2.b. shows three transitions for the hard segment, a T_g at 55°C, crystallization at 185°C and melting at 221°C. Some of the samples exhibit two endotherms due to crystalline melting, the second occurring at a higher temperature than that shown. Some difficulty was encountered in locating the T_g. It is not detected by TMA or by DSC when typical sample sizes of 10-20 mg are used. Almost 40 mg was required in order to observe the T_g. It is not observed in some samples which are linear, presumably because the hard segment is too crystalline. The hard segment weight percents are very low for these samples and this fact contributes to difficulties in observing the T_g. Note also that at a scan rate of 20°C/min., the crystallization exotherm for the hard segment did not occur and yet a melting was still visible. A slower heating rate was required in order to produce the exotherm at 185°C.

The low hysteresis values and the low temperature rises observed in these samples and discussed below are due in part to the small change in

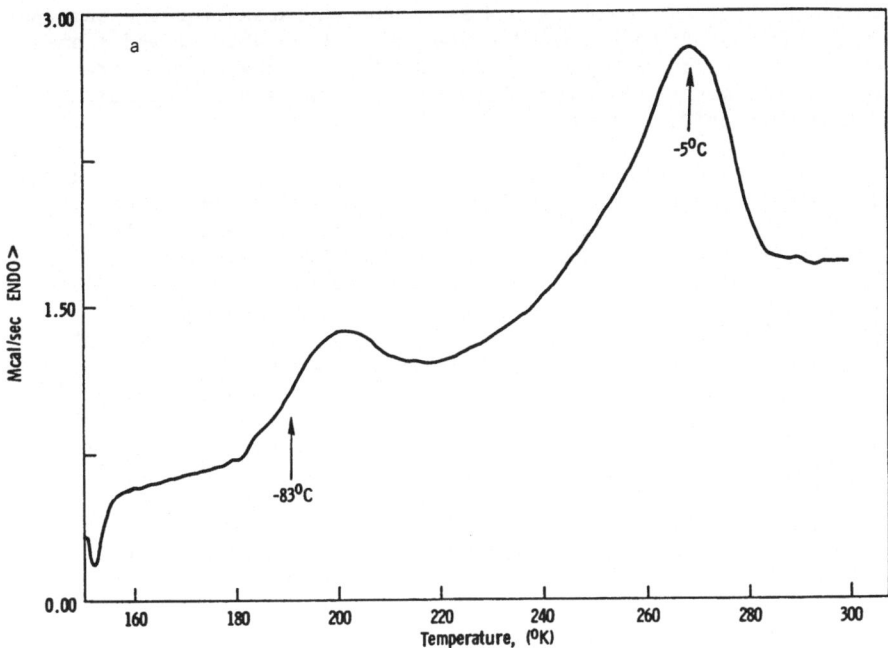

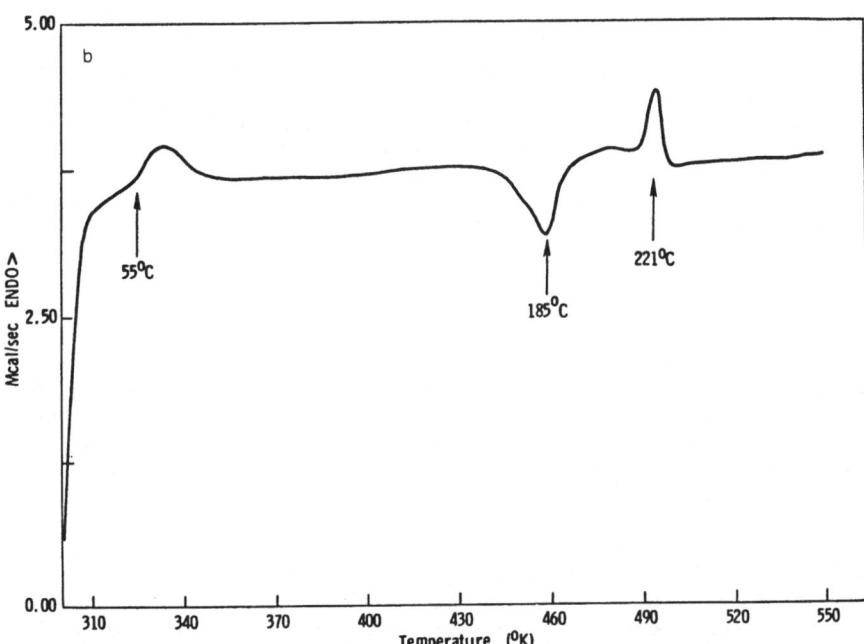

Fig. 2. a) Low temperature DSC scan for sample 1.7-2000-0.80-0.13.
Sample weight 21.14 mg. Scan Rate 20°C/min. b) High temperature DSC
scan for same sample. Sample weight 38.17 mg. Scan rate 10°C/min.

heat capacity observed at the T_g. Most of the DSC scans for the samples exhibit deviations from the baseline near 55°C which when viewed alone could not be judged to be T_g's. When scans for all the samples are compared, it becomes obvious that 80 percent of the samples exhibit transitions near 55°C.

The dynamic mechanical spectrum for 2.5-2000-1.2-0.2 is shown in Fig. 3. The curve for loss modulus (G") shows a transition due to the soft segment T_g at -80°C, a change in slope due to soft segment melting at about 0°C and possibly another transition occuring between 0° and 60-70°C. This small transition must be associated with the T_g of the hard segment. The sample retains its strength until at least 160°C, which is above the temperatures attained in compressive cycling. It had been suggested that tan δ values for these polyurethanes might be useful in predicting hysteresis. The results of dynamic mechanical analysis do not show a consistent relationship between internal temperature during compressive fatigue and tan δ. The values for tan δ were compared at 25°C, 100°C and at the temperature reached during compression testing. Some of the differences in tan δ are within the experimental error of the technique. The absence of a plateau region in the temperature range of interest in most of the samples contributes to the difficulty in analyzing the results. A possible improvement in results could be obtained by carefully controlling the thermal history of the samples tested.

Thermal analysis combined with information about the hardness values in Table 2 led to the choice of samples for initial compressive testing. Hardness values were considered important because many of the vulcani-zates of commercially important elastomers have lower hardness values than those of the polyurethanes reported here. Increasing the weight percent of soft segment by using PTMO 2695 was one way to reduce the hardness. Reducing the weight percent of hard segment by using two moles CHDI also produced softer samples. An alternative might be to prepare the polyurethanes by a one step technique rather than by the prepolymer method. This would probably lead to softer samples with more poorly organized hard segment phases. Unfortunately, increasing the softness frequently also leads to an increase in hysteresis. It should be noted that polyurethanes prepared from CHDI exhibit unusually high Shore A hardness values at low weight percent hard segment, ranging from 17.3 to 22.8 weight percent for sample compositions shown in Table 2 prepared with PTMO 2000.

Another factor which would be important in resistance to wear on rough surfaces is tear resistance. Values for tear strength are shown in Table 2. A trouser tear test was used in an effort to reduce the tensile contribution to the result. Samples containing more CHDI and exhibiting higher hardness values have higher tear resistance values. Within the groups of samples varying only in NCO/OH equivalent ratio, the samples with values of 0.95 tend to have higher tear resistance. Samples with the least amount of trimethylolpropane have the highest tear resistance. The highest value in the table is exhibited by the only linear sample which contains no crosslinker. The linear sample would be expected to possess greater order or crysallinity in the hard segment domains. The presence of crosslinker in the hard segment phase causes a reduction of order in that domain and this is manifested in lower tear strengths. The best way to evaluate the performance of the samples in use would be to measure the tear resistance at the internal temperatures reached during compressive fatigue. This set of experiments will be included in ongoing research on these materials.

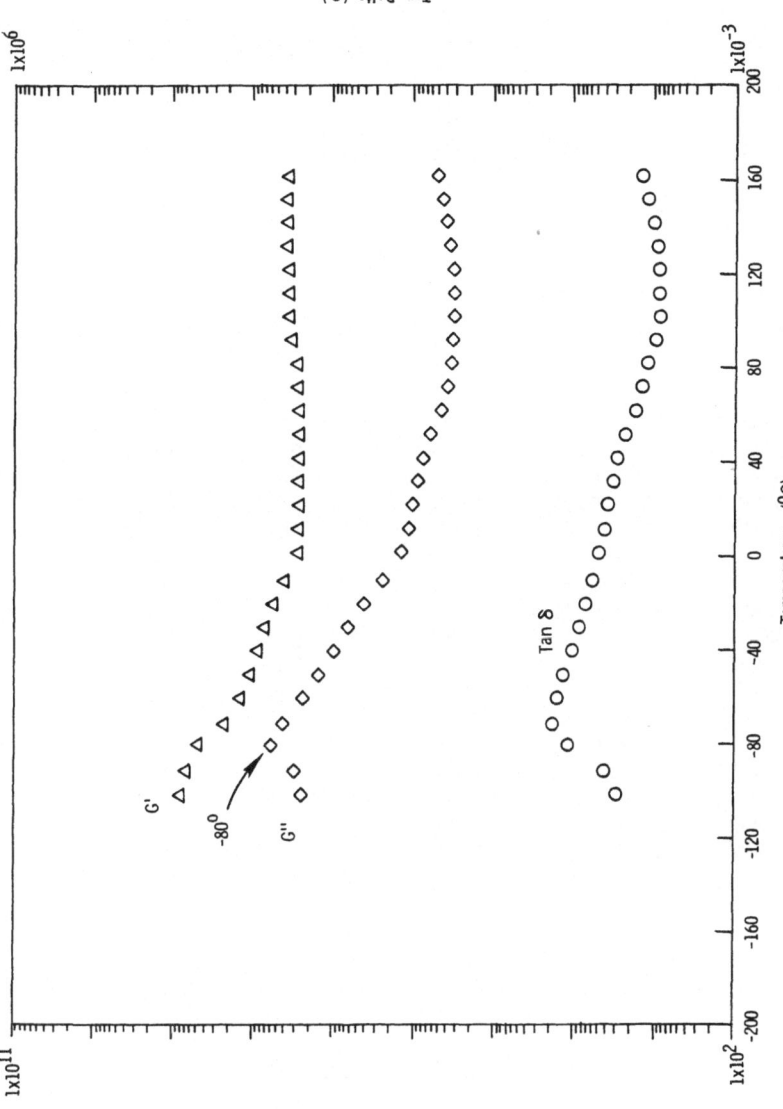

Fig. 3.   The dynamic mechanical spectrum for sample 2.5-2000-1.20-0.2.

Table 2. Shore A Hardness and Tear Strengths of Polyurethanes Prepared with CHDI and H_{12}MDI

Moles/1 Mole/Moles/Moles CHDI/ PTMO/ BD /TMP	NCO/OH	Hardness	Tear Strength (kN/m)	(pli)
1.7 - 2000 - 0.80 - 0.13	0.85	88	7.8	44.8
1.9 - 2000 - 0.80 - 0.13	0.95	90	8.3	47.6
2.0 - 2000 - 0.80 - 0.13	1.00	91	6.5	37.4
2.1 - 2000 - 0.80 - 0.13	1.05	84	10.9	62.1
1.7 - 2000 - 0.90 - 0.07	0.85	89	4.5	25.5
1.9 - 2000 - 0.90 - 0.07	0.95	91	23.3	132.9
2.0 - 2000 - 0.90 - 0.07	1.00	76[a]	---	---
2.1 - 2000 - 0.90 - 0.07	1.05	87	---	---
2.1 - 2000 - 0.60 - 0.27	1.05	87	4.0	23.0
2.1 - 2000 - 1.20 - 0.20	0.85	90	6.4	36.4
2.4 - 2000 - 1.20 - 0.20	0.95	94	9.7	55.7
2.5 - 2000 - 1.20 - 0.20	1.00	86	9.0	51.2
2.6 - 2000 - 1.20 - 0.20	1.05	94	5.8	33.4
2.8 - 2000 - 1.20 - 0.20	1.10	96	8.8	50.4
2.1 - 2000 - 1.35 - 0.10	0.85	94	26.5	151.7
2.4 - 2000 - 1.35 - 0.10	0.95	94	25.8	147.5
2.5 - 2000 - 1.35 - 0.10	1.00	95	12.8	73.4
2.6 - 2000 - 1.35 - 0.10	1.05	96	12.4	70.8
2.8 - 2000 - 1.35 - 0.10	1.10	95	11.2	63.9
2.6 - 2000 - 0.90 - 0.40	1.05	90	3.8	21.8
2.5 - 2000 - 1.50 - 0.00	1.00	95	30.6	175.1
2.1 - 1000 - 0.80 - 0.13	1.05	97	14.0	79.7
2.6 - 1000 - 1.20 - 0.20	1.05	95	13.5	77.3
2.0 - 2695 - 0.80 - 0.13	1.00	81	---	---
2.5 - 2695 - 1.20 - 0.20	1.00	73[a]	---	---

Moles
H_{12}MDI

3.5 - 2000 - 2.0 - 0.33	1.00	74[a]	---	---
3.0 - 2000 - 1.6 - 0.27	1.00	69[a]	---	---
2.5 - 2000 - 1.2 - 0.2	1.00	77[a]	---	---
2.0 - 2000 - 0.8 - 0.13	1.00	66[a]	---	---

[a]measured on blocks, since no sheets were available

Table 3. The Results of Compression Testing of Polyurethanes

Moles/1 Mole/Moles/Moles CHDI/ PTMO / BD /TMP	Temp. at End[a] of Experiment(°C)	Hysteresis[b] (in-lb)	(N-m)	Time to[a] Temp. (min)
1.7 - 2000 - 0.80 - 0.13	74	20.8	3.6	15[c]
1.9 - 2000 - 0.80 - 0.13	162	60.0	10.0	34[d]
Run 2	101	16.4	1.8	144[e]
2.0 - 2000 - 0.80 - 0.13	57[f]	9.9	1.1	85
Run 2	122[g]	8.2	0.9	45
Sample 2	89	1.1	0.1	45
2.1 - 2000 - 0.80 - 0.13	86	14.2	1.6	177
1.7 - 2000 - 0.90 - 0.07	102	39.4	4.4	5[c]
1.9 - 2000 - 0.90 - 0.07	89	15.6	1.7	180
2.0 - 2000 - 0.90 - 0.07	87	14.0	1.6	180
2.1 - 2000 - 0.90 - 0.07	72[g]	12.0	1.4	137
2.0 - 2000 - 0.60 - 0.27	82[g]	13.5	1.5	103
2.1 - 2000 - 1.20 - 0.20	110	33.6	3.7	7[c]
2.4 - 2000 - 1.20 - 0.20	69	11.6	1.3	135
2.5 - 2000 - 1.20 - 0.20	57	6.4	0.7	185
Sample 2	78	12.1	1.3	135
2.6 - 2000 - 1.20 - 0.20	66	9.3	1.0	108
2.8 - 2000 - 1.20 - 0.20	52	8.6	0.9	144
2.1 - 2000 - 1.35 - 0.10	55	8.5	0.9	135
2.4 - 2000 - 1.35 - 0.10	73	7.5	0.8	90
2.5 - 2000 - 1.35 - 0.10	59	9.5	1.1	180
2.6 - 2000 - 1.35 - 0.10	54	0.4	0.1	180
2.8 - 2000 - 1.35 - 0.10	51	2.7	0.3	180
2.5 - 2000 - 0.90 - 0.40	88	13.2	1.5	182
2.5 - 2000 - 1.50 - 0.00	58	7.8	0.9	90
2.0 - 2695 - 0.80 - 0.13	100	16.0	1.8	182
2.5 - 2695 - 1.20 - 0.20	100	23.8	2.7	176
Moles H_{12}MDI				
3.5 - 2000 - 2.00 - 0.33	165	36.8	6.4	23[c]
3.0 - 2000 - 1.60 - 0.27	161	37.7	6.6	10[c]
2.5 - 2000 - 1.20 - 0.20	124	41.8	7.3	6[c]
2.0 - 2000 - 0.80 - 0.13	141	42.6	7.5	12[c]

[a]This is also the maximum temperature during the experiment. After the time indicated, the temperature was constant.

[b]Hysteresis at end of experiment

[c]Damage to block, test stopped

[d]Block slipped out, test stopped

[e]Power failed at 160,000 cycles

[f]Thermocouple failed, temperature still rising

[g]Thermocouple failed, temperature levelling off

The results of compression testing of the sample blocks are shown in Table 3 and Figures 4 and 5. The poorest samples in the table are all marked with superscript c and showed damage early in the test. All four samples prepared with H_{12}MDI failed early and had high hysteresis values. Three of the samples prepared with CHDI also failed. All three had NCO/OH equivalent ratios of 0.85, the lowest tested. Failure modes were of two types, cracking in the center of the block near the thermocouple or blow out at the edges of the blocks. All other samples were undamaged at the end of the test. For the undamaged samples, the hysteresis changes initially and then levels off as the blocks attain a steady elevated temperature. The times in the last column of Table 3 require some explanation. The undamaged samples, with one exception to be mentioned below, exhibited nearly constant temperatures during the experiment after an initial break in period. Sample 2.1-2000-0.9-0.07 exhibited the same temperature from 137 to 180 minutes, the end of the test. Sample 2.0-2000-0.9-0.07 exhibited a nearly constant temperature, increasing a degree every 20,000 cycles or so up to the end of the test.

The hysteresis values are very high initially for three of the samples, exemplified by 2.0-2000-0.9-0.066 in Fig. 4. In the others the initial change is small. A nearly constant hysteresis value and temperature is attained in all cases except one before 3,000 cycles have been completed. The sample prepared with 2.5 CHDI and PTMO 2695 was the only sample tested which showed a continuing significant increase in hysteresis and temperature throughout the 180 minute experiment. Some type of change in polymer chain organization was undoubtedly continuing in that sample. Samples prepared with PTMO 1000 were not tested because they were too hard. The reasons for this are discussed in the section on hardness. The reasons why some samples show large hysteresis values initially and others do not is uncertain. In addition to the sample already mentioned, the other two compositions which show high hysteresis initially are 2.1-2000-0.8-0.13 and 2.0-2695-0.8-0.13. This behavior is probably related to the method of preparing the samples rather than to any intrinsic differences. Since these polyurethanes are prepared without a solvent, there are probably inhomogeneities in them which would not be found in solvent-cast samples. Sample 2.1-2000-0.8-0.13 was subjected to a second compressive fatigue experiment and did not show a high initial hysteresis value.

The hysteresis values in Table 3 are all values found at the end of the tests, even where thermocouple failure occurred, since the hysteresis could still be obtained after the internal temperature could no longer be measured. All of the undamaged samples exhibited low internal temperatures during compressive fatigue, illustrated in Figure 5. Temperature was measured every minute when it was changing rapidly and every ten minutes as it levelled off.

For samples which were undamaged under the conditions of this test, some generalizations can be made. An increase in weight percent of hard segment, either by increasing the amount of CHDI or by reducing the molecular weight of PTMO, leads to harder samples with lower hysteresis. In a given series of samples, increasing the NCO/OH equivalent value improved the performance of the sample in compressive fatigue. An increase in amount of crosslinking by TMP had little or no effect in the compression testing. The one sample prepared without TMP performed as well as the crosslinked samples.

The good performance of the polyurethanes prepared from CHDI is related to the morphology of the samples. They are well phase separated.

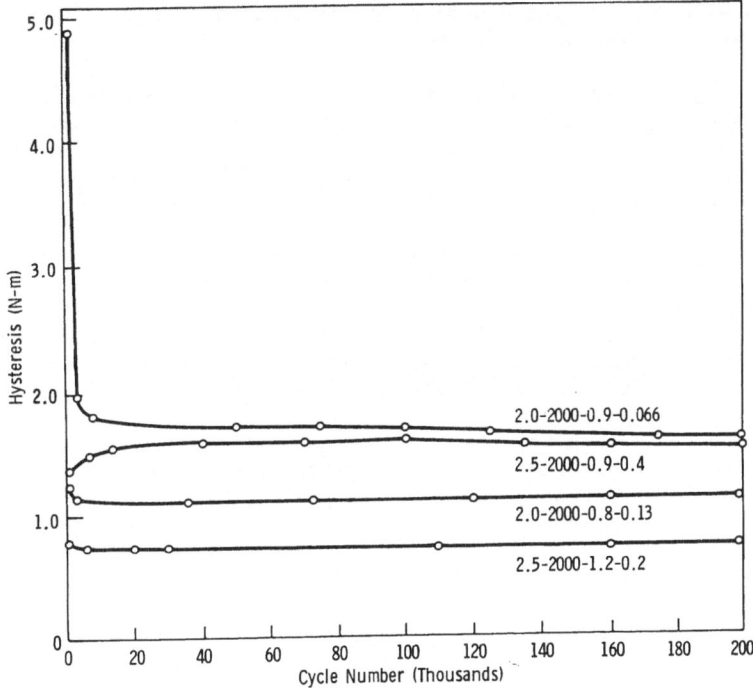

Fig. 4. The hysteresis during compressive fatigue at 18.5 Hertz.

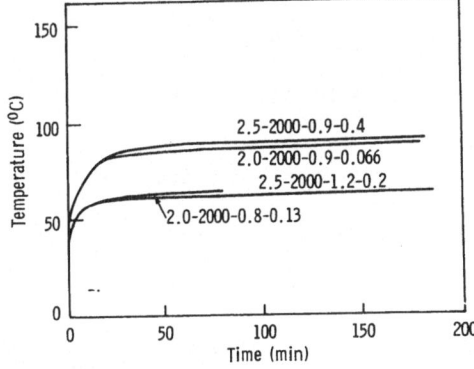

Fig. 5. The temperature at the center of the sample blocks during compressive fatigue at 18.5 Hertz.

They contain a low weight percent of hard segment. The internal temperatures generated during compressive fatigue are low. In some cases they are very close to the T_g's of the hard segments. The samples do not show a loss of properties above the T_g because the hard segments are partially crystalline. The crystalline melting is shown in the DSC curve in Fig. 2b. Since the samples are nearly transparent, it's likely that the crystallites are very small. Whereas large irregular crystals might interfere with polymer chain motion in compressive cycling, small crystals are less likely to do so.

Combining the results of compression fatigue with the results for tear resistance, it appears that the trends for improvement are reversed. The samples with the highest tear resistance at room temperature have low NCO/OH equivalent values, whereas the samples with the best performance in compressive fatigue have high NCO/OH values.

A choice of the best composition for a tire or track pad would best be delayed until some additional investigations are completed. More samples without TMP crosslinking should be examined since 2.5-2000-1.5-0.0 performed well. The effects of postcuring on properties should be determined. In addition, selected sample blocks should be subjected to repeated tests, to see the results of long term use.

Abrasion resistance has not been discussed here, but it is an important factor in performance. In limited field testing on another series of promising polyurethanes, the unfilled polyurethanes abrade at a rapid rate compared to compounded rubbers. The excellent abrasion resistance frequently shown by polyurethanes in conventional tests does not occur in field tests where the samples are track pads for tanks.

ACKNOWLEDGEMENTS

The author would like to thank Richard W. Matton for the synthesis, Elias R. Pattie for the Fatique Testing and Anthony L. Alesi and William W. Houghton for helpful discussions.

REFERENCES

1. A. L. Alesi, W. W. Houghton, M. E. Roylance and R. W. Simoneau, Proc. 42nd SPE Annual Tech. Conf. 583 (1984).

2. S. W. Wong and K. C. Frisch, Adv. Urethane Sci. Technol. 8, 75 (1981).

3. Unpublished results, Akzo Chemie America, Chicago, Illinois.

4. C. A. Byrne, D. P. Mack and J. M. Sloan, Rubber Chem. Technol., 58, 985 (1985); C. A. Byrne, E. A. McHugh, R. W. Matton, M. A. Cleaves, D. P. Mack and N. S. Schneider, Army Materials Technology Laboratory, Technical Report 83-57.

5. V. A. Grasso, paper presented at the Polyurethane Manufacturers Association Meeting, Boston, October 27-31, 1985.

6. A. J. Castro, P. Hentschel, W. Brodowski and T. Plummer, J. Elastomers Plast. 17, 238 (1985).

MORPHOLOGY AND PROPERTIES OF SEGMENTED POLYURETHANE-UREA

COPOLYMERS PREPARED VIA t-ALCOHOL "CHAIN EXTENSION"

Dinesh Tyagi and Garth L. Wilkes
Department of Chemical Engineering

Bin Lee and James E. McGrath
Department of Chemistry
and Polymer Materials & Interfaces Laboratory
Virginia Polytechnic Institute & State University
Blacksburg, VA 24061-6496

ABSTRACT

The structure-property behavior of novel tertiary alcohol-chain extended segmented polyurethane-urea copolymers was investigated. These copolymers were synthesized by utilizing the carbamate-isocyanate interaction at 150°C. Mechanical, thermal, dynamic mechanical and x-ray measurements were carried out to characterize the morphology and other characteristics of these segmented copolymers of systematically varying hard segment content, hard segment type, soft segment MW, and block length. It was observed that these properties depended primarily on the order in the hard domains and the order could be improved by increasing either the hard segment content at constant soft segment MW or soft segment molecular weight at the same hard segment content. The results obtained for these materials were compared with those from conventional polyurethanes to investigate the effect of intermolecular hydrogen bonding on molecular arrangement.

INTRODUCTION

In the past decade numerous detailed studies have been carried out for investigating the structure-property relationships in various segmented polyurethanes. A wide variety of aromatic and aliphatic diisocyanates and low molecular weight diol or amine chain extenders have been employed. The chain extenders commonly used for the hard segment have about 2 to 12 carbon atoms. The soft segments have been typically 1000 to 3000 molecular weight polyester, polyether or polybutadiene polyols. Many of these studies have established that in segmented polyurethanes, phase separation of the urethane hard segment into microdomains can take place even when the segmental length is relatively short. The primary driving force for domain formation is the strong intermolecular interaction between the urethane units which often have aromatic character and are capable of forming interurethane hydrogen bonds. Several workers have also observed the presence of hydrogen bonding between the urethane NH groups and oxygen of the macroglycol ether or ester linkage[1,2]. This observation is consistent with the postulate that some hard segments are dissolved in the soft segment matrix phase or vice versa. Polyurethanes are not high temperature polymers and cannot be used for continuous application at

103

temperatures in excess of 100°C. At high temperatures, polyurethanes begin to degrade thermally in an oxidative environment and/or may lose their two phase texture due to phase mixing[3].

The incorporation of urea linkages in polyurethane hard segments has a profound effect on the phase separation and domain structure of polyester or polyether-based polyurethane-ureas (PEUU). This is due to the high polarity differences between the hard and soft segments and the likely development of a three dimensional urea hydrogen bonding network[4-6]. These urea linkages can be introduced by allowing an isocyanate group to react with an amine group. As a result of their improved hydrogen bonding capability, good elastomeric behavior has been observed in other urea-linked segmented copolymers which contained as low as 6 % hard segment by weight[7,8].

From the literature it is obvious that a variety of polyurethane-urea copolymers have been prepared and studied. Sung et al.[5,9,10] have described the properties of segmented PEUU's based on tolylene-2,4-diisocyanate (TDI), ethylenediamine (ED) and poly (tetramethylene oxide) (PTMO), a polyether glycol also referred to as polytetramethylene glycol. The morphological investigation by small angle x-ray scattering (SAXS) has been carried out by Abouzahr et al.[11] on these materials. In contrast, Khransovskii[12] studied the PEUU composed of PTMO, TDI and 4,4'-diaminodiphenyl methane (DAM). The hard segments formed in both these polyurethane-ureas are capable of only limited crystallization because[13-15] of the unsymmetric structure of TDI units. Ishihara et al.[13-15] have reported on the synthesis and behavior of more crystallizable polyurethane-ureas based on PTMO, methylene bis(4-phenylisocyanate) (MDI), and DAM. A symmetric MDI/DAM hard segment would have better packing efficiency than a non-symmetric TDI/DAM hard segment. As a result, much stronger hard segment domains would be obtained with the former and the copolymer would be expected to display a stronger mechanical response. As compared to MDI/polybutadiene diol (BD) based polyurethanes, the thermal properties would no doubt be enhanced because of the complete aromatic nature of the hard segments.

Unfortunately, the large thermodynamic incompatibility between the two types of segments and hydrogen bonding-capability of the hard segments also make the synthesis and processing of these segmented polyurethane-ureas very difficult. The problem also arises because the reaction between amine and isocyanate groups proceeds at a much faster rate than the reaction of isocyanate and hydroxyl groups. In a one-step bulk polymerization, the high reactivity of $-NH_2$ and -NCO groups could result primarily in the formation of polyurea which can prematurely precipitate out of the reaction mixture. Some success is achieved when a strong polar group is attached to the diamine component to retard the isocyanate-amine reaction, e.g., MDI and 3,3'-dichloro-4, 4'-diaminodiphenyl methane (MOCA). Use of the latter diamine has been discontinued because of carcinogenic hazards. Solution polymerization of PEUU's is sometime possible when the reaction mixture is maintained at low temperatures to control the $-NH_2$/-NCO reaction. However, the solution polymerization approach suffers from the difficulty of obtaining a good common solvent for all the reaction components. In brief, the choice of solvent clearly affects the degree of polymerization and a mechanically weak polymer would be obtained if the overall molecular weight of the segmented copolymer is low.

Attempts have been made to synthesize the MDI/DAM-based segmented polyurethane-urea by addition of stoichiometric amounts of water to the mixture of PTMO and MDI. Adding controlled amounts of water would convert the excess MDI into DAM by the reaction scheme as illustrated on the following page.

$$RNCO + H_2O \longrightarrow [RNHCOOH] \longrightarrow [RNH_2] + CO_2$$

The water molecule causes the formation of the unstable carbamic acid which decomposes further to yield DAM and CO_2 and is given off in the process. This approach, although promising,[2] is difficult to use effectively because of water immiscibility with the reaction system. As a consequence, addition of water results in the formation of water droplets and a copolymer with heterogenous properties may thus be obtained.

Historically, in 1948, Saunders and Slocombe[16] reported that tertiary alcohols dehydrate at reasonably high temperatures in the presence of isocyanate groups. Although no mechanism was proposed, it was postulated that a tertiary alcohol would produce water in the presence of a weak acidic environment. Therefore, it may be possible to synthesize segmented polyurethane-ureas by taking advantage of such a proposed dehydration reaction. Specifically, a mixture of isocyanate-capped polyether, diisocyanate and a tertiary alcohol, when heated under inert conditions, could result in the formation of a PEUU. This could result from the fact that the tertiary alcohol would dehydrate and the water generated would cause the formation of an amine group by the reaction with isocyanate. Next, this amine-terminated prepolymer unit would react with MDI to serve as the chain extender for the MDI-based hard segments. Thus, this scheme provides a novel approach to the synthesis of segmented PEUU's as high molecular weight copolymers can be obtained by a homogeneous, in situ chain-extension step involving essentially two terminal isocyanate groups. Since the water is introduced only at molecular levels, it should be consumed immediately in the formation of amine . Quantities of water in the form of droplets are thus avoided and the resulting copolymers might be expected to be compositionally more uniform.

Using the above procedure, novel segmented polyurethane-ureas have been successfully prepared and some synthesis details have already been reported.[17] These copolymers are obtained by reacting PTMO and excess MDI (or TDI) at high temperatures. Either tertiary cumyl or dicumyl alcohol*(CA or DCA) was added for the purpose of providing the in situ conversion of isocyanate groups. Two series of PEUU were prepared using both tertiary alcohols. The molecular weight ($\overline{Mn}$) of the polyether macroglycol ranged from 650 to 2000. Each series of samples contained three levels of hard segment content chosen to maintain a comparable hard segment fraction.

We are currently[18] attempting to establish the reaction mechanism by which the isocyanate group is converted into an amine group. Dehydration of the alcohol may not be necessary. An intermediate carbamate can also explain the polymerization phenomenum. The current paper reports on our investigation of structure-property relationships in these novel segmented polyurethane-ureas. For this purpose, it was desirable to determine how the mechanical and thermal properties were affected with hard segment content, crystallization and length of the soft and hard segment. The chemical composition was thus used to control the size, shape and the degree of order in these segmented copolymers. Finally, these PEUU's with symmetric and aromatic hard segments were compared with conventional PTMO/MDI/BD polyurethane and other polyurethane-ureas reported in the literature.

*Strictly speaking, it should be named as α,α-dihydroxy-p-diisopropyl-benzene, a tertiary aromatic dialcohol. The name dicumyl alcohol is used for convenience.

EXPERIMENTAL

Segmented polyurethane-urea copolymers were synthesized by a step-growth (condensation) reaction involving the prepolymer and amine terminated groups as chain extender. The prepolymers were prepared by reacting poly (tetramethylene oxide) glycol [DuPont's Teracol[R]] with excess 4,4'-diisocyanatodiphenyl methylene (MDI) or a 80/20 mixture of 2,4- and 2,6- toluenediisocyanate (TDI) [Mobay Chemical Co.]. Macroglycol of molecular weights 650, 1000, 2000 was employed as the soft segments in these elastomers. The polyols were dried at 60°C in a vacuum oven for 48 hours prior to use. The diisocyanates were vacuum distilled and middle cuts were stored at 0°C before use.

Table 1. Characteristics of Various Polyurethane-Ureas Synthesized

Sample	Mole Ratio MDI/PTMO/CA or DCA	% Urea Content
PTMO-2000-MDI-23.3-CA	2.5/1/1.5	40
PTMO-2000-MDI-31-CA	3.7/1/2.7	57
PTMO-2000-MDI-41-CA	5.6/1/4.6	70
PTMO-2000-MDI-31-DCA	3.7/1/1.35	57
PTMO-1000-MDI-31-DCA	1.8/1/0.4	29
PTMO- 650-MDI-31-DCA	1.2/1/0.1	9
PTMO-2000-TDI-31-CA	5.3/1/2.2	69
PTMO-2000-MDI-31-BD	CONTROL	0

$$\% \text{ Urea Content} = \frac{\text{No. of all possible urea linkages}}{\text{Total urea and urethane linkages}}$$

Tertiary alcohols, dicumyl alcohol (The Goodyear Tire & Rubber Company) and cumyl alcohol (Aldrich) were used to carry out the necessary conversion of excess isocyanate groups to amine groups as mentioned earlier. Thus, these tertiary alcohols are not directly involved in the chain extension step. The amount of excess MDI and tertiary alcohols determines the hard segment content in the polyurethane-urea samples. The composition of the various copolymers synthesized is listed in Table 1.

To carry out the reaction, MDI was charged into a N_2-filled resin kettle equipped with a high torque mechanical stirrer at room temperature. PTMO was then slowly delivered by means of a syringe into the reactor at 110°C over a period of 20 minutes. The mixture was then continuously stirred at 110°C for 1 hour. The specified tertiary alcohol was then added to the mixture of prepolymer and excess MDI. The molar ratio of alcohol to isocyanate was about 1:2. The reaction was then allowed to continue for another 3 hours at 150°C under inert conditions. At temperatures below 110°C no significant change in the bulk viscosity was observed, indicating that the uncatalyzed chain extension step cannot be carried out at these low temperatures. However, at 150°C an immediate increase in the viscosity was observed. By changing the molar ratio of MDI and PTMO, a series of samples with systematically varied hard segment content and relative block length were synthesized. The resulting copolymer was extracted with THF for 24 hours to remove low molecular weight impurities, if any. The final product was dried in a vacuum oven for 36 hours at 70°C. A clear polymer solution resulted when this polymer was dissolved in N,N dimethyl formamide (DMF) at 100°C suggesting that an essentially linear polymer was obtained.

Films of the segmented PEUU copolymer and a control polyurethane (Estane[R], BFGoodrich) were prepared by compression molding the dry material at 200 to 240°C and 10,000 psi. The temperature was slightly higher when the hard segment content was increased to 41 weight % MDI. The material was first heated in the press for two to three minutes and the pressure was then raised slowly. The film was kept in the press for about 5 minutes prior to its removal. After removal from the press, all samples were immediately quenched to room temperature and placed under vacuum in a dessicator until further testing. No obvious degradation of the materials at these high temperatures was observed as all the films were transparent or translucent (for materials based on 2000 MW PTMO and MDI) after removal. The segmented polyurethane-urea copolymer obtained with 2000 molecular weight PTMO, MDI and CA is referred to as PTMO-2000-MDI-31-CA, where 31 represents the weight percent of the hard segment. A similar nomenclature is employed for all other samples.

Structural characterization of the polymers was obtained by FT-IR and NMR spectroscopy using a Nicolet MX-1 and Varian EM-390 spectrometer. The IR spectra for the oligomers and the copolymers were obtained from solution cast films on KBr discs. Dog-bone specimens were cut from the films for mechanical testing of the segmented copolymers. Mechanical measurements included stress-strain, stress relaxation, tensile hysteresis and permanent set behavior. All these tests were performed on an Instron Model 1122 at room temperature. All stress-strain and hysteresis measurements were carried out at a strain rate of 200% per minute based on the initial sample length.

Dynamic mechanical spectra were obtained with a Rheovibron Model DDV II-C at 110 Hz and the Dynamic Mechanical Thermal Analyzer (DMTA) manufactured by Polymer Laboratories, England at 1Hz. The approximate heating rate for the sample was maintained near 3°C per minute in the Rheovibron and 5°C per minute for the DMTA.

Thermal analysis (DSC, TGA, TMA) of the block copolymers was performed on a Perkin-Elmer System 2 and System 4. Experiments were carried out under a helium or nitrogen atmosphere using a heating rate of 10°C per minute. The DSC and TMA scans were started at -140°C, whereas thermal degradation studies (TG) began at 50°C. During TMA measurements, a constant load of 10 g was employed.

The formation of domain structure (microphase separation) was also verified by using small angle X-ray analysis (SAXS). An automated Kratky small angle x-ray camera was utilized for the SAXS experiments. The x-ray source was a Siemens AG Cu 40/2 tube, operated at 40 kV and 20 mA by a GE XRD-6 generator. A Cu Kα-radiation of wavelength .154 nm was obtained by Ni-foil filtering. A Philips table-top X-ray generator PW 1720 was utilized in conjunction with a standard vacuum-sealed Statton (Warhus) camera to obtain wide-angle x-ray diffraction (WAXD) patterns for the copolymers. The samples were approximately 0.5 mm thick and an exposure time ranging from 10 to 24 hours was used depending on the system.

RESULTS AND DISCUSSION

In order to characterize the reaction mechanism and to confirm the formation of urea linkages, infrared spectra were obtained from the polymer films cast on KBr disks. The FT-IR spectra for sample PTMO-2000-MDI-31-CA confirmed the incorporation of urea linkages by the presence of an absorption peak at 1640 cm^{-1} which is attributed to the hydrogen bonded urea carbonyl. No absorption peak, corresponding to this wavenumber was observed for the PTMO-2000-MDI-31-BD polyurethane (ESTANE®) control material where urea groups are absent. The IR spectra obtained on the polymerization by-product, collected from the sides of the reaction kettle, indicated the presence of an olefin absorption in the spectra. The absorption peaks appearing at 1640 cm^{-1} and 850 cm^{-1} correspond to the C=C stretch and =CH$_2$ wag, respectively, and compare well with an authentic IR spectra for α-methylstyrene. This was to be expected when cumyl alcohol was utilized. These absorption peaks in the polymerization condensate imply that there can be two possible reaction schemes by which the t-alcohols interact with isocyanate groups. These two schemes are illustrated in Figure 1. The first scheme indicates the formation of unstable carbamate which ionically dissociates to produce CO$_2$ from the unstable anion and an amine-terminated moiety. The second scheme would require dehydration of t-alcohols initially, which is then followed by formation of amine terminated groups through the more widely considered carbamic acid route. A recent detailed study[18] carried out on model reactions seem to indicate that the former mechanism is more likely to take place. However, a dual reaction mechanism where both mechanisms compete may also be possible.

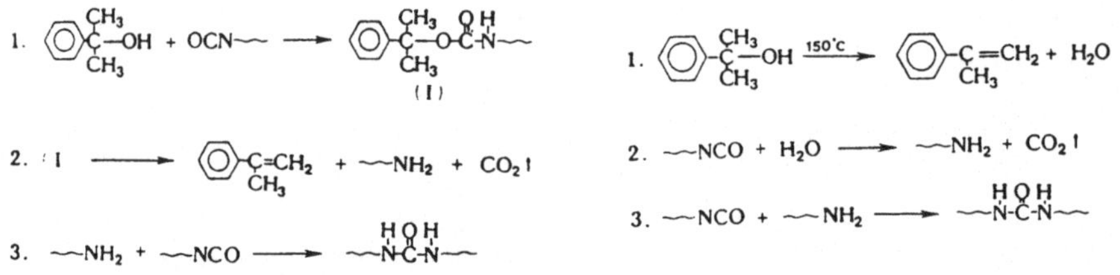

Scheme I   Scheme II

Fig. 1:   Scheme 1 - Possible urea formation mechanism via carbamate rearrangement and Scheme II - Possible urea formation mechanism via tertiary alcohol dehydration.

Mechanical Analysis

The engineering stress-strain curves for segmented MDI-based polyurethane-ureas are shown in Figure 2 through Figure 4. All curves are shown up to the fracture stress of the sample. The Young's modulus, ultimate tensile strength, and ultimate elongation were determined from these measurements and are listed in Table 2 for each sample. It is well accepted that the tensile behavior for a thermoplastic elastomer generally depends upon the size, shape and concentration of the hard segments; intermolecular bonding within the hard domains; and the molecular weight of the soft segments.[19, 20] The tensile behavior of these segmented copolymers establishes the expected correlation with these parameters.

In Figure 2, the stress-strain behavior for materials with different soft segment molecular weights but the same hard segment content are shown. The results indicate that the Young's modulus and the tensile strength in these samples increase as the hard segment length is increased. This can be explained on the basis of the increase in the hard segment length necessary to maintain the same hard segment content with increasing molecular weight of the soft segments. Increasing the block length not only increases the aspect ratio of the dispersed hard domains but also leads to a higher degree of order in the hard domain since this results in more urea linkages per hard segment unit. Although

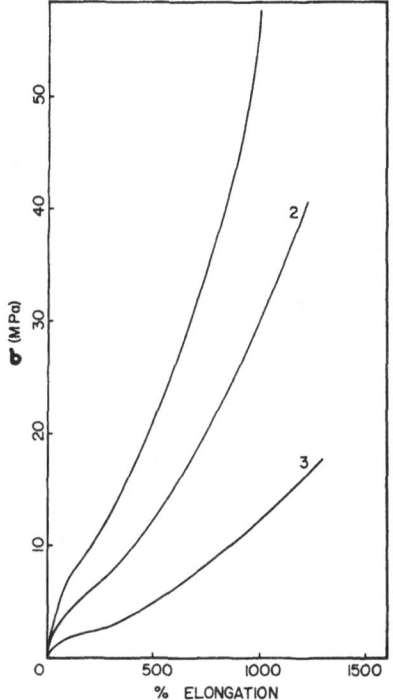

Fig. 2: Stress-strain curves for various segmented polyurethane-urea copolymers indicating the dependence of tensile behavior on the soft segment molecular weight. (1) PTMO-2000-MDI-31-DCA, (2) PTMO-1000-MDI-31-DCA, and (3) PTMO-650-MDI-31-DCA.

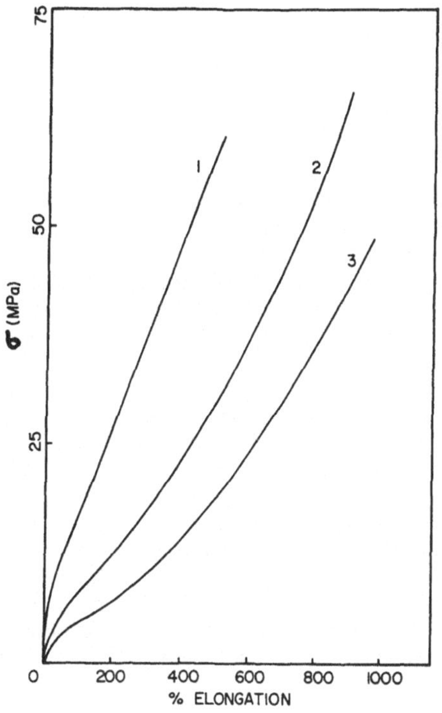

Fig. 3: Stress-strain curves showing the effect of hard segment content on the tensile behavior of various segmented polyurethane-urea copolymers. (1) PTMO-2000-MDI-41-CA, (2) PTMO-2000-MDI-31-CA, and (3) PTMO-2000-MDI-23.3-CA.

the sample PTMO-650-MDI-31-DCA has a higher density of the hard segment units, the pseudo multifunctional cross-links formed by the hard domains are shorter and likely weaker. As a consequence of these poorly defined hard domains, the sample PTMO-650-MDI-31-DCA exhibits lower mechanical strength.

Table 2. Mechanical Properties of Segmented Polyurethane-Ureas

Sample	Ultimate Tensile Strength (MPa)	Young's Modulus (MPa)	%Elongation at Break
PTMO-2000-MDI-23.3-CA	49.0	15.1	1000
PTMO-2000-MDI-31-CA	67.5	17.6	900
PTMO-2000-MDI-41-CA	59.0	104.5	600
PTMO-2000-MDI-31-DCA	59.0	11.4	1000
PTMO-1000-MDI-31-DCA	40.5	9.3	1100
PTMO-650-MDI-31-DCA	17.5	6.8	1300
PTMO-2000-TDI-31-CA	50.5	35.6	550
PTMO-2000-MDI-3-BD	41.0	13.2	1000

The stress-strain behavior for polyurethane-urea elastomers based on
2000 molecular weight PTMO is shown in Figure 3 as a function of
increasing hard segment content. It is indicated that the mechanical
response of these materials is strongly affected when hard segment
content is raised from 23.3% to 41% by weight. The observed behavior can
be explained on the basis of the introduction of a higher volume
fraction of hard segments as well as a higher degree of order in the
hard domains. The higher modulus and tensile strength with increasing
hard segment content in these samples is also consistent with their
greater urea content which results in more cohesive hard domains.

The results in Figure 4 are presented to bring out the effect of
varying intermolecular binding forces; hard segment symmetry or shape;
and finally the use of two different tertiary alcohols (namely CA and
DCA). All four samples indicated in this figure have 31% hard segment by
weight. A comparison among these curves indicates that the lowest
ultimate tensile strength is displayed by sample PTMO-2000-MDI-31-BD
(ESTANE). This sample differs from the rest in the sense that it does
not contain any urea linkages. Clearly, the weaker interdomain secondary
binding forces result in less cohesive hard domains and may limit the
development of a three dimensional hydrogen bonding network. Phase
separation is, no doubt, also affected and a possible larger interfacial
region and/or higher segmental mixing present in ESTANE could be
responsible for higher initial tensile modulus. The lack of symmetry in
the ESTANE hard domains (MDI/BD vs. MDI/DAM) may also influence the
tensile behavior as the packing of the hard segments would not be
efficient either.

When cumyl alcohol is used instead of dicumyl alcohol to carry out
in situ chain extension, some dissimilarity between the mechanical
behavior of the corresponding copolymers is observed. This disparity
presumably is a consequence of the differences in the chain extension
step caused by the nature of the two tertiary alcohols involved. This
difference may affect the distribution of hard segment lengths and would
be narrower in the copolymer where CA is employed. In either case, it
appears that the copolymer prepared with DCA apparently has less order
in the hard domains and possibly a larger interfacial region between the
two phases due to a larger distribution in the hard segment length. The
SAXS results (discussed later) have also indicated that at the same hard
segment content, the thickness of the interfacial zone is larger in the
copolymers when DCA is used for the chain extension step. The

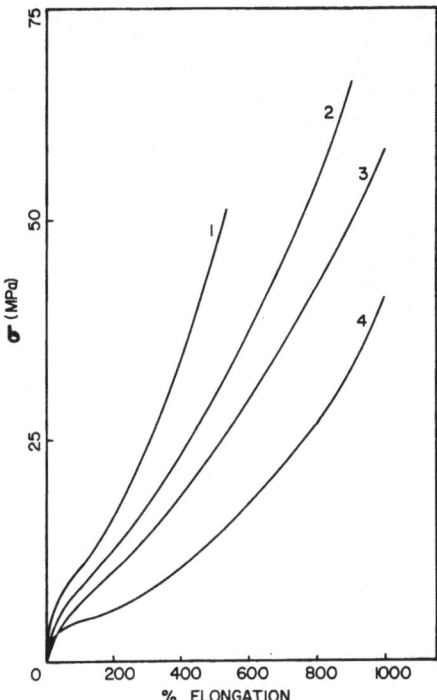

Fig. 4: Stress-strain behavior for various segmented polyurethane-urea copolymers. (1) PTMO-2000-TDI-31-CA, (2) PTMO-2000-MDI-31-CA, (3) PTMO-2000-MDI-31-DCA, and (4) PTMO-2000-MDI-31-BD.

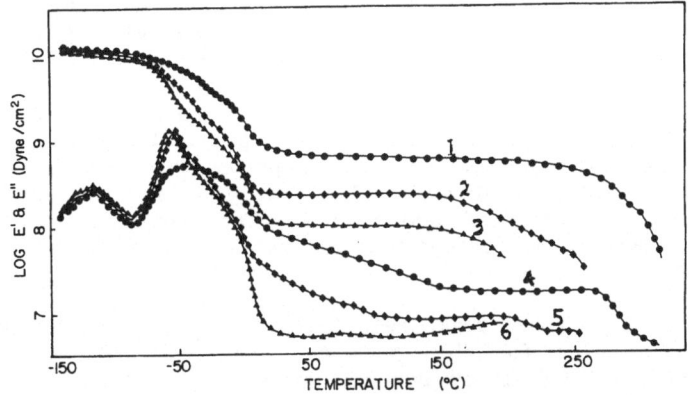

Fig. 5: Dynamic mechanical spectra for segmented poly-urethane-urea copolymers. (●) PTMO-2000-MDI-41-CA, (♦), PTMO-2000-MDI-31-CA and (▲) PTMO-2000-MDI-23.3-CA determined on the Rheovibron. 110 Hz frequency. Curves 1-3 are for storage moduli data and curve 4-6 are for loss modulus data.

differences in the mechanical properties, observed for the copolymers obtained via these two alcohols, is therefore an indication of the differences in the chain extension step caused by these two reactants. When the degree of chain extension is small (23.3% hard segment by weight), the difference between the tensile properties of the segmented copolymer prepared by the two tertiary alcohols is also small.

The difference in the TDI and MDI based polyurethane-ureas, however, is observed because of two reasons. First, for constant hard segment content by weight, TDI based copolymers have a higher concentration of urea linkages since the molecular weight of one TDI unit is less than that of a MDI unit. Hence a stronger mechanical response is obtained owing to a higher degree of cohesiveness. The second reason is interrelated with the dissimilarities observed in the orientation behavior of the soft segment chains upon deformation for the two representative copolymers. As will be discussed shortly, it was found by WAXS analysis that strain-induced crystallization of the PTMO chains occurs at lower elongations for TDI based system as compared to samples with equal weight percent of MDI-based hard segments. An earlier onset of the crystallization in TDI system is also likely to increase the stress level in the sample. The reasons for this observed difference in the crystallization behavior will be discussed in more detail when the dynamic mechanical and WAXS results are presented.

Dynamic Mechanical Analysis

To investigate the compositional dependence of the local scale motion as well as cooperative segmental motion which may exist in these segmented polyurethane-urea copolymers, dynamic mechanical spectra were obtained. It was also of interest to determine the mechanical and thermal properties in light of the role played by the variety of hydrogen bonding possibilities in these materials. Figures 5 and 6 show the overall dynamic behavior in terms of storage (E') and loss (E") modulus as a function of temperature.

The dynamic mechanical results in Figure 5 demonstrate the effect of hard segment content for a fixed molecular weight of the polyether soft segment. A rubbery plateau, from the E' curves, is observed to be composition-dependent and extends to $200^{\circ}C$ for 23.3% hard segment content material and near $300^{\circ}C$ for the sample with 41% hard segment content. The plateau modulus also increases significantly upon increasing the hard segment content in the sample. An increase in the order of hard domains arising from hard segment content results in the formation of a stiffer material which is reflected in the E' behavior. Interestingly, the plateau modulus in these polyurethane-ureas is nearly one order of magnitude higher than that reported for a MDI/ED-based PEUU with a similar hard segment content[21]. The loss modulus curves appear to be complex as a number of damping peaks (or shoulders) are observed. The various relaxations, responsible for the origin of these peaks, are now discussed.

The secondary loss peak near $-120^{\circ}C$ is designated as the γ relaxation. The value for this peak has been reported to lie between -130 and $-100^{\circ}C$ at 110 Hz by several workers[21-26]. This peak has been ascribed to the motion of methylene sequences. The magnitude and location of the γ peak is found to be nearly independent of the hard segment content. The γ relaxation temperature observed in these copolymers is in excellent agreement with the reported value ($-120^{\circ}C$) for polyolefin systems. This indirectly suggests that the degree of phase separation achieved in these copolymers may be very high. This is expected because of the highly polar nature of the aromatic urea

112

linkages that form the hard segments in these copolymers.

The primary relaxation observed near -55°C in the dynamic mechanical spectra at 110 Hz for these copolymers is designated as the α_a relaxation and has been attributed to the glass transition temperature of the soft segments.[21-23] The intensity and location of the α_a peak is strongly dependent on the chemical composition of the samples. An increase in the hard segment content at constant prepolymer molecular weight broadens the α_a relaxation and decreases its intensity. This is accompanied by a shift of the peak temperature to a higher temperature. It appears that the α_a relaxation process involves motion within the amorphous portion of the flexible prepolymer chains. Thus it may be inferred that increasing the aromatic urea content will tend to restrict the motion of the prepolymer phase and thus shift the α_a transition to a higher temperature.

On the high temperature side of the α_a peak, a mechanical dispersion is indicated as a shoulder near 0°C. This transition is designated as α_c and is related to the melting of crystallites of the soft segments. The melting point of PTMO homopolymer has been reported to lie near 35°C.[27] An even higher value (nearly 49°C) of Tm has been reported by Dreyfus[38] for extremly high molecular weight PTMO. The lower Tm of the soft segment observed here is believed to be due to the lower molecular weight of the PTMO segments in these copolymers or to the formation of small and/or imperfect crystallites.

The existence of δ relaxations in dynamic mechanical behavior has been observed for conventional MDI/BD based polyurethanes and shown to be associated with the hard segments of the copolymer.[22, 23] The first δ relaxation near 80°C was ascribed to the annealing effects associated with the poor ordering in the hard domains. The high temperature δ' relaxation near 200°C in the dynamic mechanical spectra has been attributed to the melting of the hard segment crystallites.[23] For these polyurethane-ureas copolymers the δ relaxation is barely visible in the loss modulus curve. This indicates that any further ordering of the hard segments does not take place in 2000 molecular weight PTMO-based copolymers. This is not surprising as the hard domains are more cohesive in these copolymers due to the presence of urea linkages and the large thermodynamic incompatibility between the two segment types. One other relaxation is observed over 200°C and is dependent on the hard segment content. This relaxation is assumed to be associated with the melting of the hard segment microcrystallites with urea linkages and this dispersion is designated as δ'. An endotherm at these temperatures was also observed by DSC. The melting point of the MDI/DAM "hard segment" control has been reported to be 370°C.[15] The same Tm value has also been observed in this laboratory, although we are not certain that this represents the value associated with very high molecular weight components.

Figure 6 shows the dynamic mechanical behavior of segmented polyurethane-ureas as a function of the molecular weight of the prepolymer. The results indicate that the α_a transition shifts to a lower temperature as the molecular weight of the soft segments is increased. This shift can be explained on the basis of longer, more ordered and well defined domain structure which results when the molecular weight of the macroglycol is increased at constant hard segment content. The intensity of the α_a peak is also composition-dependent and decreases with increasing segment molecular weight. This occurs because of the onset of crystallinity in the soft segments as the molecular weight of PTMO is increased. For the 650 and 1000 molecular weight polyether glycol prepolymer, the α_c relaxation was not observed. This is consistent with other observations that the soft segment blocks crystallize only when their molecular weight is in excess of 1000,[23] although higher hard/soft segment mixing present in these systems may also prevent the crystallization of the soft segments. The δ and δ' relaxation peaks are indicated for these materials by a slight deflection in the dynamic mechanical spectra. The γ relaxation peaks appear to be molecular weight-dependent as they shift to low temperatures for high MW prepolymer based materials of the same hard segment content. This may indicate a higher degree of segmental mixing as the molecular weight of the soft segment is lowered for the same hard segment content.

The length and extent of the rubbery plateau is also shown to depend on the molecular weight of the macroglycol employed. The plateau modulus is higher for samples where 2000 molecular weight soft segments are utilized as compared to samples where the soft segment molecular weight is lower. This behavior is attributed to the existence of less well defined domain structure when low molecular weight macroglycols are used. For copolymers prepared from higher molecular weight polyether glycols, high prepolymer molecular weight based copolymers, better phase separation is obtained and enhanced physical crosslinking and/or filler effects (arising from the presence of hard segment domains) dominate the properties. As a consequence, the effective degree of domain formation is improved in the high segment molecular weight materials. Additionally, for samples with the same hard segment content, the concentration of urea groups is higher for samples with higher molecular weight prepolymer whose hydrogen bonding network improves intradomain cohesion.

Thermal Analysis

The thermomechanical analysis (TMA) spectrum was obtained for these polyurethane-ureas to observe the transitions associated with the hard and soft segments. Figure 7 compares the TMA measurements of three representative copolymers with different hard segment contents. The

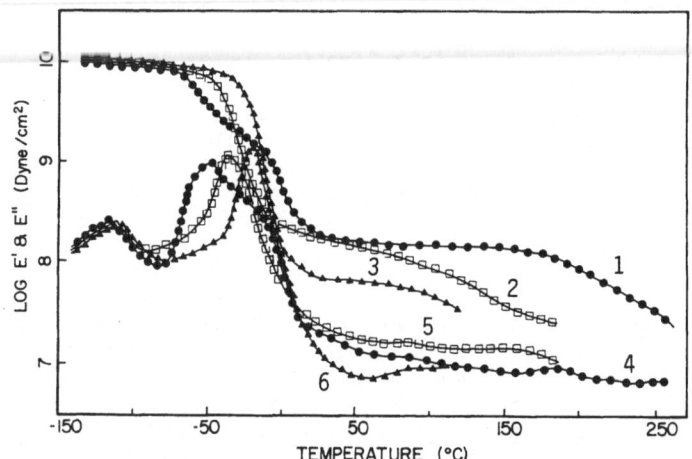

Fig. 6: Dynamic mechanical spectra for (●) PTMO-2000-MDI-31-DCA, (□) PTMO-1000-MDI-31-DCA and (▲) PTMO-650-MDI-31-DCA. Frequency 110 Hz. Curves 1-3 are for storage moduli data and curves 4-6 are for loss modulus data.

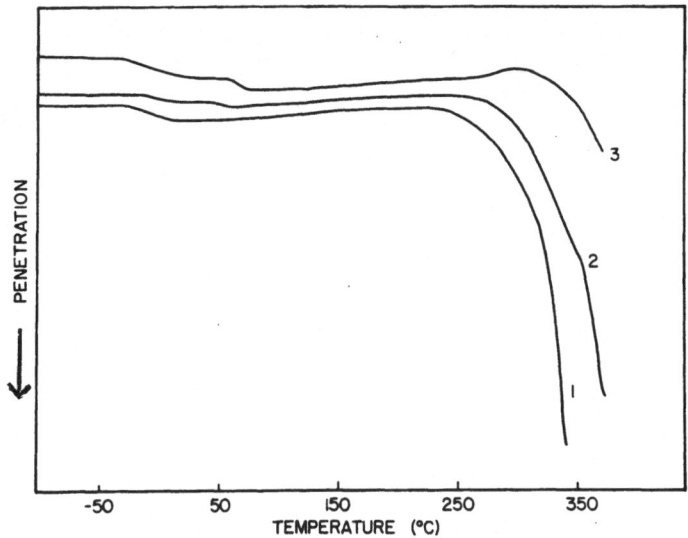

Fig. 7: Thermomechanical penetration curves for various samples with different hard segment content. (1) PTMO-2000-MDI-41-CA, (2) PTMO-2000-MDI-31-CA, and (3) PTMO-2000-MDI-23.3-CA.

primary transition near -50°C is ascribed to the soft segment glass transition. The hard segment transition is indicated to be composition dependent and varies from 240 to 300°C as the hard segment content is increased from 23.3 to 41% by weight. The nature and length of the rubbery plateau depends on the hard segment content and is higher for segmented copolymers with higher hard segment content. The TMA response is also affected when the molecular weight of the soft segment is varied at a constant hard segment content (see Figure 8). These results are consistent with the fact that the sequence length also increases as a function of hard segment content. In Figure 8, these MDI-based polyurethane-ureas are compared with Estane, a conventional diol-extended polyurethane. One significant feature observed in the ESTANE is its relatively low softening temperature (120°C). In contrast, the strength and service temperature have been dramatically improved with the incorporation of 29 and 59 weight percent aromatic urea groups in polyurethane-ureas. In general, the results obtained from TMA penetration curves are in agreement with the dynamic mechanical behavior discussed earlier.

Thermogravimetric analysis on these materials provide strong evidence of the somewhat improved thermal stability due likely to the existence of a 3-dimensional urea hydrogen bonding network as well as the complete aromatic nature of the hard segments. The TGA behavior was observed to be nearly independent of the composition and hence only one curve is indicated for MDI based copolymers. These materials exhibit little weight loss up to 300°C, but lose 50% of their weight by about 400°C.

As discussed in the preceding section, a number of different transitions were observed with dynamic mechanical measurements. In order to verify these transitions or relaxations, differential scanning calorimetry (DSC) thermograms were obtained for these polyurethane-urea

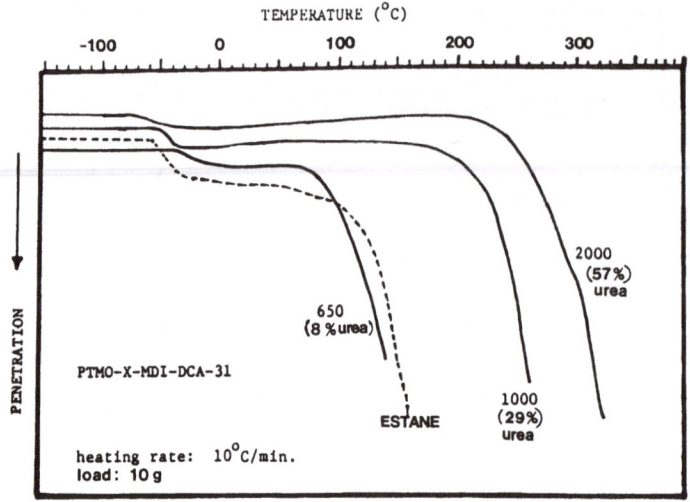

Fig. 8: TMA penetration curves of the PEUU's with different soft segment lengths.

copolymers. This study also allowed the determination of composition dependence of various transitions associated with hard and soft segments. Table 3 presents a summary of the low temperature DSC results for the segmented copolymers. Data on various thermal events occurring in the component prepolymers are also indicated in this table. The transitions observed by DSC are in good agreement with the α_a, α_c, δ, and δ' relaxations obtained by dynamic mechanical spectroscopy. No signs of the γ relaxation were indicated by the DSC thermograms but a transition associated with the melting of the crystallites was observed in the thermal scans.

Table 3. Results From Low Temperature DSC Scans ($^{\circ}$C) For Segmented Polyurethane-Urea Copolymers

Sample	Tg			Tc	Tm
	From	To	Midpoint		
PTMO-2000-MDI-23.3-CA	-83	-61	-75.2	-43.1	2.5
PTMO-2000-MDI-31-CA	-86	-62	-74.7	-34.1	3.5
PTMO-2000-MDI-41-CA	-84	-54	-72.0	-35.1	11.7
PTMO-2000-MDI-31-DCA	-89	-57	-73.7	-35.7	8.8
PTMO-1000-MDI-31-DCA	-59	-29	-47.9	ND	ND
PTMO- 650-MDI-31-DCA	-58	-19	-38.3	ND	ND
PTMO-2000-MDI-31-CA	-84	-62	-76.3	ND	ND
PTMO-2000-MDI-31-BD	-64	-29	-49.0	ND	ND
PTMO-2000	--	--	-78.0	--	24,27
PTMO-1000	--	--	-82.0	--	28,37
PTMO-650	--	--	-84.0	--	23,36

a: all temperatures in $^{\circ}$C
ND: not determinted

Annealing studies were also carried out to investigate the morphological changes in segmented elastomers induced by annealing and quenching as a function of both time and temperature. For unannealed samples, the low temperature thermograms contain the transitions associated with the soft segments and are presented in Figures 9 and 10. The thermograms in Figure 9 show the low temperature transitions for various copolymers prepared with different molecular weight polyether glycol but the same hard segment content. For PEUU based on the polyether glycol 2000 molecular weight, the soft segment glass transition occurs at -75°C and is followed by an in situ crystallization exotherm near -40°C and melting endotherm at about 10°C. Crystallization and melting transitions were not observed for copolymers derived from the lower molecular weight polyether glycols.

For polyurethane-urea samples prepared with 650 or 1000 molecular weight PTMO, the glass transition temperatures were observed at -38 and -48°C, respectively. These Tg's are substantially higher than the glass transition temperature reported in Table 3 for pure PTMO (-84 and -82°C, respectively). This indicates either a significant amount of interfacial mixing or an existence of mixed hard segments in the soft matrix of these materials, the extent of which is likely determined by the amount and length of both segments. In addition, the anchoring of soft segments at the phase boundaries of a domain structure would also raise the Tg due to the restrictions imposed by the coupling at the interface. The width of the glass transition zone is also longer for lower molecular weight polyether glycol-based copolymer, suggesting a higher degree of segment mixing. The partial mixed segment morphology indicated for samples with 650 or 1000 molecular weight PTMO can be described by two models.[6] The first would consider the presence of polyether segments in the hard domains and/or hard segments dispersed in the polyether matrix. The second model acknowledges the existence of a broad interfacial region between the relatively pure hard and soft segment domains. The DSC and SAXS analyses (discussed later) on these materials seem to suggest a model which combines the basic aspects of both models. However, Cooper's[21] IR results on polyurethane-ureas have indicated that urea carbonyls are completely hydrogen bonded. If one accepts this point of view, then only a few individual hard segments should be dissolved in the soft segment matrix, although small hard domains (a few small coupled hard segments) could no doubt still be miscible.

All low temperature DSC thermograms on copolymers with 2000 molecular weight PTMO (see Figure 10) indicate the presence of crystallization and melting transitions, above the glass transition. The presence of a melting endotherm is observed, as expected, by virtue of the greater molecular weight of soft segments involved. The crystallization exotherm is thus due to the soft segment. At room temperature, these segments are just above their melting temperatures so that rapid cooling below room temperature is effectively a melt quench to an amorphous glassy state below Tg. When reheated, the soft segments crystallize above their glass transition temperature.

From Table 3 and Figure 10, it is to be noted that for constant soft segment molecular weight, the Tg of the soft segment increases with hard segment content. This may suggest that an increase in the aromatic urea content tends to restrict the motion of the soft segments thus shifting the Tg to higher temperatures. However, the presence of a larger interfacial region in the high hard segment-content-material may also raise the Tg. The SAXS results (discussed later) on these materials tends to favor the second postulate.

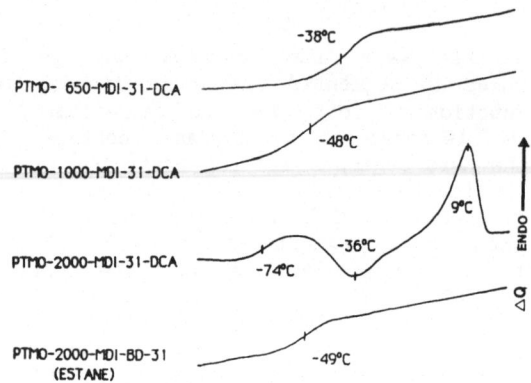

Fig. 9: Low-temperature DSC curves of the
PEUU's with different soft segment
lengths.

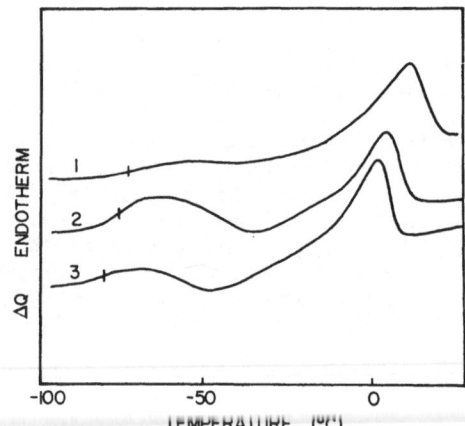

Fig. 10: DSC Thermograms for various samples with
different hard segment content.
(1) PTMO-2000-MDI-41-CA, (2) PTMO-2000-
MDI-31-CA and (3) PTMO-2000-MDI-23.3 CA.

Interestingly, the DSC results in Figure 10 also reveal that a lower
glass transition also promotes a lower soft segment crystallization
temperature (Tc). The melting endotherms in these copolymers are observed
to be shifted to lower temperatures as the hard segment content is
decreased. This implies that the proportion of hard segments present in
the soft segment matrix decreases as the hard segment content is
increased. As more hard segments are present in the soft matrix, the
melting point would be depressed because of a dilution effect. On
comparing the areas under the crystallization and melting peaks, it is

found that the difference is largest for the sample with the highest hard segment content. These results indicate that the soft segments in the sample with highest hard segment content crystallize faster and can be explained by two separate phenomena. The first one has to do with the larger crystallization window (Tg-Tm) available for the fast crystallization of soft segments. Faster crystallization could also occur due to the presence of less hard segment impurities in the soft segment matrix. When significant amounts of hard segment impurities are present in the matrix, as may be the case for samples with lower hard segment content, then they could interfere with the crystallization of the soft segments. As a result when the sample is quickly quenched, only a small proportion of the soft segments would crystallize. Therefore it is hypothesized that the difference between the areas under the crystallization and melting peaks may be correlated to the purity of the soft phase.

The Tg exhibited by the 2000 molecular weight PTMO based-copolymer is much lower (near -75°C) as compared to those of copolymers with lower soft segment molecular weight. This indicates that the mixing of the hard and soft segments is minimal when higher molecular weight polyethers are employed and that the Tg value can also be used as an indication of the relative purity of the soft segment regions.

From these DSC results it also appears that the Tg for polyurethane-urea copolymers also depends on the type of tertiary alcohol used to promote the chain extension step. When DCA is used, the Tg is higher as compared with those of the copolymers where CA was utilized. These results are consistent with the differences observed in the mechanical behavior. In DCA-based PEUU's, a less well ordered domain structure is obtained, probably because of a larger hard segment length distribution as discussed earlier. DSC curves also suggest that hard segment/soft segment mixing is affected by the hard segment type and the affinity of one segment for the other. For MDI/BD based polyurethanes, a relatively high value of Tg is indicated implying the existence of large scale mixing of hard and soft segments. This mixing in conventional MDI/BD based polyurethanes may also be caused by higher compatibility between the two types of segment compared to MDI/DAM based PEUU's.

While the sample PTMO-2000-MDI-31-CA displays the existence of both crystalline and melting temperature transitions, no such activity is observed in PTMO-2000-TDI-31-CA. The Tg of the soft segments in TDI-based materials is lower than that of MDI-based materials, suggesting better phase separation in the latter. It appears that the unsymmetric configurations of TDI units cause the the formation of random hard segment structures which inhibit, but not necessarily eliminate, the crystallization of soft segments under identical thermal histories.

The DSC scans were also obtained for two polyurea systems which form the hard segments in the MDI-based copolymers. These polyureas were obtained by reacting MDI with DAM at low temperatures for 2 hours or MDI with CA at high temperatures (150°C) for 15 hours. For both materials a Tm is observed at about 370°C. The same melting temperature has also been reported for MDI/DAM polyureas by Ishihara et al.[15] A heat of fusion of 110 cal/gm was determined for the MDI/DAM-based polyureas assuming no degradation of the sample. For MDI/CA-based polyureas, a lower heat of fusion is observed because a completely crystalline structure is not obtained. This hypothesis is confirmed by the existence of glass and crystallization transitions in the thermograms. Partially amorphous polyureas are obtained because the material likely undergoes some branching, but short of a 3-dimensional network formation, during synthesis at high temperatures. A diffuse amorphous halo was also observed in the powder diffraction patterns obtained for the MDI/CA polyurea. The high Tm and ΔH_m are, no doubt, responsible for the well defined crystalline structure which is formed mainly through hydrogen bonding and the stacking interactions of aromatic rings. The presence of strong and high temperature melting hard domains in these polyurethane-ureas is also likely to be responsible for the their superior mechanical properties and high temperature performance in comparison to diol extended conventional polyurethanes.

Wide-angle X-ray Diffraction

The differences in the onset of strain-induced crystallization behavior of MDI- and TDI- based polyurethane-ureas copolymers was investigated with the aid of wide-angle x-ray diffraction (WAXD). Flat plate diffraction patterns were obtained from two representative copolymers at various elongations (Figure 11). In the undeformed state, both samples show an isotropic arrangement of the amorphous segments and only a broad diffuse halo is observed. At higher elongations the anisotropy in the sample is indicated by the azmuthal dependence of the diffraction patterns. When strain induced crystallization of the soft segments occurs, the reflections corresponding to the oriented crystallites may be observed as sharp reflections along the equator. For MDI-based PEUU's these spots are observed when the sample elongation is more than 300%. The TDI-based system, however, develops these spots at only 110% elongation. This indicates that for MDI-based copolymers, the phenomenon of strain induced crystallization takes place at a much higher elongation than the TDI-based materials. The reason for this unusual behavior may lie in the non-symmetric configuration of of the TDI hard segments, but this is only a speculation. It may appear that the WAXD results are in direct conflict with the DSC and dynamic mechanical results discussed earlier. Those results had suggested that although the soft phase in TDI based copolymers was more pure as compared to the MDI-based materials, the random hard segment structures developed by the non-symmetric TDI units may have been responsible for the absence of cold crystallization behavior. If more hard segments impurities are present in the soft matrix for MDI-based materials, then on deformation the soft segment chains should experience less orientation effects as compared to the TDI-based materials. If this is true, then MDI based PEUU's might show strain induced crystallization at a much higher overall elongation in contrast to the TDI-based systems. The authors again point out that these latter statements are only speculations at this time.

Small-angle X-ray Analysis

In order to more directly determine the differences in the structural arrangement as a function of sample composition, the morphology in these segmented polyurethane-urea copolymers was investigated via small-angle x-ray scattering (SAXS). To accomplish this, the interdomain spacing, interfacial boundary thickness and extent of segmental mixing were determined as a function of composition variables. The analysis of small-angle x-ray scattering data is relatively straightforward and detailed information on the determination of various morphological parameters has been discussed elsewhere.[28-35]

The slit smeared x-ray data was obtained on a Kratky small angle camera and subsequently corrected for parasitic and background (thermal)

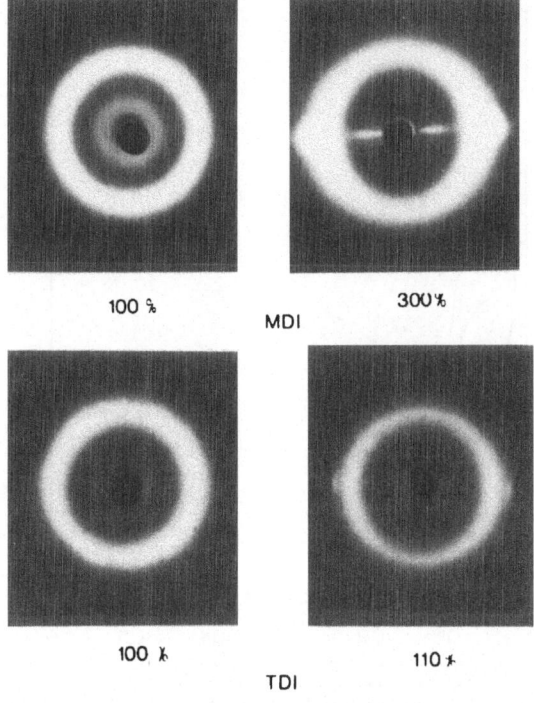

100 %   300 %

MDI

100 %   110 %

TDI

Fig. 11: **Wide-angle diffraction patterns for PTMO-2000-MDI-31-CA and PTMO-2000-TDI-31-CA at different elongations.**

Table 4.   Interdomain or D-Spacing in nm for Segmented Polyurethane-Urea Copolymers

Sample	by **Bragg's** law (nm)	From I-D Correlation Function (nm)
PTMO-2000-MDI-23.3-CA	17.5	12.0
PTMO-2000-MDI-31-CA	16.5	11.0
PTMO-2000-MDI-31-DCA	15.5	10.5
PTMO-1000-MDI-31-DCA	12.4	9.3
PTMO- 650-MDI-31-DCA	10.5	--
PTMO-2000-MDI-31-BD	11.8	9.2

scattering. The resulting **x-ray** scattering profiles for three polyurethane-urea copolymers at constant hard **segment** content are shown in Figure 12. All curves show the presence of a shoulder or **maximum** followed by a gradual fall off in the intensity. The location of the maximum **(or shoulder)** depends on the composition and appears **at** higher angles when either hard segment content (at constant soft segment molecular weight) or soft segment length (at the **same** hard segment content) are reduced. A more defined peak occurs in place of a shoulder when the hard **segment** content for a high molecular weight polyether

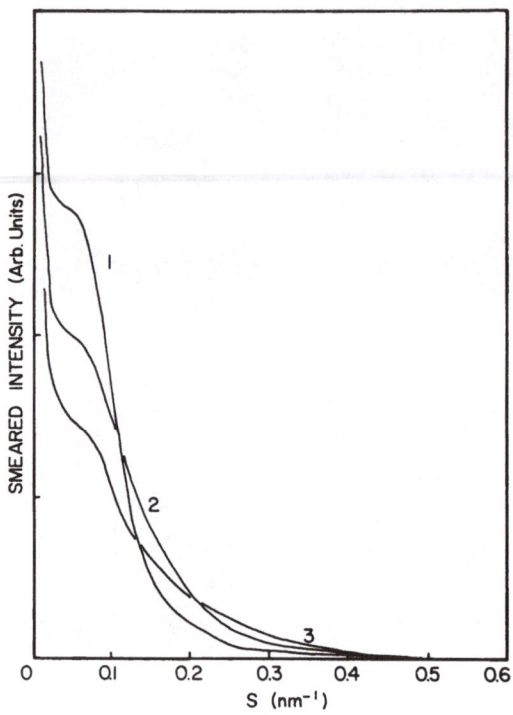

Fig. 12: Smeared SAXS scattered intensity
profiles for (1) PTMO-2000-MDI-
31-DCA, (2) PTMO-1000-MDI-31-
DCA and (3) PTMO-650-MDI-31-DCA.

glycol is increased. The breadth of the scattered intensity profile is
also larger for lower segment molecular weight materials suggesting a
larger distribution of domain sizes and/or interdomain spacings. For
PTMO-2000-MDI-41-CA the scattering profile is rather sharp indicating a
narrow distribution of interdomain spacings. The same trends are seen
in Figure 13 for the collimation-corrected intensity profiles as well.
For sample PTMO-650-MDI-31-CA, the lack of a distinct phase separated
structure causes the scattered intensity to be lower and the position of
the shoulder is also not well defined. The mixing of the two different
segments apparently leads to the lowering of the mean square electron
density fluctuation in the system and the scattered intensity is lower
as a result of lower electron density contrast. This causes a large
statistical error in the intensity profile and presents problems during
data analysis.

The interdomain spacing or d-spacings were determined from the
location of the peak in the collimation corrected intensity profiles using
Bragg's law and are listed in Table 4. Values of the interdomain spacings
were also determined by 1-D correlation function analysis. The values of
the interdomain spacing obtained by the two approaches are in reasonable
agreement with each other. The results indicate that the interdomain
spacing is affected by changes in the composition as would be expected.
For example, sample PTMO-2000-MDI-23.3-CA displays a spacing of 12 nm
while a value of 11 nm is observed for PTMO-2000-MDI-31-CA.
At constant hard segment content of 31 percent by weight, the
interdomain spacing increases from 9.3 to 11 nm as the

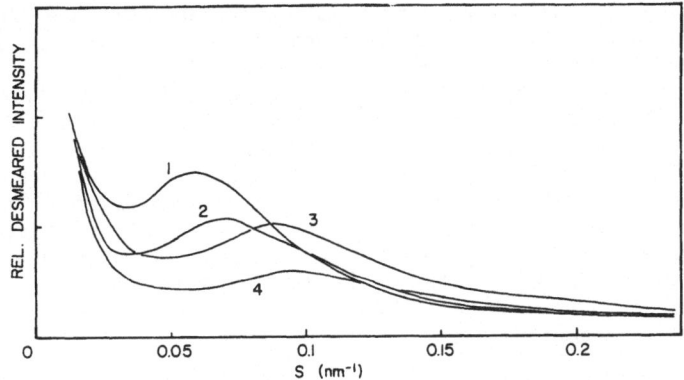

Fig. 13: Desmeared SAXS scattered intensity profiles for
(1) PTMO-2000-MDI-31-CA, (2) PTMO-2000-MDI-23.3
CA, (3) PTMO-2000-MDI-31-BD and (4) PTMO-1000-
MDI-31-DCA.

molecular weight of the soft segment is increased from 1000 to 2000,
respectively. As the molecular weight of the soft segment is increased,
the hard segment length also increases at constant hard segment content
and as a result, not only the order in hard segment domains is
increased, but the interdomain spacing increases as well. The d-spacing
increases as well when the higher molecular weight polyethers are
employed. These results, along with the morphological insight obtained
from other techniques, indicate that hard segment ordering is reflected
in the values of the interdomain spacing.

It has been suggested by Porod that for an ideal two phase system,
the desmeared scattered intensity at high angles varies with the
reciprocal of fourth power of the scattering vector. A reciprocal third
power dependence is observed when slit smeared data is used. This
implies that the product $I.s^4$ or $I.s^3$ reaches a constant value at large
angles for collimation corrected and slit smeared scattered intensities
respectively. The variable s is called the scattering vector and is
related to the angle of observation of the scattered intensity from the
center of the beam. A negative slope in a plot of $I.s^3$ vs s^2 indicates
the existence of diffuse phase boundaries. All but one sample indicated
the presence of a diffuse two phase structure in the materials. The TDI
based copolymers, however, display a near zero slope which is indicative
of sharp phase boundaries. This observation is consistent with the DSC
results where the same copolymer showed the lowest Tg for the soft
segments. However, these results only indicate that the thickness of
the interfacial zone present in the material may be small and does not
rule out the solubilization of one component in other. An estimate of
the thickness of the diffuse interfacial boundaries, as indicated by the
negative slopes, can be determined by various graphical methods.[35]

A detailed investigation of the diffuse boundary thickness was
carried out using the slit smeared scattered intensities. The
interfacial thickness parameter, σ, was estimated by several graphical
procedures and the results are reported in Table 5. The value of σ is
related to the interfacial boundary thickness assuming a sigmoidal
gradient of electron density across the interface. The results for
interfacial thickness parameters are comparable with previously reported
values of σ for segmented polyether polyurethanes.[28,29] Interestingly,
the diffuse boundary zone is found to be larger for samples containing

123

the higher hard segment content. A comparison of the interfacial boundary thicknesses for copolymers prepared with CA and DCA shows that the thickness is greater in copolymers prepared with the latter. This result is consistent with the earlier rationalization and differences in thermal and mechanical properties observed for the two copolymers. However, at constant hard segment content, the interfacial thickness determined for low molecular weight PTMO is smaller in comparison to materials based on higher soft segment molecular weight. This result would appear to be in conflict with most observations which indicate better phase separation for cases where longer segments are employed. First it must be clarified that a larger thickness of the interfacial regions <u>does not</u> necessarily point to a poor phase separation or a higher degree of segmental mixing in the material.

Table 5. Interfacial Thickness Parameter (in nm) by Various Methods

Sample	Koberstein	Ruland[1]	Bonart	Exponential	Ruland[2]
PTMO-2000-MDI-23.3-CA	0.205	0.196	0.263	0.186	0.182
PTMO-2000-MDI-31-CA	0.335	0.283	0.426	0.301	0.280
PTMO-2000-MDI-31-DCA	0.264	0.234	0.339	0.240	0.224
PTMO-1000-MDI-31-DCA	0.109	0.100	0.160	0.113	0.110
PTMO-2000-MDI-31-BD	0.130	0.145	0.179	0.127	0.126

1. Using a plotting routine of $sI(s)$ vs. $1/s^2$.
2. Using a plotting routine of $s^3I(s)$ vs. s^2.

The parameter indicating the degree of phase separation, σ, is also dictated by the relative lengths of the hard and the soft segments. For example, for a hypothetical sample A where the interfacial thickness is small, the fraction of the interfacial region could be large as compared to another sample B which may have higher interfacial thickness but the fraction of the interfacial region is small. Sample B, in such a case, would show a better overall degree of phase separation as compared to sample A. The reason for this unexpected observation may also lie in the molecular arrangement of the copolymer chains obtained with low molecular weight prepolymer. It has been shown that the presence of this type of intrasegmental mixing would cause a positive deviation in the Porod's plot.[35] If this occurs then the obtained value of σ is an artifact representing the net result of two opposing phenomena. Based on these arguments, it can also be explained why polyurethane elastomers show lower interfacial thickness by x-ray analysis as compared to these PEUU elastomers despite repeated observations by other techniques that polyether polyurethanes are not as well phase separated in comparison. Hence, the values of σ do not give any direct indication of the degree of phase separation but only an estimate of the interfacial thickness. A better indication of the degree of phase separation is obtained by the determination of the electron density variance.

Electron density variance was estimated from the scattering intensity profiles corrected for background scattering as discussed elsewhere.[31,35] These variances or mean square electron density fluctuations, $\overline{\Delta\rho^2}'$ and $\overline{\Delta\rho^2}_c$, can provide valuable insight into the state and degree of phase separation as shown by Bonart.[36,37] The term $\overline{\Delta\rho^2}'$ is determined through the calculation of the invariant while the term $\overline{\Delta\rho^2}_c$ is calculated assuming complete phase separation. The ratio

$$\overline{\Delta\rho^2}'/\overline{\Delta\rho^2}_c$$

on a semi-quantitative basis, reflects the overall degree of phase separation. A value of unity obtained for the ratio indicates a completely phase separated material while a lower value would indicate a reduced amount of segmental mixing. The term $\overline{\Delta\rho^2}_c$ is the electron density variance for an ideal two-phase system. This ideal variance in electron density can be calculated from the knowledge of the chemical composition and mass density of the component phase by assuming complete phase separation with sharp phase boundaries. The results of the electron density variance obtained for different copolymers, are presented in Table 6. In all cases, the $\overline{\Delta\rho^2}{}'$ is smaller than the ideal variance $\Delta\rho^2_c$. This indicates that some phase mixing is present in all samples. The ratios of the two variances were also calculated and are reported in the last column of Table 6. The degree of phase separation, as indicated by this ratio, is found to be higher for copolymers with higher hard segment content.

Table 6. Degree of Phase Separation as Determined by Electron Density Variance

Sample	$\overline{\Delta\rho^2}{}' \times 10^4$ (mol elec/cc)^2	$\overline{\Delta\rho^2}_c \times 10^3$ (mol elec/cc)^2	$\overline{\Delta\rho^2}{}'/\overline{\Delta\rho^2}_c$
PTMO–2000–MDI–23.3–CA	7.654	1.664	0.46
PTMO–2000–MDI–31–CA	10.175	2.098	0.56
PTMO–2000–MDI–31–DCA	7.978	1.664	0.48
PTMO–1000–MDI–31–DCA	8.373	2.032	0.43
PTMO– 650–MDI–31–DCA	6.873	1.964	0.35
PTMO–2000–MDI–31–BD	5.324	1.331	0.40
PTMO–2000–TDI–31–CA	8.412	2.103	0.41

When presented in this fashion, even the sample PTMO–1000–MDI–31–DCA indicates a less phase separated structure in comparison to PTMO–2000– MDI–31–DCA. Polyether polyurethane based on MDI/BD (ESTANE®), which had indicated a smaller interfacial thickness, shows a considerable amount of mixing between the two types of segments. In general, a good correlation can be found between the degree of phase separation as determined by x-ray analysis and mechanical and thermal characteristics observed earlier.

CONCLUSIONS

It has been shown that it is possible to synthesize novel polyurethane-ureas by utilizing either the rearrangement characteristics of branched carbamates and/or the dehydration characteristics of tertiary alcohols at a sufficiently high temperature. Mechanical, thermal, dynamic mechanical and x-ray experiments were carried out to characterize the morphology and properties of polyether polyurethane-urea copolymers of systematically varying hard segment type, hard segment content, soft segment molecular weight, and block length. The results obtained for these materials were compared with those from conventional polyurethanes to investigate the effect of intermolecular hydrogen bonding on molecular arrangement.

The tensile behavior was observed to depend primarily on the degree of order in the hard domains. It was shown that this order can be improved by increasing either the hard segment content at constant molecular weight of the soft segment or soft segment molecular weight at the same hard segment content. Both approaches increase the concentration of urea linkages in the material and hence the tensile properties. In addition to the differences in the orientation behavior, the TDI-based materials were stiffer than those based on MDI because of higher urea content at the same weight percent hard segment content. This suggests that the cohesion in hard segment domains is improved through the formation of 3-dimensional urea hydrogen bonding in the polyurethane-urea copolymers. The filler effects of the hard domains are enhanced as well because of higher hard segment content. At comparable hard segment content, all materials indicated enhanced mechanical strength over polyether polyurethanes chain extended with butanediol.

From dynamic mechanical measurements on these copolymers, the extent and nature of the rubbery plateau was shown to be composition dependent. The results also indicated a higher degree of phase separation in the samples when order in the hard segment domains was increased by altering the composition. As many as five different types of relaxation mechanisms were observed over a wide temperature range. The δ relaxation associated with annealing effects was not observed when the concentration of urea linkages was increased. TMA and TGA results also supported the earlier conclusions drawn from the mechanical and dynamic mechanical studies. Introduction of polar urea segments were shown to somewhat enhance the thermal stability of these materials.

Good agreement with dynamic mechanical measurements was observed for DSC results as well. The glass transition temperature was determined to be lower when either the molecular weight of the soft segment was increased or hard segment content was decreased. The location of the soft segment Tg was shown to be correlated with the crystallization and melting temperatures as well. DSC behavior also provided the correlation between the degree of phase separation and higher mechanical strength.

Differences in strain induced crystallization behavior were observed for MDI vs TDI based hard segments in the copolymers by wide angle x-ray diffraction. The small-angle x-ray results provided valuable insight to the morphological architecture of the materials. Values of interdomain spacing were shown to reflect the extent of hard segment ordering in the sample. Except for the TDI based system, negative deviations from Porod's law were observed and the thickness of the interfacial zone was estimated based on these deviations. Calculation of the SAXS invariant was used to examine the overall degree of phase separation in the sample and was found to be consistent with earlier observations.

ACKNOWLEDGEMENTS

It is a pleasure to acknowledge the financial support provided by TACOM through research contracts DAAE 07-83-C-R117 and DAAE 07-82-C-4094. The authors also express their thanks to the Dow Chemical Company, Freeport TX and to the Exxon Foundation for partial support of this work.

REFERENCES

(1) G. A. Senich and W. J. MacKnight, Macromolecules, 13, 106 (1980).

(2) V. W. Srichatrapimuk and S. L. Cooper, J. Macromol. Sci. Phys., B15, 267 (1978).

(3) G. L. Wilkes, S. Bagrodia, W. Humphries and R. Wildnauer, J. Polym. Sci. Polymer Letters Ed., 13, 321 (1975).

(4) R. Bonart, L. Morbitzer and E. H. Muller, J. Macromol. Sci. Phys., B9, 447 (1974).

(5) C. S. Paik Sung, T. W. Smith and N. H. Sung, Macromolecules, 13, 117 (1980).

(6)   K. Knutson and D. J. Lyman, Adv. in Chem., American Chemical Society, 199, (1982).

(7)   D. Tyagi, I. Yilgor, J. E. McGrath and G. L. Wilkes, Polymer, 25, 1807 (1984).

(8)   D. Tyagi, I. Yilgor, J. E. McGrath and G. L. Wilkes, Polym. Bull., 8, 543 (1982).

(9)   C. S. Paik Sung, C. B. Hu and C. S. Wu, Macromolecules, 13, 111 (1980).

(10) C. S. Paik Sung and C. B. Hu, Macromolecules, 14, 212 (1981).

(11) G. L. Wilkes and S. Abouzahr, Macromolecules, 14, 458 (1981).

(12) V. A. Khransovskii, Doklady Akademii Nauk SSSR, 244(2), 408 (1979).

(13) I. Kimura, H. Ishihara, H. Ono, N. Yoshihara, S. Nomura and H. Kawai, Macromolecules, 7, 355 (1974).

(14) H. Ishihara, I. Kimura, K. Saito and H. Ono, J. Macromol. Sci. Phys., B10, 591 (1974).

(15) H. Ishihara, I. Kimura, K. Saito and H. Ono, J. Macromol. Sci. Phys., B22, 713 (1983).

(16) J. H. Saunders and R. J. Slocombe, Chem. Rev., 43, 203 (1948).

(17) B. Lee, D. Tyagi, G. L. Wilkes and J. E. McGrath, Paper presented at the Annual ACS Meeting at Philadelphia, August 1984; Rubber Division Meeting, Los Angeles, April 1985; Sagamore Conference on Elastomers, July 1985.

(18) B. Lee, J. E. McGrath, D. Tyagi and G. L. Wilkes, Polym. Prepr., American Chemical Society, Div. Polym. Chem., 27(1), 100 (1986).

(19) L. E. Nielsen, Rheo. Acta, 13, 86 (1974).

(20) L. R. G. Treloar, The Physics of Rubber Elasticity, 3rd Ed., Clarendon, Oxford, 1975.

(21) C. B. Wang and S. L. Cooper, Macromolecules, 16, 775 (1983).

(22) H. N. Ng, A. E. Allegrezza, R. W. Seymour and S. L. Cooper, Polymer, 14, 255 (1973).

(23) D. S. Huh and S. L. Cooper, Polym. Eng. Sci., 11(5), 369 (1971).

(24) T. Kajiyama and W. J. MacKnight, Macromolecules, 2, 254 (1969).

(25) T. F. Schatzky, J. Polym. Sci., 57, 496 (1962).

(26) A. H. Willbourn, J. Polym. Sci., 34, 569 (1959).

(27) N. G. McCrum, B. E. Read and G. Williams, in Anelastic and Dielectric Effect in Polymeric Solid, John Wiley, New York (1967).

(28) Z. Ophir and G. L. Wilkes, J. Polym. Sci. Polym. Phys. Ed., 18, 1469 (1980).

(29) J. T. Koberstein and R. S. Stein, J. Polym. Sci. Polym. Phys. Ed., 21, 1439 (1983).

(30) J. T. Koberstein and R. S. Stein, J. Polym. Sci. Polym. Phys. Ed., 21, 2181 (1983).

(31) J. W. C. Van Bogart, Ph D Thesis, department of Chemical Engineering, University of Wisconsin, 1981. (University Microfilms 81-7545).

(32) A. Guinier and G. Fournet in Small Angle Scattering of X-rays, J. Wiley & Sons, London, 1955.

(33) L. E. Alexender in X-ray Diffraction Methods in Polymer Science, Wiley, New York, 1969.

(34) O. Glatter and O. Kratky in Small Angle X-ray Scattering, Academic Press, New York, 1982.

(35) D. Tyagi, Ph.D. Dissertation, Department of Chemical Engineering , Virginia Polytechnic Institute and State University, 1985.

(36) R. Bonart and E. H. Muller, J. Macromol. Sci. Phys., B10, 177 (1974).

(37) R. Bonart and E. H. Muller, J. Macromol. Sci. Phys. B10, 345 (1974).

(38) P. Dreyfus, Polytetrahydrofuran, Gordon & Breach, New York, 1982.

DIENE TRIBLOCK POLYMERS WITH

STYRENE-ALPHA-METHYLSTYRENE COPOLYMER END BLOCKS

L. H. Tung and G. Y. Lo

The Dow Chemical Company
Central Research-Polymeric Materials Laboratory
1702 Building
Midland, Michigan 48674

ABSTRACT

Diene triblock copolymers with styrene-alpha-methylstyrene copolymer (SAMS) end blocks have been prepared by two procedures: (1) by preparing the center diene block in cyclohexane and then the end blocks in cyclohexane-THF mixture; (2) by preparing the triblock in alpha-methylstyrene itself as the solvent. Several variations of procedure (2) were used. When all the monomers were added in the reaction mixture at the same time and a dilithium initiator was used, the product was a tapered triblock. When styrene was withheld until the diene block was formed, the triblock was untapered. The triblock was also prepared by forming the end block first using s-butyllithium as the initiator. The diene monomer was then added to the reaction mixture to form the diblock. The reactive diblock was coupled to form a triblock and also a radial block. Triblock was also formed by adding more styrene monomer instead of coupling.

The polymers prepared with a diene content between 55 to 70 weight percent were thermoplastic elastomers. The properties of the polymers prepared by the several different procedures were comparable and also comparable to conventional styrene-diene block polymers with the exception that the softening temperature as measured by thermomechanical analysis was 20°C higher than that of the corresponding styrene-diene block polymers. The SAMS-based triblock polymers prepared with a diene content below 40% were clear impact resisting plastics. The Vicat softening points of the clear plastics samples were also about 30°C higher than those of similar polymers based on styrene and butadiene.

INTRODUCTION

Triblock polymers and radial block polymers of styrene and butadiene (or isoprene) are successful commercial thermoplastic elastomers pioneered by the Shell Chemical Company and the Phillips Petroleum Company. The low diene and high styrene block polymers are clear impact resisting plastics marketed by the Phillips Petroleum Company under the trade name K-Resins. One common deficiency of these polymers is the relatively low glass temperature (T_g) of the polystyrene end blocks. For the thermoplastic elastomers the service temperatures are limited to below 65°C.[1] The Vicat softening

point of K-Resins is at 93°C.[2] Attempts have been made to replace the
polystyrene end blocks with end blocks of higher T_g. Cunningham[3] prepared
diene triblock polymers with end blocks of poly(alpha-methylstyrene) which
has a T_g of 170°C. The end block of poly(alpha-methylstyrene) demands a
fabrication temperature which is too high for the polydiene center block
to survive. Poly(t-butylstyrene) has a T_g midway between those of polystyrene and poly(alpha-methylstyrene) and appears to be high enough to
impart a meaningful increase of the service temperature of the triblock
polymer and yet not too high to endanger the integrity of the polydiene
block during fabrication. The poly(t-butylstyrene) block, however, was
reported to have too close a solubility parameter to that of the polydiene
center block for phase separation to take place in a reasonable molecular
weight range.[4] Phase separation of the blocks into microdomains is necessary for the polymers to have good elastomeric properties. Copolymers of
styrene and alpha-methylstyrene (SAMS) also have the desirable T_g. Neumann[5]
prepared a butadiene triblock with SAMS at the ends (SAMS-B-SAMS), but the
vinyl-1,2 content of his polybutadiene center block was about 45%.

In this paper we describe the preparation and the properties of the
title triblock with a low vinyl-1,2 (or 3,4 in the case of polyisoprene)
polydiene center block. Two different solvent systems were used as the
media of polymerization. In the first system, the polydiene center block
was prepared in cyclohexane. Alpha-methylstyrene (AMS) and a polar solvent
tetrahydrofuran (THF) were then added. This was followed by a slow and
continuous styrene addition to complete the end block preparation. In the
second system, AMS itself was used as the solvent with no other solvent
added. The second solvent system enabled us to use several different polymerization schemes. The center block could be prepared first to form a
tapered or untapered triblock. The end block copolymer also could be prepared first and then the diblock and then coupled to form a tri- or a
radial block polymer. Instead of coupling, more styrene could be added to
complete the triblock. All these different routes of preparation were
used in this work.

EXPERIMENTAL

Materials

Rubber grade 1,3-butadiene was purchased from the Phillips Petroleum
Company. Two columns, one packed with the potassium form of a sulfonic
acid ion exchange resin (Dowex MSC-1-K) and the other packed with an activated alumina, were used to purify the butadiene. Isoprene was purchased
from Aldrich Chemical Company. Stabilizer-free styrene monomer was obtained directly from the Dow styrene monomer plant. Alpha-methylstyrene
(AMS) was purchased from U.S. Steel Corporation. The last three monomers
were all purified by passing through a column packed with activated alumina and then vacuum distilled over calcium hydride.

Toluene, cyclohexane, and THF, distilled-in-glass grade, were obtained
from Burdick and Jackson Laboratory. Toluene was dried over sodium ribbon
before use. Cyclohexane was dried by passing through an activated alumina
column. THF was distilled over sodium-naphthalene complex in vacuum.

Phenyl benzoate (99%) was obtained from Aldrich; silicon tetrachloride
was from Matheson, Coleman and Bell. Both reagents were dissolved in purified toluene first and then degassed just before use.

s-Butyllithium, 14% in cyclohexane, was purchased from Foote Mineral
Company. It was diluted with cyclohexane to about 0.5 N before use. The
concentrations of the final s-butyllithium solutions were determined by the

double titration method of Gilman.[6] The dilithium initiators used in the triblock polymerization were prepared by reacting in a 100 mL flask a toluene solution of 1,3-bis-(1-phenylethenyl)benzene or 1,3-di[1-(4-methyl-phenyl)ethenyl]benzene with s-butyllithium at room temperature.[7] Stoichiometrical amounts of the reactants were used and the reaction completed in about 1 hour. The dark red initiator solutions in concentrations of about 0.08 millimol per mL were usually prepared just before use, although they could be kept for several weeks at room temperature without any loss of activity.

Polymerization

Anionic polymerization was carried out in a three-neck round bottom flask under a nitrogen atmosphere. A detailed description of the apparatus was given in an earlier report.[8] Two solvent systems were used in the preparation of block polymers.

The procedure for the first system is as follows. Cyclohexane was siphoned into the flask. After degassing, the flask was cooled by an ice bath while a dry ice-methylene chloride mixture was used as the coolant for the reflux condenser. Butadiene in liquid form was discharged directly from the purification columns to the reaction flask. The amount of butadiene delivered was measured volumetrically by a scale attached to the side of the flask. The ice bath was removed after butadiene was introduced. AMS and the dilithium initiator were then added separately by means of a syringe through a septum on a side arm of the reaction flask. The reaction mixture was then brought to a temperature between 50 to 60°C by a hot water bath. The dark red color of the initiator then gradually faded to yellow. An increase of polymerization temperature usually occurred as the solution became viscous at the 40 to 60 minute mark. Another hour was allowed to complete the butadiene polymerization. The reaction mixture was then cooled with an ice bath to below 30°C for the polymerization of SAMS end block. Two mL of the styrene monomer were added to the reaction mixture, then the required amount of THF. A reduction of viscosity and a change of color from yellow to red followed. The remaining amount of styrene was then added slowly at a constant rate by means of a syringe pump. In about 5 minutes after all the styrene monomer was added, methanol was added to terminate the living anions. The resulting block polymer was coagulated in methanol and dried in vacuum at room temperature.

In the second system, AMS, which was used as the solvent, was added first. After degassing, a small amount of butyllithium was used to react with the remaining impurities in AMS until a faint red color appeared. Butadiene (or isoprene) and styrene were then added. The solution was then brought to about 50°C and the dilithium initiator was introduced. The strong reddish color from the initiator then faded to yellow and the temperature rose due to the exothermic polymerization reaction. About 10 to 20 minutes after the temperature peaked, the color turned red again indicating that the SAMS end block polymerization was taking place. A second temperature rise was usually observed. The temperature of the polymerization was kept above 60°C for about another 30 minutes before methanol was added to terminate the living chains.

The end block of SAMS copolymer was polymerized first in some preparations in AMS solvent. In this case, AMS was added and blanked with butyllithium as before. Styrene monomer and s-butyllithium for initiation were then added to start the SAMS polymerization at temperatures above 50°C. At the time designated for butadiene addition, a small amount of the butadiene monomer was bled in first. The remaining butadiene was added after the reaction solution was cooled to near 5°C. The reaction solution was then brought back to about 50°C for a period of 70 to 80 minutes for the butadiene to polymerize. A coupling agent or additional styrene was then added. Isopropyl alcohol was used as the final terminating agent when phenyl benzoate was used as the coupling agent.

Characterization

Beckman AccuLab 3 spectrometer was used for IR analysis of polybutadiene microstructure. The absorption coefficients reported by Morero, et al.[9] were used for the calculation of the microstructure. Varian EM 360 spectrometer was used for the proton NMR composition analysis.

GPC measurements were made on a modified Waters Model 200 instrument equipped with a UV photometer as the second detector. The molecular weights of SAMS-B-SAMS triblock polymer were calculated from the following equations:

$$\log M = (1 - x)\log M_A + x \log M_B \tag{1}$$

$$M_A = M_S[104(1 - y) + 118y]/104 \tag{2}$$

$$M_B = M_S/1.75 \tag{3}$$

where   x = weight fraction of butadiene
             y = mole fraction of alpha-methylstyrene in the end block
             M_S = molecular weight at the peak elution volume, polystyrene scale
             M_A = molecular weight at the peak elution volume, SAMS copolymer scale
             M_B = molecular weight at the peak elution volume, polybutadiene scale

Equations (1) and (3) were established in a previous work.[10] Equation (2) is an assumption. Weight fraction of butadiene, x, and mole fraction of AMS in the end block, y, were determined from NMR composition analysis; M_S was based on polystyrene calibration in GPC.

The weight percentages of SAMS, SAMS-B diblock, and SAMS-B-SAMS triblock in the product prepared by coupling were calculated from the UV absorption peak areas of the respective species in the GPC chromatogram. The composition for the entire sample, determined by proton NMR, was used to assign the weighting factor of the peaks.

RESULTS AND DISCUSSION

Preparation of SAMS-B-SAMS in Cyclohexane

The procedure that we used to prepare SAMS-B-SAMS triblock was similar to that used by Neumann.[5] The dilithium initiator used by him was prepared in the presence of diethyl ether which was responsible for the high 1,2-vinyl microstructure in his polybutadiene center block. In our experiment the first stage of butadiene polymerization was carried out in the absence of any polar solvent. A set of preliminary experiments of SAMS copolymerization in benzene and cyclohexane were made to determine the conditions required for the second stage polymerization. Since styrene polymerizes much faster, the concentration of styrene should be kept at a low level at all times. This was accomplished by adding styrene continuously during the run. Several common polar solvents were tried as the additive to promote the formation of copolymer. Among them THF, hexamethylphosphoramide and triethylene glycol dimethyl ether were found to be more effective. Of these three, THF was found to show the least influence in broadening the

Table 1. Microstructure (IR) and Glass Transition Temperature
of Polybutadiene Prepared in Cyclohexane

AMS (vol %)	Microstructure (%)			T_g (°C) (DSC)
	Cis-1,4	Trans-1,4	1,2-vinyl	
0	36	55	9	-95
10.6	36	54	10	-96

molecular weight distribution of the copolymer. The optimum concentration of THF was found to be between 5 and 10 percent of the total volume of the solvents. To avoid broadening the distribution at this level of THF, the copolymerization temperature should be kept below 40°C. Based on the above results, the procedure described in the experimental section was adopted.

In adopting the procedure, we had the option of adding AMS early in the polymerization or at the start of the second stage polymerization. The advantage of adding it early was to minimize the introduction of impurities in the second stage. The impurities introduced early can be reacted out by blanking with butyllithium before polymerization starts. We examined, therefore, the microstructure of polybutadiene produced in cyclohexane with and without a 10.6 volume % of alpha-methylstyrene present. Table 1 shows the comparison. The infrared spectrum showed no aromatic component in the polybutadiene prepared in the presence of AMS. The GPC chromatogram also showed no UV absorption.

Table 2. Conditions of SAMS-B-SAMS Polymerization in Cyclohexane

Run No.	Solvents (mL) Cyclo-hexane	THF	Monomers (g) B	S	AMS	Initiator (millimol)	End Block Polym. Temp. (°C)	S Addition Time (min)	% AMS Converted
1	350	40	20	36.0	54.5	1.17	27	170	66
2	360	40	21	36.3	54.5	1.23	27	170	74
3	360	40	21	35.4	59.0	0.84	26	170	75
4	350	40	30	30.8	49.9	1.15	20	185	66
5	370	32	30	43.5	36.3	1.16	31	135	54
6	360	40	30	16.3	68.1	1.14	30	180	47
7	370	40	39	29.0	49.0	1.26	28	180	72
8	350	40	43	27.2	46.3	1.16	26	185	83
9	353	40	32	10.0	22.7	0.64	5	170	50
10	360	40	33	10.0	22.7	0.64	20	170	62
11	367	40	34	9.1	19.1	0.67	30	150	72
12	900	100	110	17.2	38.1	1.86	26	175	77

In the headings: B is butadiene; S is styrene.

Table 2 shows the polymerization conditions of the SAMS-B-SAMS preparation. The second stage polymerization would stop when all the styrene monomer in the reaction solution was consumed, as the residual alpha-methylstyrene would then be below the ceiling concentration at the polymerization temperature.[11] Thus the amount of AMS incorporated in the end blocks depended on its concentration, the amount of styrene added, the rate of styrene addition, and the polymerization temperature. High conversion of AMS was obtained when the end block polymerization was carried out at 20 to 30°C; the styrene addition time was 150 to 185 minutes; and the total monomer ratio of AMS to styrene was 1.5 to 2.5. Quantitative effects of these variables on the end block composition were not determined. A typical GPC chromatogram of the triblock prepared is shown in Figure 1.

The first 8 runs in Table 2 led to SAMS-B-SAMS with butadiene content below 40 weight %. These were rigid polymers with good clarity. Their compositions and properties are shown in Table 3. In Table 3 the end block T_g values were measured by differential scanning calorimetry (DSC). Both the Vicat softening temperatures and end block T_g's were above the boiling point of water for these block polymers. The softening points increase with AMS/S molar ratio in the end block as expected. Also, as expected, the tensile strength decreased and elongation and Izod impact strength increased with increasing butadiene content in the block polymer.

An attempt was made to measure the sequence distribution of AMS and styrene in the block polymers by ^{13}C NMR. Because of the interference of the tacticity peaks, the diads and triads of AMS could not be identified. The styrene diads were detected in sample no. 4 which had an AMS to styrene ratio of 0.94, but not in sample no. 6 which had a higher AMS content.

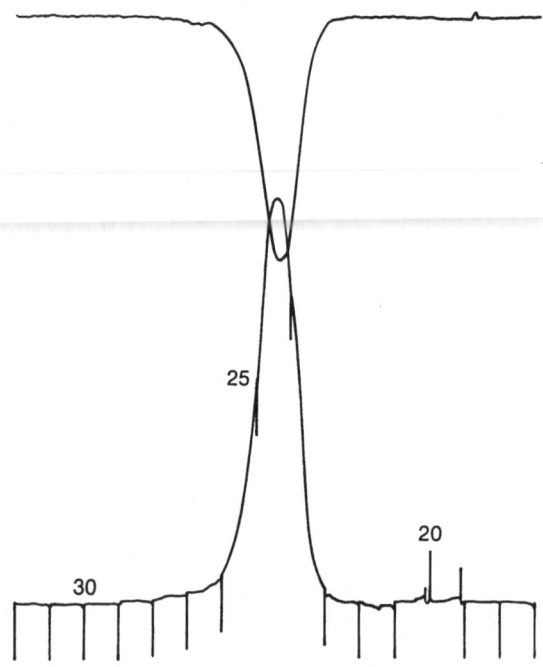

Fig. 1. GPC chromatogram for run no. 10, a SAMS-B-SAMS triblock thermoplastic elastomer prepared in cyclohexane using a dilithium initiator (upper curve from UV detector).

Table 3. Properties of Rigid SAMS-B-SAMS
(SAMS-B-SAMS Prepared in Cyclohexane)

Run No.	Polymer Composition B (wt. %)	AMS/S (molar ratio)	Mol Wt at GPC Peak	Tensile Yield (psi)	Elonga-tion (%)	Izod Impact Str (ft-lb/in)	Vicat Softening Temp. °C	T_g,°C (DSC)
1	19.3	0.88	94000	4000	8.1	0.55	127	119
2	20.3	0.98	92000	4280	8.2	0.43	122	115
3	24.9	1.10	156000	4330	4.3	0.61	118	125
4	29.5	0.94	98000	4130	14.1	0.58	120	122
5	31.9	0.40	90000	3500	13.7	0.61	102	109
6	34.3	1.72	100000	3430	11.7	0.76	128	132
7	36.4	1.07	122000	2970	22.7	1.05	124	–
8	38.4	1.25	112000	2470	>350	1.39	120	124

This suggested that AMS sequence longer than 2 was absent in all the samples. Reinforcing this view was the observation that in run no. 6 the AMS to styrene mole ratio was only 1.72 in the polymer, although the ratio of these monomers was close to 4 in the feed. All these observations suggested that the SAMS copolymer end blocks may indeed be close to a random sequence distribution.

The last 4 runs in Table 2 led to SAMS-B-SAMS with butadiene contents higher than 55 weight percent. These were thermoplastic elastomers and their compositions, molecular weights, and stress-strain properties are listed in Table 4. The tensile strength and elongation at break are similar to those reported for styrene-butadiene-styrene triblock (S-B-S) thermoplastic elastomers.[12] The T_g of the end blocks, however, could not be determined unequivocally by differential scanning calorimetry. Hence, they are not reported. The softening temperature of a typical SAMS-B-SAMS tapered triblock was compared with that of a commercial S-B-S thermoplastic elastomer on a DuPont thermomechanical analysis (TMA) instrument. At 5% probe penetration, the temperature for the SAMS-based triblock was 20°C higher. The curves of the measurement are shown in Figure 2.

Table 4. Properties of SAMS-B-SAMS Thermoplastic Elastomers
(SAMS-B-SAMS Prepared in Cyclohexane)

Run No.	Polymer Composition B (wt %)	AMS/S (molar ratio)	Peak Mol. Wt.	Tensile Strength at Break (psi)	Elongation at Break (%)
9	56.7	1.0	94000	3500	790
10	57.1	1.23	113000	3970	780
11	58.5	1.33	86000	3100	790
12	69.5	1.50	94000	3930	930

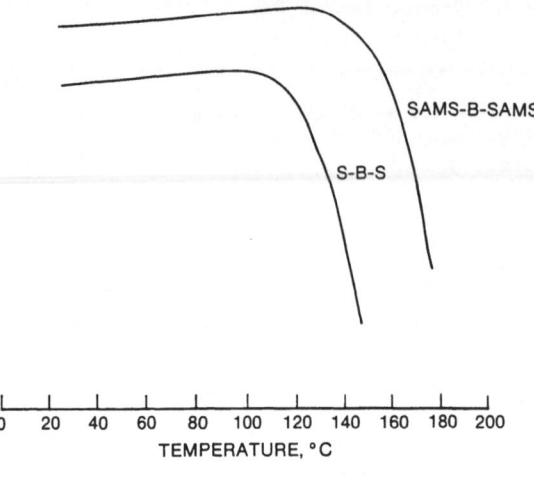

Fig. 2. Comparison of thermomechanical behavior of a typical SAMS-B-SAMS
thermoplastic elastomer with that of a commercial S-B-S
(Kraton D1101 Thermoplastic Rubber).

Table 5. Polymerization Conditions and Some Properties of Two Other
AMS-Containing Triblock Polymers

	Run 13	Run 14
Initiator (millimol)	0.71	0.70
Cyclohexane (mL)	370	360
THF (mL)	40	40
Diene (g)	isoprene 34	butadiene 44
AMS (g)	13.6	15.3
3rd monomer (g)	styrene 5.9	vinyltoluene 7.4
End block polym temp. (°C)	27	30
3rd monomer add. time (min.)	160	180
Diene (wt. %) in polymer	71.6	74
AMS/3rd monomer (molar ratio)	1.14	0.91
Mol wt. x 10^{-3}	129	116
Tensile strength at break (psi)	3900	3400
Elongation at break (%)	1325	975

Two other triblock polymers were prepared in cyclohexane. Run no. 13
used polyisoprene in place of polybutadiene as the center block. Run
no. 14 used vinyltoluene in place of styrene in the SAMS end block copolymer. The polymerization conditions and results are shown in Table 5.

Polymerization in Alpha-methylstyrene

Bates and Alfrey[13] prepared SAMS copolymer by using AMS itself as the
solvent at temperatures above 60°C, the ceiling temperature of poly(AMS).
The normally slow AMS copolymerization was accelerated by the large excess
of AMS. At the same time, long sequences of AMS could not enter the copol-

ymer chains because the temperature was above the ceiling temperature of poly(AMS). Cunningham[3] used AMS as the solvent in the preparation of a triblock with poly(AMS) end blocks. He kept the temperature of end block polymerization below the ceiling temperature of poly(AMS) to allow the formation of AMS homopolymer. At 40°C, end blocks with molecular weights of 15,000 to 20,000 were obtained in 4 to 8 hours. The mass action was responsible for rate increase of AMS polymerization in the absence of a polar solvent.

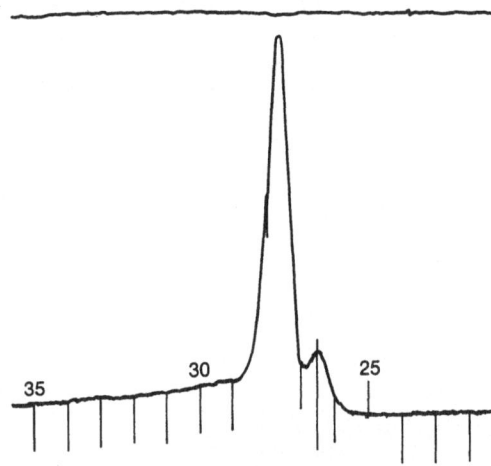

Fig. 3. GPC chromatogram of a polybutadiene prepared in AMS (upper curve from UV detector, lower from differential refractive index detector).

A set of preliminary experiments were conducted to explore whether SAMS-B-SAMS could also be prepared in AMS as the only solvent. In the preliminary experiments of preparing SAMS copolymers, the styrene used in the feed was between 3 to 5%. The other 97 to 95% was AMS. The initiator was s-butyllithium. At 27°C, the polymerization rate was found to be very slow and the amount of AMS incorporated in the copolymer was very small. At 42-45°C, conversion of styrene was complete within 35 minutes. The proportion of AMS copolymerized was exceptionally high, indicating that at this temperature some AMS homopolymerization took place. The GPC chromatograms of all the runs were narrow. In the runs made at 52-70°C, the AMS/S molar ratios in the copolymers were about 1. The molecular weight distributions remained narrow. At temperatures above 70°C, the distribution became broad indicating that self termination or chain transfer reactions had taken place. Another preliminary experiment was conducted to examine the polybutadiene prepared in AMS. In this polymerization, styrene monomer was not used. Butadiene dissolved in AMS was initiated by s-butyllithium at about 50°C. Figure 3 shows that there is no detectable amount of UV absorption for AMS in the GPC chromatogram of the polybutadiene produced. The 1,2-vinyl content of the polymer was 14.3% by IR, slightly higher than the values shown in Table 1. Polyisoprene prepared similarly in AMS showed some UV activity in GPC chromatograms indicating that some copolymerization had taken place. Proton NMR showed that the amount of AMS incorporated in the polyisoprene was about 2%.

Several different schemes were used in the preparation of triblock polymers in AMS. The preparation using dilithium initiators was the simplest.

In this scheme, all monomers were added to the reaction mixture before initiation. After initiation at 50°C, the center diene block polymerized first. Depending on the amount of the diene monomer used, the temperature of polymerization could reach a maximum of 60 to 80°C during the 70 to 90 minute polymerization time. After the diene block formation was completed, the end block polymerization began as indicated by a change of color from yellowish to dark red in the polymerization mixture. A second exothermic temperature rise then took place. The end block polymerization required about another 10 to 30 minutes, and the total time for a run was less than 2 hours. The polymers prepared have a tapered section between the end block and the center diene block. The styrene monomer could also be withheld until the diene block was polymerized. The product then was a conventional untapered triblock. Figure 4 shows the proton NMR spectra of tapered and untapered SAMS-B-SAMS prepared by these two procedures. A sharp spike on the shoulder of the styrenic peak was observed for the tapered triblock. A similar observation for tapered and untapered S-B-S has been reported.[14] Table 6 shows the feed compositions, and Table 7 shows some of the properties of the triblocks prepared. The properties of both tapered and untapered SAMS-B-SAMS were comparable to those prepared in cyclohexane as shown in Table 4. The GPC chromatogram of a typical triblock prepared by the above method is shown in Figure 5.

Table 6. Feed and Polymer Compositions of SAMS-B-SAMS Prepared in AMS Using a Dilithium Initiator

| Run No. | Monomers in Feed | | | Polymer Composition | | Peak Mol Wt |
	B (g)	S (g)	AMS (g)	B (wt %)	AMS/S (molar ratio)	
15	42	10	430	61.1	1.50	139000
*16	38	11	410	58.1	1.42	107000
*17	32	41	470	28.2	0.80	122000
18	34	43	470	31.1	0.63	88000

*Indicates tapered triblock.

Table 7. Properties of SAMS-b-SAMS Prepared in AMS Using a Dilithium Initiator.

Run No.	Tensile Strength at Break (psi)	Elongation at Break (%)	Izod Impact Strength (ft-lb/in)	Vicat Softening Temp. (°C)
15	3860	750	–	–
*16	3080	750	–	–
*17	3980	33	0.90	128
18	3250	8.3	0.80	112

*Indicates tapered triblock.

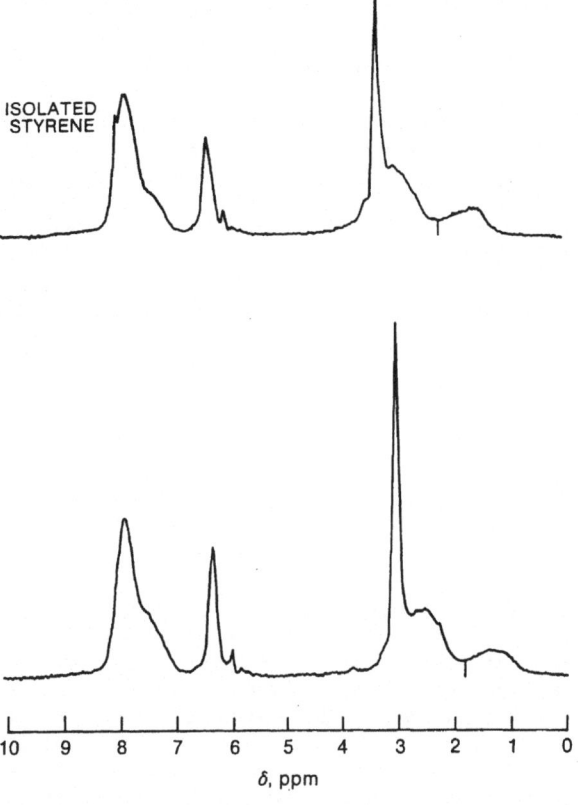

ISOLATED
STYRENE

δ, ppm

Fig. 4.   Proton NMR spectra of SAMS–B–SAMS triblock.   Upper spectrum is
from a tapered triblock, run no. 17, and lower spectrum is from
an untapered triblock, run no. 18.

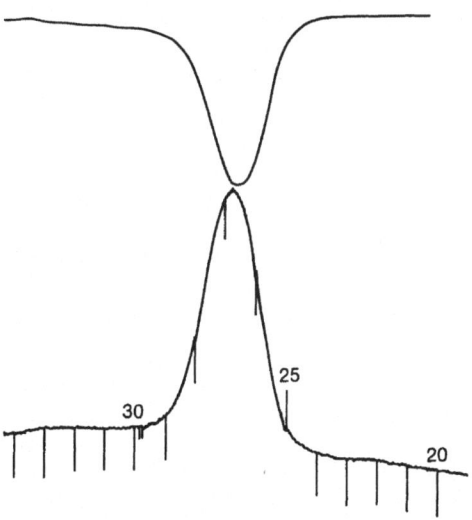

Fig. 5.   GPC chromatogram for run no. 16, a tapered SAMS–B–SAMS triblock
thermoplastic elastomer prepared in AMS using a dilithium
initiator.

Table 8. Conditions of SAMS-B-SAMS Polymerization in AMS using s-Butyllithium Initiation

Run No.	Monomers AMS (g)	S (g)	B (g)	s-BuLi mmol	1st Block Polym. Temp. (°C) Start	Max	Method of triblock or multi-block formation
19	410	16.3	45	1.69	50	67	Coupling by phenyl benzoate
20	400	12.7	45	1.81	50	61	Coupling by SiCl$_4$
21	400	12.7	45	0.96	54	59	Styrene monomer added with B
22	400	11.8	48	0.96	53	60	Styrene monomer added later
23	500	40	27	2.05	50	76	Coupling by phenyl benzoate
24	510	40	27	2.05	41	61	Coupling by phenyl benzoate

Several triblock polymers were also prepared by using s-butyllithium to polymerize the SAMS copolymer end block first. The diene monomer was withheld until the end block polymerization was near completion. The timing in the addition of butadiene was found to be critical. In the polymerization of SAMS copolymer end block, an exotherm was observed. When the addition was made several minutes past the exothermic peak of the temperature rise, a large number of the anions had already been terminated. Only when butadiene was added just after the exothermic peak did the majority of the chains form the desired diblock. The amount of premature termination was indicated by the SAMS copolymer peak in the GPC chromatogram of the final polymer. After the diene polymerization, a coupling agent was added. The timing for adding the coupling agent was far less critical. In another polymerization run, styrene monomer was added after the diblock formation for the polymerization of the second end block. Table 8 shows the preparation conditions of some of the successful runs.

In run no. 20, the polymer was a 4-arm radial block. The GPC chromatogram of a typical phenyl benzoate-coupled triblock (run no. 24) is shown in Figure 6. The major peak at the right is the triblock; the peak to the left of the major peak is the uncoupled diblock; and the peak further to the left is the SAMS copolymer formed by premature termination. The GPC chromatogram for run no. 20 showed only the major peak and the SAMS copolymer peak. The coupling efficiency of SiCl$_4$ apparently was high. In run no. 21, additional styrene monomer was added with butadiene, and the triblock produced was tapered at one end and untapered at the other end. Compositions of the block polymers are shown in Table 9.

In Table 9, the AMS/S molar ratios were measured by proton NMR on the entire sample. The values also represent the compositions of the end blocks except for run no. 21 in which styrene was added in the second stage with butadiene. The sample was half tapered, and the AMS/S ratio in the table would not be the same as that in the end blocks. The next three columns in Table 9 are the relative amounts of SAMS copolymer, diblock, and triblock calculated from GPC chromatograms in combination with NMR results. The butadiene content for the triblock peak listed in the table was calculated similarly, and the butadiene content for the entire sample, though not explicitly shown, can be easily obtained from the numbers given in the table. Of the six block polymers listed in Table 9, four (no. 19-22) were thermoplastic elastomers, and two (no. 23, 24) were rigid polymers. Some of their properties, shown in Table 10, are in general comparable to those made by other methods.

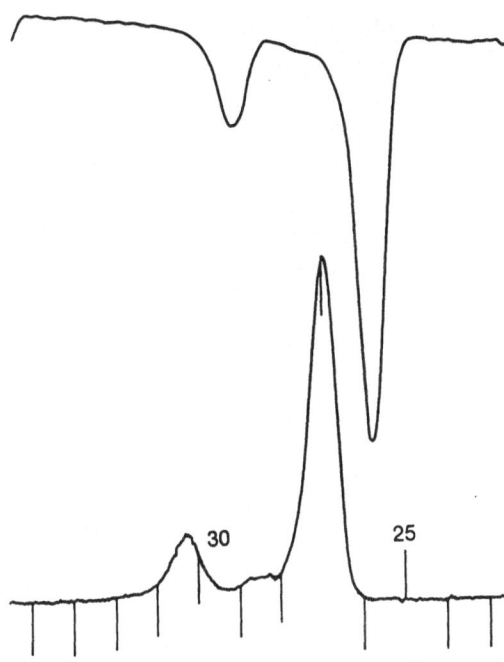

Fig. 6. GPC chromatogram for run no. 24, a SAMS-B-SAMS clear impact plastic prepared in AMS by the coupling of diblock anions with phenyl benzoate.

CONCLUSIONS

Diene triblock polymers with SAMS copolymer end blocks have been prepared by several polymerization procedures. A radial block of similar composition has also been prepared. The most attractive way of preparing the triblock polymers appeared to be by initiating with a dilithium initiator in AMS as the solvent. In this case, all the monomers were mixed at the start, and the polymerization was completed in less than 2 hours. The structure of the triblock was tapered. A convenient variation of the method consisted of withholding the styrene monomer at the start and adding it after the diene block was formed, and in this case a pure triblock was obtained in about the same polymerization time.

Table 9. Compositions of SAMS-B-SAMS Prepared with s-Butyllithium in AMS

Run No.	AMS/S (molar ratio)	SAMS copo. (wt. %)	Diblock (wt. %)	Triblock (wt. %)	Triblock Peak mol wt.	Triblock Peak B (wt. %)
19	1.26	17.0	10.4	72.6	148000	70.4
20	1.43	10.5	–	89.5	160000	65.8
21	1.34	1.9	–	98.1	89000	61.7
22	1.96	7.8	–	92.2	130000	60.3
23	0.83	37.6	← 62.4 →		134000	46.3
24	0.83	16.8	6.6	76.6	112000	35.6

Table 10. Properties of SAMS-B-SAMS Prepared with s-Butyllithium in AMS

Run No.	Tensile Strength at Break (psi)	Elongation at Break (%)	Izod Impact Strength (ft-lb/in)	Vicat Softening Temp. (°C)
19	3760	780	–	–
20	3980	570	–	–
21	3070	740	–	–
22	2930	510	–	–
23	3010	1.4	0.2	119
24	3240	10	0.4	115

When the SAMS-based triblock was prepared in AMS by using conventional butyllithium to polymerize the end block first, the timing for the addition of diene monomer was found to be critical. When the addition was too early, the desired end block would not be formed; when too late, premature termination of the chains gave a large amount of SAMS copolymer.

The diene triblock polymers with end blocks of SAMS copolymer were found to have properties similar to those of conventional S-B-S. The softening points for these triblocks were, however, 20 to 40°C higher.

ACKNOWLEDGEMENT

Our sincere thanks to J. P. Heeschen of the Dow Analytical Laboratory for his help in proton and ^{13}C NMR determinations to R.A. McDonald of Dow Systems Research for providing us with the scanning differential calorimetry results, and to F. E. Towsley, our colleague, for providing us with the results of thermal mechanical analysis.

REFERENCES

1. "Kraton Thermoplastic Rubber," Shell Chemical Company publication SC:68-81 (1981).
2. Technical Service Memorandum TSM-245, Phillips Petroleum Company (1974).
3. R. E. Cunningham, J. Appl. Polym. Sci., 22:2907 (1978).
4. L. J. Fetters, E. M. Firer, and M. Dafauti, Macromolecules, 10:1200 (1977).
5. F. E. Neumann, British Pat. 1,264,741 (1972).
6. H. Gilman and A. H. Haubein, J. Am. Chem. Soc., 66:1515 (1944).
7. L. H. Tung, G. Y. Lo, and D. E. Beyer, U.S. Pat. 4,172,100 (1979).
8. L. H. Tung, G. Y. Lo, and J. A. Griggs, J. Polym. Sci. Polym. Chem. Ed., 23:1551 (1985).
9. P. Morero, A. Santambrogio, L. Porri, and F. Ciampelli, Chim. Ind. Milan, 41:758 (1959), cited in "Polymer Handbook, Section V-2, 2nd Ed.," J. Brandrup and E. H. Immergut, eds., John Wiley & Sons (1975).
10. L. H. Tung, J. Appl. Polym. Sci., 24:953 (1979).
11. H. W. McCormick, J. Polym. Sci., 25:488 (1957).
12. G. Holden and R. Milkovich, U.S. Pat. 3,265,765 (1966).
13. S. I. Bates and T. Alfrey, Jr., unpublished information (1979).
14. X. W. Chang, S. N. Zhu, and X. Z. Yang, Gaofenzi Tongxun, 72 (1981).

PHASE-SELECTIVE CURING OF POLY(p-METHYLSTYRENE-b-BUTADIENE-b-p-METHYLSTYRENE)

Roderic P. Quirk* and Michael T. Sarkis

Institute of Polymer Science
The University of Akron
Akron, Ohio 44325
        and
Dale J. Meier

Michigan Molecular Institute
1910 West St. Andrews Drive
Midland, Michigan 48640

ABSTRACT

The preparation, properties and phase-selective crosslinking reactions of poly(p-methylstyrene-b-butadiene-b-p-methylstyrene), PMS-BD-PMS, have been examined. The PMS-BD-PMS triblock copolymer was synthesized anionically by sequential monomer addition. The ability to selectively crosslink the poly(p-methylstyrene) end-block segments was examined by curing with various polymers containing hydroperoxide end groups. Both polystyrene hydroperoxide, PSO_2H, and poly(p-methylstyrene) hydroperoxide, $PMSO_2H$, were found to significantly increase tensile stress at break and the percent elongation-at-break of the PMS-BD-PMS triblock copolymer. In addition, the modified triblock copolymer was insoluble in toluene, heptane and methyl ethyl ketone. When unfunctionalized styrene homopolymer was added the effects were quite different. These results have been compared to the unselective, non-phase-selective peroxide, dicumyl peroxide, which dramatically decreases the elastomeric properties of the triblock copolymer after curing. In addition, the specificity of this effect has been investigated by carrying out equivalent experiments with an analogous poly(styrene-b-butadiene-b-styrene), PS-BD-PS, triblock copolymer. No enhancement of tensile properties was

---

* Author to whom correspondence should be addressed.

observed on the treatment of the PS-BD-PS triblock copolymer with PSO_2H, followed by curing. The phase-selective curing of PMS-BD-PMS resulted in a product with better tensile properties at 70°C than that observed for the corresponding uncured PMS-BD-PMS and PS-BD-PS triblock copolymers.

INTRODUCTION

Alkyllithium-initiated, anionic polymerization of vinyl monomers is a very useful synthetic method since the major variables affecting polymer properties can generally be controlled, e.g., molecular weight, molecular weight distribution, copolymer structure and composition, configurational microstructure, molecular architecture, and chain-end functionality.[1-5] This control is a direct consequence of the fact that in the absence of reactive impurities many of these polymerizations are termination and chain transfer free; therefore, the products of these polymerizations are "living polymers"[6] with carbanionic chain ends.

One of the most important and unique aspects of anionic polymerization is the ability to prepare heterophase, multiblock copolymers (e.g., A-B, A-B-A, A-B-C, etc.) by sequential monomer addition with control of the molecular weight, molecular weight distribution and diene microstructure.[7,8] For example, thermoplastic elastomeric compositions of poly(styrene-b-diene-b-styrene) with high 1,4-polydiene microstructure can be prepared by sequential monomer addition using an alkyllithium initiator in hydrocarbon media.[7,9] One of the challenges to improve the properties of such block copolymers is to prepare materials which have a greater latitude in their useful temperature range by using hard-phase end blocks which have higher glass transition temperatures.[8] For example, Fetters and Morton[10] reported that well-defined poly(α-methylstyrene-b-isoprene-b-α-methylstyrene) copolymers exhibit both higher room temperature tensile strength and higher temperature capability compared to the analogous poly(styrene-b-isoprene-b-styrene) copolymers with the same molecular weights and compositions. The glass transition temperature of poly(α-methylstyrene) is 173°C.[10] Other monomers which can be utilized as hard-phase end block components to increase the useful temperature range of triblock thermoplastic elastomers include p-hydroxystyrene (after hydrolysis of the t-butyldimethylsilyl ether protecting group)[11] and 2-isopropenylnaphthalene.[12]

Poly(p-methylstyrene) is an interesting polymer in this regard, since it not only has a higher glass transition temperature (T_g=113°C) compared to that of polystyrene (T_g=102°C) but it also undergoes cross-linking much more readily than polystyrene.[13] Herein we report the results of a study of preparation, properties, and selective crosslinking reactions of triblock copolymers based on poly(p-methylstyrene) end blocks for evaluation as thermoplastic elastomers for applications at elevated temperatures.

EXPERIMENTAL

p-Methylstyrene (99.7 wt. % vinyl; 97% para, 3% meta, Mobil Chemical Company) was purified by initial stirring and degassing over freshly-crushed CaH_2 on a high-vacuum line, followed by distillation onto dibutyl-magnesium (Lithium Corporation). Final purification involved distillation from this solution directly into a calibrated ampoule. 1,3-Butadiene (99.0%, Air Products) was condensed directly from the storage cylinder into a flask containing freshly crushed calcium hydride on the vacuum line. After degassing and stirring for several hours at -78°C, 1,3-butadiene was distilled directly into a calibrated ampoule. Benzene was purified as described previously.[14] Solutions of sec-butyllithium (Lithium Corporation, 12.0 wt. % in cyclohexane) were analyzed by the double titration method with 1,2-dibromoethane.[15]

2,6-Di-t-butyl-p-cresol (BHT, Eastman) and dicumyl peroxide (Di-Cup®, Hercules) were used as received.

Polymerizations were carried out in benzene at 30-35°C in all-glass, sealed reactors using breakseals and standard high-vacuum techniques.[7] The block copolymer was synthesized by sequential monomer addition. Polystyrene hydroperoxide and poly(p-methylstyrene) hydroperoxide were prepared by oxidation of the corresponding poly(styryl)lithiums in the solid state after complexation with N,N,N',N'-tetramethylethylenediamine as described previously.[16] The total peroxide content of polystyrene hydroperoxide [$\overline{M}_n$(GPC)=4200] was 0.95 equivalents per mole as determined by iodometric titration.[17] The analogous total peroxide content of poly(p-methylstyrene) hydroperoxide [$\overline{M}_n$(GPC)=3700] was 0.93 equivalents per mole.

Size exclusion chromatographic analyses of 0.025g/10mL concentrations of polymers were performed by HPLC (Perkin-Elmer 601 HPLC) using five μ-Styragel columns (10^6, 10^5, 10^4, 10^3, and 500 Å) with tetrahydrofuran as solvent at 25°C and a flow rate of 1mL/min. after calibration with standard polystyrene samples. Glass transition temperatures were determined by differential scanning calorimetry (DSC) using either a Dupont Instruments 1090 Thermal Analyzer operating at a heating rate of 10°C/min. or a Polymer Laboratories Dynamic Mechanical Thermal Analyzer

(DMTA) operating in the shear mode at a frequency of 1Hz and a heating rate of 4°C/min.

The transmission electron micrograph was obtained with a JEOL Model JEM 120 Transmission Electron Microscope using a sample microtomed at -160°C and stained with osmium tetroxide.

Test specimens (dumbells 71mm long, 8mm wide at clamp and 2mm wide at neck) for tensile property measurements (Instron Universal Testing Instrument Model 1130; load cell, 5kg full scale; strain rate of 5cm/min.) were prepared by solvent casting from dilute (<10%) solutions in benzene on mercury; the films were then dried to constant weight in a vacuum oven. The films were cured by placing them in a Pasadena Press at 2500 psi at 160°C for varying time periods.

RESULTS AND DISCUSSION

A triblock copolymer, poly(p-methylstyrene-b-butadiene-b-methyl-styrene), PMS-BD-PMS, was prepared by sequential monomer addition using sec-butyllithium as the initiator in benzene solution. The resulting polymer exhibited an $\overline{M}_n$(GPC)=115,000 [$\overline{M}_n$(stoichiometric)=110,000], $\overline{M}_w/\overline{M}_n$(GPC)=1.10; the stoichiometrically expected block segment composition was 14,000-82,000-14,000.

By analogy with the theoretical and experimental results for poly-(styrene-b-diene-b-styrene) block copolymers[8,20-30] it was anticipated that the PMS-BD-PMS block copolymers would undergo microphase separation because of the thermodynamic incompatibility of the poly(p-methylstyrene) and polybutadiene segments.[31] Experimental evidence for the heterophase nature of the PMS-BD-PMS triblock copolymer was obtained by DSC measurements which showed two discrete transitions, one at -83°C (corresponding to the polybutadiene phase) and one at 92°C [corresponding to the poly-(p-methylstyrene) phase] when films cast from methyl ethyl ketone were used. Methyl ethyl ketone should be a selective solvent for the poly-(p-methylstyrene) phase.[32] Only one DSC peak was observed at -90°C when the casting solvent was benzene. Benzene should be a good solvent (i.e., non-selective) for both block segments.[33] These results are a further confirmation of the fact that the morphology of heterophase polymers can be affected by the casting solvent.[8,22,33]

The heterophase nature of the PMS-BD-PMS triblock copolymer was also confirmed by dynamic mechanical analysis (DMTA). DMTA analysis of the film cast from methyl ethyl ketone exhibited two distinct transitions at -65°C and +100°C as shown in Figure 1.

146

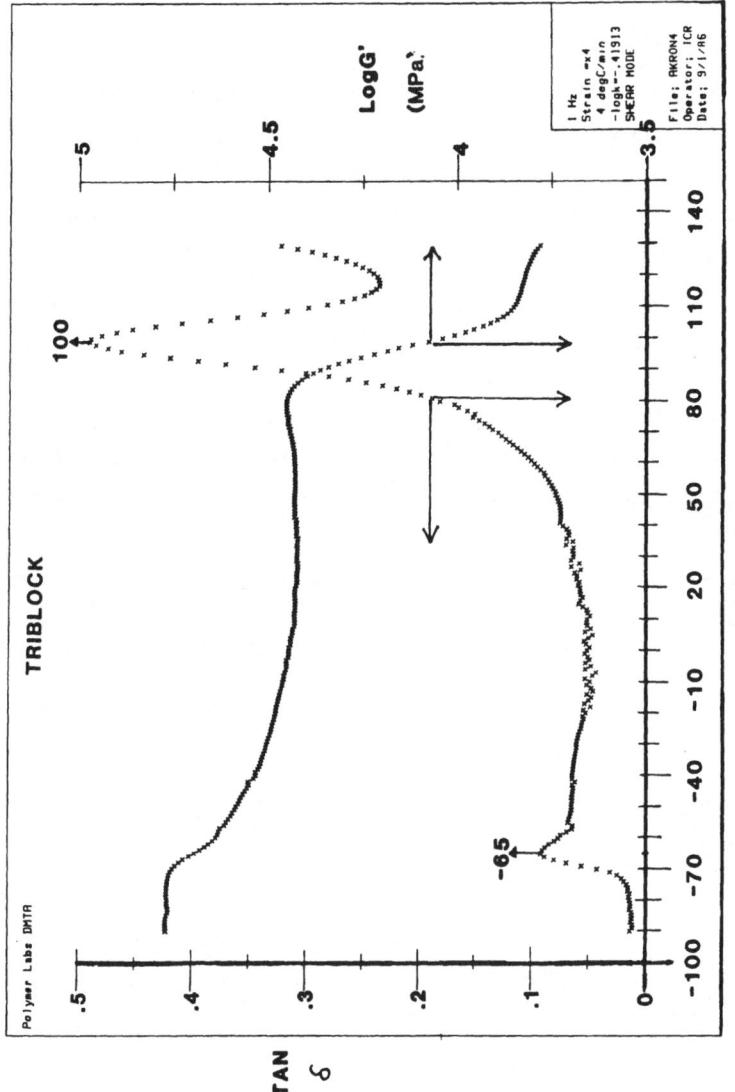

Figure 1. Dynamic Mechanical Thermal Analysis of PMS-BD-PMS Triblock Copolymer Film Cast From Methyl Ethyl Ketone Solution.

By analogy with the corresponding poly(styrene-_b_-diene-_b_-styrene) triblock copolymer, the domain morphology of the PMS-BD-PMS triblock copolymer (25% _p_-methylstyrene) might be expected to exhibit either spherical or cylindrical domain morphology for poly(_p_-methylstyrene). Preliminary electron microscopic examination of a film sample cast from benzene and stained with osmium tetroxide is shown in Figure 2. This is consistent with a phase separated, spherical domain morphology of the PMS-BD-PMS triblock copolymer.

Although poly(_p_-methylstyrene) exhibits a slightly higher glass transition temperature than polystyrene, no major improvement in higher-than-room-temperature properties would be expected for PMS-BD-PMS versus the corresponding polystyrene analog. However, it was envisioned that significant improvement in higher-temperature properties could be realized if one could selectively crosslink the poly(_p_-methylstyrene) hard phase without effecting crosslinking of the polybutadiene matrix. It has been reported that poly(_p_-methylstyrene) undergoes crosslinking much more readily than polystyrene.[13] Therefore, if free radicals could be selectively generated in the poly(_p_-methylstyrene) phase of PMS-BD-PMS, these radicals could abstract hydrogen atoms from the reactive benzylic hydrogens of the methyl groups in poly(_p_-methylstyrene). The resulting benzylic radicals would be expected to terminate by coupling reactions which would effectively crosslink the hard phase as shown in Scheme 1.

(Scheme 1)

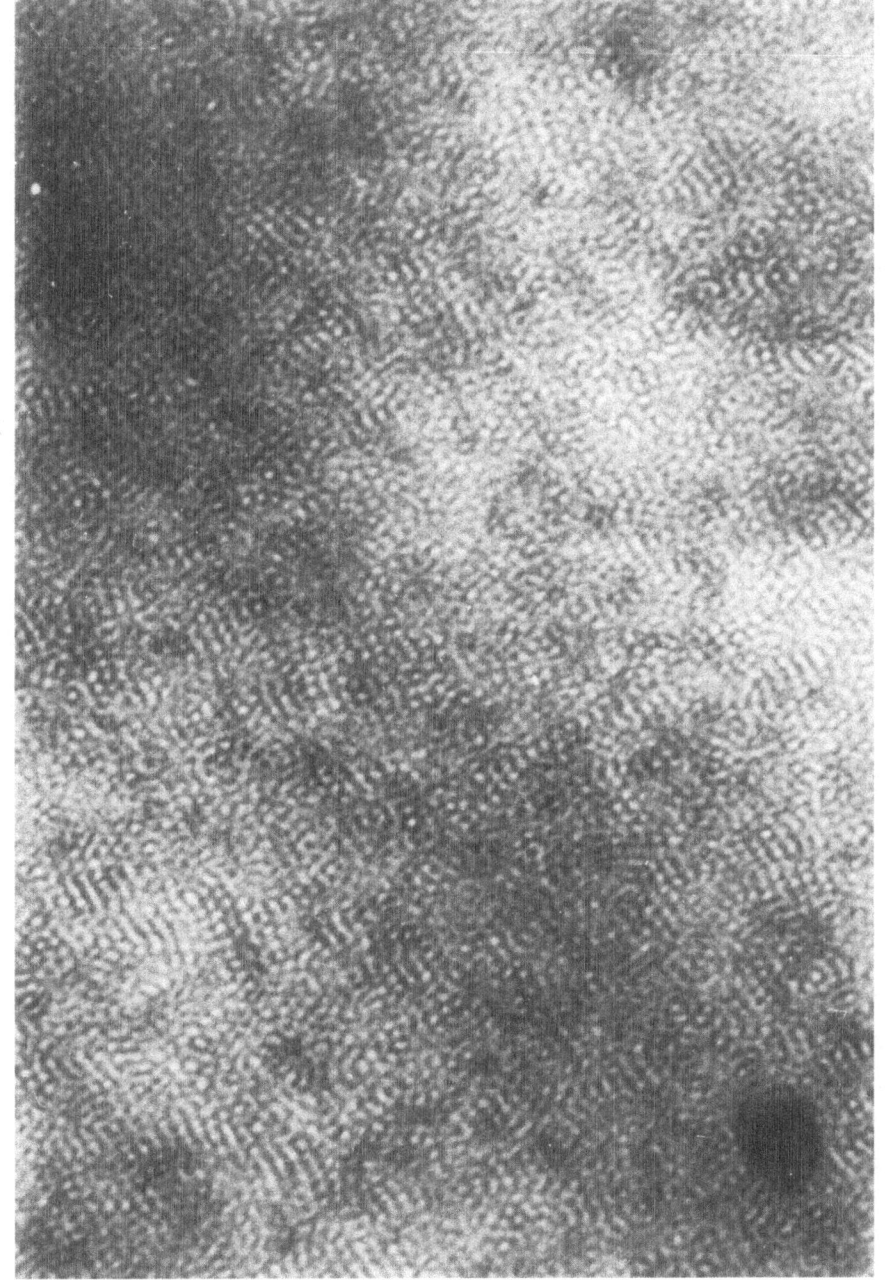

Figure 2. Electron Micrograph of PMS-BD-PMS Triblock Copolymer sample C, Table 1).

Table 1. Effect of Polystyrene Hydroperoxide and Di-Cup® on Stress-Strain Properties of PMS-BD-PMS at 25°C.

Sample[a]	Stress at Break (MPa)	% Elongation at Break
A (Triblock + PSO$_2$H)	10.4	294
B (Triblock + Di-Cup®)	7.3	12
C (Triblock + BHT)	2.2	197

[a] Films were heated in a press for 40 minutes at 160°C.

Polystyrene hydroperoxide, PSO$_2$H, ($\bar{M}_n$=4200)[16] was chosen as a suitable source of free radicals which would be compatible with the poly(p-methylstyrene) phase but incompatible with the polybutadiene phase in PMS-BD-PMS triblock copolymers. In order to test this hypothesis, three samples were prepared and compared: (A) 5g triblock (0.043mmole) plus 0.9g PSO$_2$H (0.2mmole peroxide); (B) 5g triblock plus 0.058g Di-Cup® (0.21mmole); and (C) 5g triblock plus 0.005g BHT. Each of these samples was dissolved in benzene and then cast into films by evaporation on mercury. The resulting films were heated in a press at 160°C for 40 minutes and then their tensile properties were measured at 25°C. The results are shown in Table 1. The tensile properties are consistent with unselective non-phase specific crosslinking in sample B (DiCup®), compared to the control sample C (BHT), i.e., the tensile strength is increased but the elongation is decreased dramatically. The properties of sample A (PSO$_2$H) are consistent with selective crosslinking of the hard-phase segments of poly(p-methylstyrene) with PSO$_2$H compared to the sample with antioxidant (BHT) only, i.e., an increase in tensile strength with no significant effect on the elastomeric polybutadiene center block. Further evidence for phase-specific crosslinking was obtained from swelling experiments as shown in Table 2. Although the sample C containing BHT dissolved in n-heptane or toluene, both the Di-Cup® and PSO$_2$H samples were only swollen in these solvents and in methyl ethyl ketone. Furthermore, the PSO$_2$H-treated sample exhibited much more swelling than the Di-Cup®-treated sample in heptane, a selective solvent for the polydiene phase,[33] and in toluene, a non-specific solvent.[33]

Since the morphology and properties of the poly(styrene-b-diene-b-styrene) block copolymer depend on the volume fractions of the block segments,[8,20-30] part of the effect of added PSO$_2$H could be associated with a change in the volume fraction of the hard-phase from 25% in the base triblock copolymer to 36% when mixed with PSO$_2$H. This change in

Table 2. Swelling Measurements on Cured and Uncured PMS-BD-PMS at 25°C.

Sample	Solvent	% Weight Increase
A. (Triblock + PSO$_2$H)	Heptane	861
	Toluene	3680
	MEK	106
B. (Triblock + Di-Cup®)	Heptane	64
	Toluene	188
	MEK	39
C. (Triblock + BHT)	Heptane	Soluble
	Toluene	Soluble
	MEK	250

volume fraction could also shift the morphology from spherical to lamellar or cylindrical.[8] In order to evaluate this factor, control experiments were performed by measuring the tensile properties of film samples analogous to samples B & C in Table 1, but which had been mixed with unfunctionalized polystyrene ($\overline{M}_n$=4200) in an amount equivalent to the weight of PSO$_2$H added for sample A in Table 1. The results of these experiments are shown in Table 3.

These data indicate that indeed part of the effect of added polystyrene hydroperoxide (Table 1) may be to alter the morphology of the system (e.g., compare sample E with sample C), since both the tensile strength and elongation at break are increased by simply adding an equivalent amount of polystyrene. However, the tensile strength of sample A is twice that of sample E. Therefore, there is another factor, presumably crosslinking of the hard phase, which enhances the tensile strength of sample A versus sample E.

Another variable which we have examined is the compatibility of the polymeric peroxide crosslinking agent with the triblock copolymer. For

Table 3. Effect of the Addition of Polystyrene on Stress-Strain Properties of Cured and Uncured PMS-BD-PMS at 25°C.

Sample	Stress at Break (MPa)	% Elongation at Break
D (Triblock + PSH + Di-Cup®)	12.7	35
E (Triblock + PSH + BHT)	5.8	340

PSH = Polystyrene sample isolated prior to oxidation of PSLi ($\overline{M}_n$=4200).

Table 4. Effect of Varying Amounts of Poly(p-methylstyrene) Hydroperoxide on Tensile Properties of Cured PMS-BD-PMS at 25°C.

Sample	Stress at Break (MPa)	% Elongation at Break
F (Triblock + 0.9g PMSO$_2$H)	8.12	333
G (Triblock + 1.35g PMSO$_2$H)	12.0	1497

this purpose we have compared the effect of poly(p-methylstyrene) hydroperoxide, PMSO$_2$H, with an equivalent amount of polystyrene hydroperoxide. Although the properties of PMS-BD-PMS cured samples using poly(p-methylstyrene) hydroperoxide (sample F) are similar to those using polystyrene hydroperoxide (sample A), utilization of a larger amount of poly(p-methylstyrene) hydroperoxide (sample G) gave the best enhancement of properties after curing as shown in Table 4.

If the enhancement of tensile properties in the polymer hydroperoxide treated PMS-BD-PMS samples is due to selective crosslinking of the poly(p-methylstyrene) phase, then a corresponding polystyrene triblock copolymer would not be expected to exhibit analogous enhancements. This conclusion is based on reports that polystyrene is very difficult to crosslink by electron beam radiation or by peroxide.[13,34] In order to compare the effect of phase-selective peroxide curing agents on polystyrene versus poly(p-methylstyrene) triblock copolymers, precipitated samples (5g each) of Kraton D 1102 Thermoplastic Rubber[35] [a poly(styrene-b-butadiene-b-polystyrene) triblock copolymer with 28% styrene and molecular weight of approximately 78,000]* was mixed with 0.005g BHT (sample H), 0.9g PSO$_2$H (sample I), 0.9g PSH and 0.005g BHT (sample J), and 0.058g Di-Cup® (sample K). Films were cast from a benzene solution of each mixture on mercury. The resulting films were subjected to a heat treatment identical to that used for the corresponding poly(p-methylstyrene) triblock copolymer. The tensile properties of these films are shown in Table 5.

It is noteworthy that no evidence for phase-selective crosslinking of the polystyrene block is deduced from these experiments, i.e., samples I and J exhibit similar properties and both have poorer physical properties than sample H. The effect on non-phase-selective crosslinking can be seen in sample K which exhibits very poor elongation, analogous to sample B, Table 1. These results are consistent with the

_____

* It is reported[36] that Kraton D 1102 Thermoplastic Rubber is actually a mixture of 84% triblock and 16% diblock.

Table 5. Effect of Polystyrene Hydroperoxide on Stress-Strain Properties of PS-BD-PS. (Kraton D 1102 Thermoplastic Rubber) at 25°C.

Sample	Stress at Break (MPa)	% Elongation at Break
H (Kraton D 1102 + BHT)	31.4	1670
I (Kraton D 1102 + PSO$_2$H)	27.8	1400
J (Kraton D 1102 + PSH + BHT)	26.4	1530
K (Kraton D 1102 + Di-Cup®)	7.6	25.6

expectation that phase-selective crosslinking is more efficient for poly-(p-methylstyrene) blocks compared to polystyrene blocks. Another factor is that this polystyrene triblock copolymer has volume fractions of the two blocks which optimize physical properties.[8] Therefore, added polystyrene does not enhance these properties (compare samples J and H) as observed for the p-methylstyrene triblock polymer as shown in sample E, Table 3, versus sample C, Table 1.

The stress-strain properties of these polymers have been briefly examined at 70°C. The results are shown in Table 6. It is apparent that both the PMS-BD-PMS and PS-BD-PS triblock copolymers have poorer properties at 70°C than at 25°C (compare Table 6 with Tables 1 and 5). However, the tensile properties of the PMS-BD-PMS triblock copolymer (sample C) are equivalent to those of the PS-BD-PS triblock copolymer (sample H) at 70°C in contrast to the results at 25°C (compare the results for these samples in Tables 1 and 5). It is significant that the partially crosslinked PMS-BD-PMS triblock copolymer (sample A) exhibits enhanced higher temperature tensile properties compared to both the corresponding uncrosslinked polymer (sample C) and to the analogous polystyrene triblock copolymer (sample H). These results indicate that phase-selective crosslinking may be a useful method for enhancing the higher-temperature properties of thermoplastic triblock copolymers. They also indicate the unique behavior of poly(p-methylstyrene) triblock copolymers compared to the corresponding polystyrene triblock copolymers.

Table 6. Tensile Properties of Cured and Uncured Thermoplastic Elastomers at 70°C.

Sample	Stress at Break (MPa)	% Elongation at Break
A (Triblock + PSO$_2$H)	1.75	490
B (Triblock + Di-Cup®)	2.02	3
C Triblock + BHT)	0.51	352
H (Kraton D 1102 + BHT)	0.50	363

We are continuing to explore the scope and limitations of phase-selective crosslinking reactions in thermoplastic elastomers and related materials. The results of these investigations will be reported in a future publication.

ACKNOWLEDGEMENT

The authors would like to thank the Mobil Chemical Company for their generosity in supporting this research at the University of Akron and for providing samples of p-methylstyrene.

REFERENCES

(1) M. Morton, Anionic Polymerization: Principles and Practice, Academic Press, New York, 1983.

(2) R.N. Young, R.P. Quirk and L.J. Fetters, Adv. Polym. Sci., 56, 1 (1984).

(3) M. Szwarc, Adv. Polym. Sci., 49, 1 (1983).

(4) S. Bywater, Prog. Polym. Sci., 4, 27 (1974).

(5) S. Bywater, Anionic Polymerization in Encyclopedia of Polymer Science and Engineering, 2nd Ed., Wiley-Interscience, New York, Vol. 2, 1985, p. 1.

(6) M. Szwarc, Carbanions, Living Polymers and Electron Transfer Processes, Interscience, New York, 1968.

(7) M. Morton and L.J. Fetters, Rubber Chem. Technol., 48, 359 (1975).

(8) A. Noshay and J.E. McGrath, Block Copolymers: Overview and Critical Survey, Academic Press, New York, 1977.

(9) J.C. Falk and M.A. Benedetto in Macromolecular Synthesis, E.M. Pearce, Ed., Wiley-Interscience, 1982, Vol. 8, p. 61.

(10) L.J. Fetters and M. Morton, Macromolecules, 2, 453 (1969).

(11) A. Hirai, K. Yamaguchi, K. Takenaka, K. Suzuki and S. Nakahama, Makromol. Chem., Rapid Commun., 3, 941 (1982).

(12) D. Engel and R.C. Schulz, Eur. Polym. J., 19, 967 (1983).

(13) W.W. Kaeding and G.C. Barite in New Monomers and Polymers, B.M. Culbertson and C.U. Pittman, Eds., Plenum, New York, 1984, p. 223.

(14) R.P. Quirk and W.-C. Chen, Makromol. Chem., 183, 2071 (1982).

(15) H. Gilman and F.K. Cartledge, J. Organomet. Chem., 2, 447 (1964).

(16) R.P. Quirk and W.-C. Chen, J. Polym. Sci. Polym. Chem. Ed., 22, 2993 (1984).

(17) A.J. Martin, Determination of Organic Peroxide in Organic Analysis, J. Mitchell, Ed., Interscience, New York, 1960, Vol. IV, pp. 1-64.

(18) P.A. Small, J. Appl. Chem., $\underline{3}$, 77 (1953).

(19) A. Ahmad and M. Yaseen, Polym. Eng. Sci., $\underline{19}$, 858 (1979).

(20) S.L. Aggarwal, Polymer, $\underline{17}$, 938 (1976).

(21) D.J. Meier, J. Polym. Sci. Part C, $\underline{26}$, 81 (1969).

(22) D.J. Meier in Block and Graft Copolymers, J.J. Burke and V. Weiss, Eds., Syracuse University Press, Syracuse, New York, 1973, Ch. 6, p. 105.

(23) D.J. Meier, Polym. Prepr. Am. Chem. Soc., Div. Polym. Chem., $\underline{15(1)}$, 171 (1974).

(24) E. Helfand, Macromolecules, $\underline{8}$, 552 (1975).

(25) E. Helfand and Z.R. Wasserman, Macromolecules, $\underline{9}$, 879 (1976).

(26) E. Helfand, Accounts Chem. Res., $\underline{8}$, 295 (1975).

(27) M. Morton in Encyclopedia of Polymer Science and Technology, Wiley, New York, 1971, Vol. 15, p. 508.

(28) S. Krause, J. Polym. Sci. Polym. Lett. Ed., $\underline{7}$, 249 (1969).

(29) S. Krause, Macromolecules, $\underline{3}$, 84 (1970).

(30) D.J. Meier, Block Copolymers: Science and Technology, MMI Symposium Series, Harwood Academic, 1983.

(31) K. Solc, Polymer Compatibility and Incompatibility, Principles and Practice, MMI Symposium Series, Harwood Academic, 1982.

(32) S.L. Aggarwal, R.A. Livigni, L.F. Marker and T.J. Dudek in Block and Graft Copolymers, J.J. Burke and V. Weiss, Eds., Syracuse University Press, Syracuse, New York, 1973, Ch. 9, p. 157.

(33) R. Seguela and J. Prud'homme, Macromolecules, $\underline{11}$, 1007 (1978).

(34) A.E. Platt and H. Keskkula in Encyclopedia of Polymer Science and Technology, Wiley-Interscience, New York, 1970, Vol. 13, p. 236.

(35) "Typical Property Guide, Kraton D and G," Brochure #SC 68-85, Shell Development Company, Houston, Texas, 1985.

(36) L.J. Fetters, B.H. Meyer and D. McIntyre, J. Appl. Polym. Sci., $\underline{16}$, 2079 (1972).

A NEW FREE RADICAL APPROACH TO THE SYNTHESIS OF
POLYDIMETHYLSILOXANE-VINYL MONOMER BLOCK POLYMERS

James V. Crivello, Julia L. Lee and David A. Conlon

General Electric Corporate Research and Development
Post Office Box 8
Schenectady, New York   12301

ABSTRACT

The platinum catalyzed condensation of bis(silylpinacolate) free radical initiators bearing vinyl groups attached to silicon with α,ω-hydrogen functional polydimethylsiloxane oligomers, i.e., oligomers containing terminal Si-H bonds, leads to the preparation of high molecular weight macroinitiators. The thermolysis of these macroinitiators in the presence of various vinyl monomers provides a direct synthesis of block polymers. Depending on the monomer chosen, simple triblock and/or multisequence block polymers can readily be prepared. Analysis of the products of the block polymerizations using styrene monomer shows that only block polymers are formed. These block polymers display unusual properties such as intense iridescence, reversible stress crazing and solvent dependent mechanical properties. The stress-strain properties of the block polymers have been measured and found to be related to both the relative proportions of the hard and soft blocks and to their respective block lengths. At hard block contents of less than approximately 50%, the block polymers are thermoplastic elastomers while at compositions greater than 50% the block polymers are rubber modified thermoplastics.

INTRODUCTION

Block polymers containing polydimethylsiloxane soft blocks have been the subject of considerable recent synthetic activity. In particular, polydimethylsiloxane-b-polystyrene polymers have received considerable attention as thermoplastic elastomers. For the most part, anionic polymerization methods have provided the most successful routes to the preparation of these block polymers. Among the most notable papers in this field are those of Dean,[1] Saam[2] et al, Juliano,[3] and Bajaj and coworkers.[4] Recently Chaumont and his coworkers[5] have prepared polydimethylsiloxane-b-polystyrene polymers by the platinum catalyzed condensation polymerization of α,ω-vinyl terminated polystyrene oligomers with α,ω-hydrogen terminated polydimethylsiloxanes.

While anionic and condensation methods have been used in the preparation of polydimethylsiloxane-b-vinyl monomer block polymers, there has been no mention in the literature of free radical routes to these same polymers. It was the goal of this investigation to explore the possibility of utilizing free radical methodology for the preparation of these block polymers.

157

Since polydimethylsiloxanes (PDMS) are not polymerized by free radical techniques, it is necessary either to start with preformed blocks of this type and add the vinyl monomer blocks to them or to preform the vinyl monomer blocks and then find some method of attaching them to the PDMS blocks. The simplest method of preparing the desired block polymers is to synthesize a PDMS oligomer having initiator groups attached as end groups and then to grow the vinyl polymers off both ends as shown in equation 1 where I-I represents an attached free radical initiator group.

$$I-I\left[\begin{array}{c}CH_3\\ |\\ Si-O\\ |\\ CH_3\end{array}\right]_n I-I \xrightarrow[\substack{CH_2=CH\\ |\\ X}]{\Delta} \left[\begin{array}{c}CH-CH_2\\ |\\ X\end{array}\right]_m I\left[\begin{array}{c}CH_3\\ |\\ Si-O\\ |\\ CH_3\end{array}\right]_n\left[\begin{array}{c}CH_2-CH\\ |\\ X\end{array}\right]_q$$

$$+\ 2\ I\left[\begin{array}{c}CH_2-CH\\ |\\ X\end{array}\right]_r \quad (1)$$

Similar free radical approaches have been employed by the research groups of Bamford,[6,7] Smets,[8,9] Heitz,[10] Piirma[11] and Tobolsky[12-14] to prepare block polymers from vinyl monomers. The difficulty with this method is that in addition to the generation of macrodiradical species which leads to the formation of the desired block polymer, unattached free radicals are also generated which result in the formation of homopolymers of vinyl monomers. Homopolymers are removed only with difficulty from block polymers and often contribute to reduced mechanical properties. In order to circumvent these difficulties, we decided to prepare instead macroinitiators consisting of linear PDMS polymers having initiator groups along the backbone. As depicted in equation 2, if the molecular weight of the macroinitiators is sufficiently high, the end groups may be neglected and only PDMS macrodiradical species will be formed which should lead to the predominant formation of block polymers.

$$\left[I-I\left[\begin{array}{c}CH_3\\ |\\ Si-O\\ |\\ CH_3\end{array}\right]_n\right]_m \xrightarrow{\Delta} m\ \cdot I\left[\begin{array}{c}CH_3\\ |\\ Si-O\\ |\\ CH_3\end{array}\right]_n I\cdot \quad (2)$$

EXPERIMENTAL

Materials and Characterization Methods

All of the aromatic ketones and silanes were used as purchased from commercial sources. Tetrahydrofuran (THF) was purified prior to use by first drying over calcium hydride, then fractionally distilling. Alternatively, THF was distilled under nitrogen from sodium naphthalene complex. N,N,N',N'-Tetramethylurea (TMU) was dried over calcium hydride and then fractionally distilled, while hexamethylphosphoramide was used as purchased. Magnesium metal was used in the form of 30 mesh shot. 2,3,5,6-Tetrachloro-1,4-benzoquinone used in the kinetic studies was recrystallized from acetone prior to use. Toluene, p-xylene, and benzene were dried by refluxing over calcium hydride followed by fractional distillation. Octamethylcyclotetrasiloxane (D_4) was dried and purified by distillation at atmospheric pressure. The first and last fractions were discarded and the center fraction, boiling at 175–176°C, was used. Karstedt's catalyst was obtained from General Electric Silicone Products Business Division.[15] Filtrol-20 was used as purchased from the Filtrol Company. Styrene monomer was freed from

inhibitors by washing with 10% aqueous NaOH, dried over MgSO$_4$, then fractionally distilled. Melting points were determined on a Thomas-Hoover
capillary melting point apparatus. Molecular weight measurements were
made in CHCl$_3$ at 25°C with the aid of a Waters 244 gel permeation chromatograph equipped with μ-styragel columns and are based on polystyrene standards. Osmotic pressure molecular weight measurements were made in toluene
using a Hewlett Packard 502 Membrane Osmometer. A Varian EM 390 90 MHz
Nuclear Magnetic Resonance Spectrometer was used to obtain the ^{1}H NMR spectra, while the ^{29}Si and ^{13}C NMR spectra were recorded on a Varian XL 200-
MHz spectrometer. The T$_g$ measurements of the vinyl blocks were performed
on a Perkin-Elmer DSC-2 Differential Scanning Calorimeter. The samples
were first annealed at 120°C, allowed to cool and then scanned at 20°C/min.
The determinations of the low temperature T$_g$ of the PDMS blocks were carried out on a DuPont 1050 Differential Scanning Calorimeter equipped with
a liquid nitrogen cooling system. The samples (10 mg) were first heated
to 120°C, held at this temperature for several minutes, and then quenched
at -180°C. The data were collected at a heating rate of 10°C/min.
Throughout, the samples were blanketed with an atmosphere of dry helium.
The mechanical properties of the block polymers were obtained from films
cast either from n-hexane or CH$_2$Cl$_2$. The films were dried in vacuo at
60-80°C for 2 days. At the end of this time, the films were examined for
the presence of residual solvent by running their ^{1}H NMR spectra. The
block polymers showed the absence of bands due to the presence of solvent
in their spectra. Using a die, dogbone specimens were cut from the films
and stress-strain measurements made in accordance with ASTM Method D412
using an Instron 4204 Tensile Tester.

Although a number of bis(silylpinacolate) initiators, α,ω-hydrogen
functional PDMS oligomers, and block polymers were prepared during the
course of this work, only representative examples of these syntheses will
be presented here. For a more complete experimental description of this
work, the reader is directed to reference 22.

Synthesis of Bis(dimethylvinylsilyl)benzopinacolate [benzopinacole bis-(dimethylvinylsilyl)ether] (II)

Combined into a 250 mL round-bottomed flask were 18.2 g (0.1 mol)
benzophenone, 1.2 g (0.05 mol) magnesium metal, 50 mL THF, 12.1 g (0.1 mol)
dimethylchlorovinyl silane, and 5 mL hexamethylphosphoramide. After stirring for 2 minutes, an exotherm was noted with the temperature rising to
50°C. When the exotherm had subsided, the reaction mixture was heated to
reflux for 2 h. Then the reaction mixture was stirred at room temperature
overnight. After removing the THF under reduced pressure, a yellow oil was
obtained which was slurried with approximately 100 mL CHCl$_3$ and filtered.
The solid MgCl$_2$ was washed with a little fresh CHCl$_3$ and the washings combined with the filtrate and then the solvent removed yielding a white solid.
The product was recrystallized from absolute ethanol to give a 50.3% yield
in two crops of pure bis(silylpinacolate) (II), having a m.p. of 135-137°C
(dec.). Elemental analysis; Calc., %C, 76.35; %H, 7.16; %Si, 10.50.
Found, %C, 76.33; %H, 7.27; %Si, 10.7.

Determination of Initiator Half-lives

The spectrophotometric method of Ziebarth and Neumann[17] was modified for
use in the determination of the half-lives of the thermolysis of the bis-
(silylpinacolate) initiators. A 3 x 10^{-4} M stock solution of p-chloranil and
a 1 x 10^{-1} M stock solution of the bis(silylbenzopinacolate) were prepared in
freshly dried and distilled toluene. For the half-life determination of the
bis(silylfluorenonepinacolate)s, p-xylene was used as the solvent. A 0.9 mL
aliquot of the p-chloranil solution was transferred by syringe into a
standard quartz micro UV spectrophotometer cell (1 mL capacity). The cells

were placed into a Shimadzu UV-240 spectrophotometer equipped with a Perkin Elmer thermostated cell holder and the solution allowed to equilibrate at the desired temperature. A 0.1 mL aliquot of the initiator stock solution was added by syringe and the absorbance at 410 nm determined at periodic intervals. The data were collected and plotted as the change in the absorbance versus time. Only those data which fell on the linear portion of the resulting curve were used in the calculation of the rate constants and half-lives.

Model Polymerization of Styrene Using bis(trimethylsilyl)benzopinacolate (I)

There was prepared a stock solution of freshly distilled styrene containing 0.1 M of the initiator, bis(trimethylsilyl)benzopinacolate. Ten mL aliquots of the stock solution were transferred to clean, dry vials with polyethylene screw caps. The vials were purged thoroughly with dry nitrogen gas and then placed into a thermostated oil bath at 100°C. Sample vials were withdrawn at periodic intervals and immediately quenched in an ice water bath. The polymers were recovered by first dissolving the samples in CH_2Cl_2, precipitating them into methanol, filtering, and finally drying them to constant weight in a vacuum oven at 60°C.

Preparation of an α,ω-Hydrogen Functional Polydimethylsiloxane Oligomer

Into a dry three-necked 3000 mL round-bottomed flask fitted with a magnetic stirrer, thermometer with Thermo-Watch temperature controller, efficient long path reflux condenser, and nitrogen inlet were placed 2177 g (9.806 mol) freshly dried and distilled D_4, 33.5 g (0.25 mol) tetramethyldisiloxane, and 12 g Filtrol-20 acid treated bentonite clay. The reaction mixture was heated with stirring under a nitrogen atmosphere using a mantle heater at 60°C for 20 h. After this time, the mixture was cooled and filtered through a sintered glass funnel to remove the catalyst. Next, the reaction mixture was returned to the reaction flask and stripped of starting materials and other volatiles at 150°C under approximately 0.1 Torr for several hours. The ^{29}Si-NMR was then run and the molecular weight determined using the following equation

$$N_{av} = 2\left[\frac{Si_{CH_3}}{Si_H} + 1\right]$$

where N_{av} is the average degree of polymerization, Si_{CH_3} is the area of the band at -21.996 ppm due to the silicon atoms in the main chain bearing methyl groups, and Si_H is the area of the band at -7.007 ppm assigned to the silicon atoms bearing a hydrogen atom located at the ends of the polymer chain.

Preparation of a Polydimethylsiloxane-b-polystyrene Block Polymer

To 24.73 g (0.0025 mol) of an α,ω-hydrogen functional polydimethylsiloxane having a number average molecular weight of 9894 g/mol (determined by ^{29}Si NMR) in a 100 mL micro resin kettle, there were added 4 mL dry toluene. The reaction flask was fitted with a four-armed head to which were attached a paddle stirrer, a reflux condenser and a nitrogen inlet. The reaction flask was immersed in a thermostated oil bath. Between the condenser and the head there was placed a trap filled with CaH_2 and dry toluene. The trap was constructed in such a manner that the refluxing toluene passed through the trap and was further dried prior to its return to the reaction flask. The contents of the reaction flask were then heated to reflux under a nitrogen atmosphere and drying continued for a period of 2 h. After cooling to 40°C, there were added 1.369 g (0.0025 mol) bis(dimethylvinylsilyl)benzopinacolate (II) followed by 8 µL of Karstedt's catalyst. Hydrosilylation proceeded rapidly, and within 10 min. the reaction mixture had become too viscous to stir. The reaction was allowed to proceed for a total of 1.5 to 2 h.

Next, there were added 50 g styrene monomer and the reaction mixture stirred to dissolve the macroinitiator. Then, the flask was heated to 100°C to initiate the polymerization and held at that temperature for 12 h. The block polymer was obtained as a solid plug and was removed from the reaction flask and dissolved in CH_2Cl_2 and precipitated into methanol. The polymer was dried overnight at 60°C in vacuo. There were obtained 73.1 g of the block polymer amounting to a yield of 95%. Molecular weight determination by GPC using polystyrene standards gives a $\overline{M}_n$ = 96,400 and a $\overline{M}_w$ = 290,000 g/mol.

RESULTS AND DISCUSSION

The Design of Vinyl Functional Free Radical Initiators

In equation 3 is shown the general scheme which was employed for the preparation of polydimethylsiloxane macroinitiators.

$$H\left[\begin{array}{c}CH_3\\|\\Si-O\\|\\CH_3\end{array}\right]_n\begin{array}{c}CH_3\\|\\Si-H\\|\\CH_3\end{array} + CH_2{=}CH{-}I{-}I{-}CH{=}CH_2 \xrightarrow[\text{catalyst}]{Pt}$$

$$\left[\!\!\left[\begin{array}{c}CH_3\\|\\Si-O\\|\\CH_3\end{array}\right]_n\begin{array}{c}CH_3\\|\\Si-CH_2{-}CH_2{-}I{-}I{-}CH_2{-}CH_2\\|\\CH_3\end{array}\right]_m \qquad (3)$$

This approach utilizes the platinum-catalyzed hydrosilylation reaction and requires both α,ω-hydrogen functional PDMS oligomers and free radical initiators bearing two olefinic double bonds. A series of such initiator compounds can be prepared in good to excellent yields (40-80%) by the condensation of aryl ketones with vinyl substituted chlorosilanes in the presence of magnesium and a promoter such as N,N,N',N'-tetramethylurea or hexamethylphosphoramide.[16]

$$Ar_2CO + \begin{array}{c}R_2SiCl\\|\\CH{=}CH_2\end{array} + Mg \xrightarrow[TMU]{THF} \begin{array}{c}Ar_2C{-\!-\!-\!-}CAr_2\\|\qquad\quad|\\O\qquad\quad O\\|\qquad\quad|\\R_2Si\qquad SiR_2\\|\qquad\quad|\\CH_2{=}CH\quad CH{=}CH_2\end{array} \qquad (4)$$

The resulting benzopinacole bis(dialkylvinylsilyl) ethers, which we have termed bis(silylpinacolate)s in this paper, are crystalline compounds which dissociate reversibly as shown in equation 5 on heating to give the resonance and inductivity stabilized diarylsilylketyl radicals.

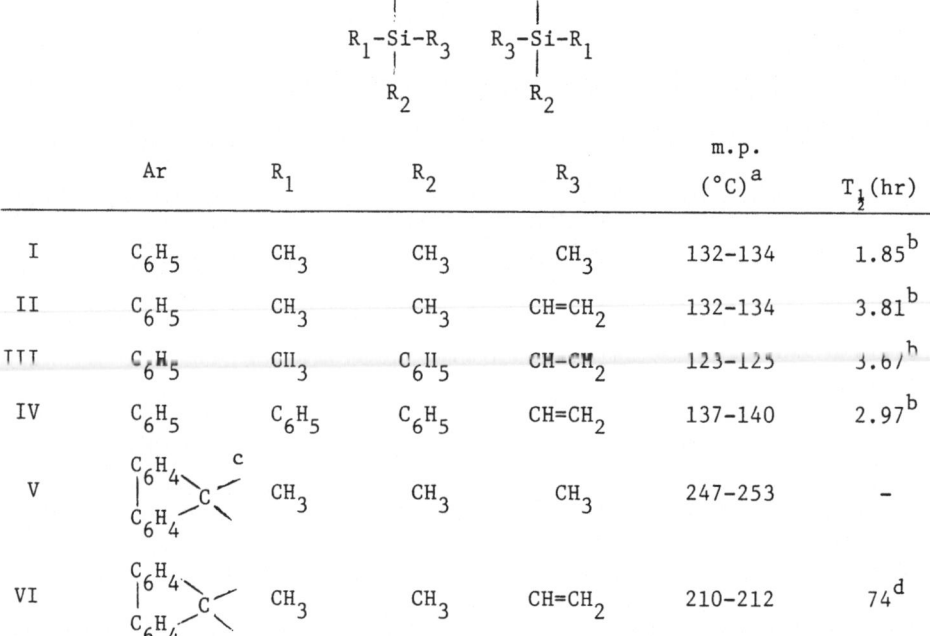

$$\begin{array}{ccc} \text{Ar}_2\text{C}\!\!-\!\!\!-\!\!\text{CAr}_2 & & \cdot\text{CAr}_2 \\ \mid \quad\quad \mid & & \mid \\ \text{O} \quad\quad \text{O} & \xrightarrow{\Delta} & \text{O} \\ \mid \quad\quad \mid & & \mid \\ \text{R}_2\text{Si} \quad \text{SiR}_2 & & \text{SiR}_2 \\ \mid \quad\quad \mid & & \mid \\ \text{CH}_2\!\!=\!\!\text{CH} \quad \text{CH}\!\!=\!\!\text{CH}_2 & & \text{CH}\!\!=\!\!\text{CH}_2 \end{array} \quad (5)$$

In Table 1 are given the structures and properties of some of the bis(silyl-pinacolate)s prepared during the course of this work, along with their half-lives for dissociation. The half-lives were determined spectrophotometrically using p-chloranil as a radical trap.[17]

The Preparation of α,ω-Hydrogen Functional PDMS Oligomers

To complete the scheme shown in equation 3 we required a series of perfectly difunctional α,ω-hydrogen terminated PDMS oligomers. After some experimentation, it was found that the desired oligomers could be readily

Table 1. Characteristics of Bissilylpinacolate Initiators

$$\begin{array}{ccc} \text{Ar}_2\text{C}\!\!-\!\!\!\!-\!\!\!\!-\!\!\text{CAr}_2 \\ \mid \quad\quad\quad \mid \\ \text{O} \quad\quad\quad \text{O} \\ \mid \quad\quad\quad \mid \\ \text{R}_1\!\!-\!\!\overset{}{\underset{}{\text{Si}}}\!\!-\!\!\text{R}_3 \quad \text{R}_3\!\!-\!\!\overset{}{\underset{}{\text{Si}}}\!\!-\!\!\text{R}_1 \\ \mid \quad\quad\quad \mid \\ \text{R}_2 \quad\quad\quad \text{R}_2 \end{array}$$

	Ar	R_1	R_2	R_3	m.p. (°C)[a]	$T_{\frac{1}{2}}$(hr)
I	C_6H_5	CH_3	CH_3	CH_3	132–134	1.85[b]
II	C_6H_5	CH_3	CH_3	$CH=CH_2$	132–134	3.81[b]
III	C_6H_5	CH_3	C_6H_5	$CH=CH_2$	123–125	3.67[b]
IV	C_6H_5	C_6H_5	C_6H_5	$CH=CH_2$	137–140	2.97[b]
V	$\begin{smallmatrix}C_6H_4\\ \;\;\;\;\;\;C<\\ C_6H_4\end{smallmatrix}$[c]	CH_3	CH_3	CH_3	247–253	–
VI	$\begin{smallmatrix}C_6H_4\\ \;\;\;\;\;\;C<\\ C_6H_4\end{smallmatrix}$	CH_3	CH_3	$CH=CH_2$	210–212	74[d]

a) Melting is accompanied by decomposition

b) Determined spectrophotometrically at 40°C

c) Fluorenylidine

d) Determined spectrophotometrically at 120°C

prepared in high yields by the controlled acid catalyzed ring opening polymerization of octamethylcyclotetrasiloxane using tetramethyldisiloxane as a chain stopper. This reaction is shown in equation 6. An acid treated bentonite clay, Filtrol-20, was employed as the catalyst.

$$\left[\begin{array}{c} CH_3 \\ | \\ Si-O \\ | \\ CH_3 \end{array}\right]_4 + \left[\begin{array}{c} CH_3 \\ | \\ H-Si \\ | \\ CH_3 \end{array}\right]_2 O \xrightarrow[60^\circ C]{H^+} H\left[\begin{array}{c} CH_3 \\ | \\ Si-O \\ | \\ CH_3 \end{array}\right]_n \begin{array}{c} CH_3 \\ | \\ Si-H \\ | \\ CH_3 \end{array} \tag{6}$$

During the reaction, it is important to scrupulously exclude water to avoid the formation of OH end groups. Using this reaction, a range of oligomers were prepared ranging in molecular weight from 1,300 to 29,000 g/mol.

The Synthesis and Characterization of PDMS Macroinitiators

The preparation of high molecular weight polymers using the hydrosilylation reaction has, to our knowledge, never been accomplished. The chief reason is the presence of complicating side reactions. Specifically, in the presence of platinum catalysts, water and other hydroxyl-containing impurities react with silicon-hydrogen bonds to give Si-OH or Si-OR compounds. If precautions are taken to carefully dry the reaction mixture, the polymerization of α,ω-hydrogen functional PDMS oligomers with vinyl substituted bis-(silylpinacolate)s proceeds rapidly and gives high molecular weight macroinitiators. To further eliminate the sources of side reactions, instead of the usual hydrosilylation catalyst, a solution of chloroplatinic acid in isopropanol, we employed the alcohol-free catalyst described by Karstedt[15] as a catalyst for this polymerization.

Initial attempts to characterize those PDMS macroinitiators having bis-(silylbenzopinacolate) groups in the backbone were frustrated. These polymers gave irreproducible viscosity measurements and GPC molecular weights. Furthermore, the molecular weights appeared to decrease with storage time and solvent dilution. We ascribe these changes to a mobile equilibrium between linear PDMS macroinitiators and macrocycles which takes place to some extent even at room temperature due to the facile rupture of the strained carbon-carbon bond of bis(silylbenzopinacolate) groups within the chain. This process is shown in equation 7.

Stable, characterizable, PDMS macroinitiators were prepared using the more stable bis(dimethylvinylsilyl)fluorenone pinacolate (VI) instead of the corresponding benzopinacolate derivative (II). The half-life for VI is 74 hours at 120°C, whereas the half-life for the analogous benzopinacolate compound II is 3.8 hours at 40°C. Figure 1 shows the ^{13}C NMR spectrum of this macroinitiator in CDCl$_3$. Unequivocal assignments could be made for each of the peaks appearing in the spectrum through comparison with the spectra of linear PDMS and the model compound, bis(trimethylsilyl)fluorenone pinacolate (V). Measurement of the molecular weight by GPC for this macroinitiator having PDMS blocks of $\overline{M}_n$ = 9894 g/mol gave a $\overline{M}_w$ = 342,800 g/mol based on polystyrene standards. Applying a correction factor for PDMS polymers, a $\overline{DP}$ = 45 was calculated for the macroinitiator. By analogy, it was concluded that the macroinitiators based on bis(silylbenzopinacolate)s are initially obtained in similarly high molecular weights. Although ring-chain equilibration does occur in these latter macroinitiators, their ability to form block polymers is unimpaired since the macrocyclic PDMS initiators which are formed also give macrodiradicals on thermolysis.

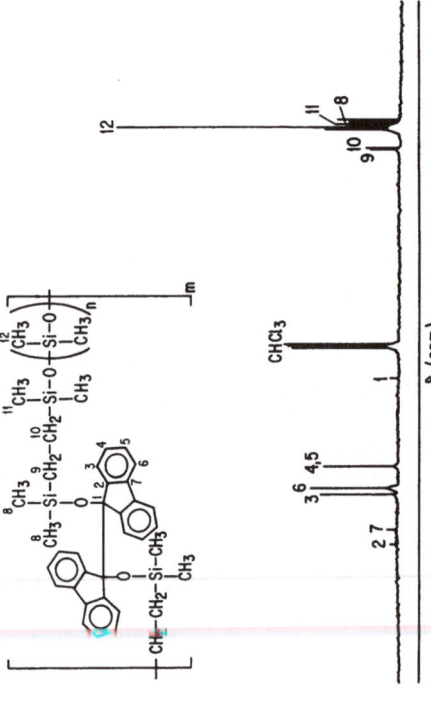

Figure 1. 13C NMR spectrum in CDCl₃ of a macroinitiator derived from bis(dimethylvinylsilyl)fluorenone pinacolate and an α,ω-hydrogen functional polydimethylsiloxane (M_n = 9894 g/mol). Peak assignments:

1 – 88.3 ppm   4⎤
2 – 146 ppm   5⎦ 118.5 ppm
3 – 128.4 ppm   6 – 126.1 ppm

7 – 140.6 ppm   10 – 9.06 ppm
8 – 0.723 ppm   11 – 0.436 ppm
9 – 9.40 ppm   12 – 0.965 ppm

$$
\left[\left[\begin{array}{c}CH_3\\|\\Si-O\\|\\CH_3\end{array}\right]_n\begin{array}{c}CH_3\\|\\Si-CH_2CH_2\\|\\CH_3\end{array}\begin{array}{cc}CH_3 & Ph \ \ Ph\\| & | \ \ \ |\\Si-O-C----C-O-\\| & | \ \ \ |\\CH_3 & Ph \ \ Ph\end{array}\begin{array}{c}CH_3\\|\\Si-CH_2CH_2\\|\\CH_3\end{array}\right]_m
$$

$\Big\downarrow\ \Big\uparrow$

$$
\begin{array}{cc}Ph & CH_3\\| & |\\\cdot C-O-Si-CH_2CH_2\\| & |\\Ph & CH_3\end{array}\left[\begin{array}{c}CH_3\\|\\Si-O\\|\\CH_3\end{array}\right]_n\begin{array}{c}CH_3\\|\\Si-CH_2CH_2-Si-O-C\cdot\\|\\CH_3\end{array}
$$

$\Big\downarrow\ \Big\uparrow$

$$
\begin{array}{c}
\quad\quad Ph\ \ Ph\\
\quad\quad |\ \ \ |\\
O-----C--C-----O\\
\quad\quad |\ \ \ |\\
\quad\quad Ph\ \ Ph\\
\\
CH_3-Si-CH_3 \quad\quad\quad CH_3-Si-CH_3\\
|\quad\quad\quad\quad\quad\quad\quad\quad\quad |\\
CH_2\quad\quad\quad\quad\quad\quad\quad\quad CH_2\\
|\quad\quad\quad\quad\quad\quad\quad\quad\quad |\\
CH_2-\left[\begin{array}{c}CH_3\\|\\Si-O\\|\\CH_3\end{array}\right]_n\begin{array}{c}CH_3\\|\\Si\\|\\CH_3\end{array}-CH_2
\end{array}
$$

(7)

The Synthesis of Block Polymers Using PDMS Macroinitiators

Before embarking on the synthesis of block polymers using PDMS macro-initiators several polymerizations were conducted using monomeric bis(silyl-pinacolate)s as initiators. In Figures 2 and 3 are shown the curves for the molecular weight and conversion as a function of time for the polymerization of styrene at 100°C using bis(trimethylsilyl)benzopinacolate (I) as the initiator. Instead of the formation of high molecular weight polymers at short reaction times usually observed in free radical polymerizations, the molecular weight was found to increase as a function of time. This behavior is more typical of condensation, living, or pseudo-living polymerizations. Otsu and Yoshida[18] have described analogous observations in free radical polymerizations using certain photoinitiators and have termed such polymerizations "living free radical polymerizations". In Scheme 1 we propose a mechanism which explains these results.

The first step involves the thermolytic dissociation of the initiator groups within the chain. Although the diarylsilylketyl radicals (I·) produced are resonance and inductively stabilized, they undergo slow addition to monomer (M) to initiate polymerization (step 2). Propagation (step 3) then proceeds as usual with chain growth. However, since the rate of initiation by the diarylsilylketyl radicals is relatively slow, there is initially a high concentration of these ketyl radicals present which can couple with the growing chains to terminate their growth. If the monomer employed is unsymmetrical, such as MMA or styrene, the new bond (M-I')

165

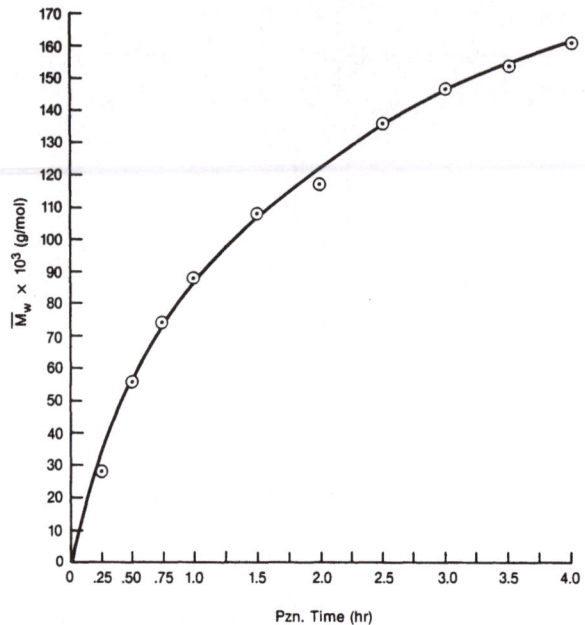

Figure 2. Plot of molecular weight as a function of polymerization time for the bulk polymer of styrene obtained at 100°C in nitrogen using 0.1 bis(trimethylsilyl)benzopinacolate, I.

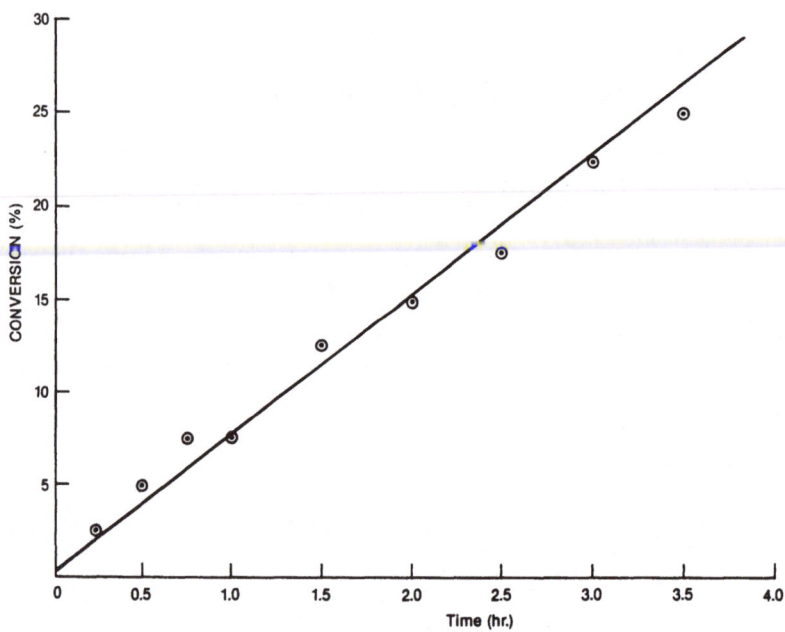

Figure 3. Plot of conversion as a function of polymerization time for the bulk polymer of styrene obtained at 100°C in nitrogen using 0.1 M bis(trimethylsilyl)benzopinacolate, I.

Scheme 1. Basic Reactions in "Long Lived" Radical
Block Polymerizations

$$I\text{-}I \longrightarrow 2\ I\cdot \qquad\qquad \text{Dissociation} \qquad (1)$$

$$I\cdot\ +\ M \longrightarrow I\text{-}M\cdot \qquad\qquad \text{Initiation} \qquad (2)$$

$$I\text{-}M\cdot\ +\ nM \dashrightarrow I\text{-}(M)_n^{\cdot} \qquad\qquad \text{Propagation} \qquad (3)$$

$$I\text{-}M\cdot\ +\ I\cdot \longrightarrow I\text{-}(M)\text{-}I' \qquad\qquad \text{Termination} \qquad (4a)$$

$$I\text{-}(M)_n^{\cdot}\ +\ I\cdot \longrightarrow I\text{-}(M)_n\text{-}I' \qquad\qquad \text{Termination} \qquad (4b)$$

$$I\text{-}(M)\text{-}I' \longrightarrow I\text{-}M\cdot\ ,\ +\ I\cdot \qquad \text{Redissociation} \qquad (5a)$$

$$I\text{-}(M)_n\text{-}I' \longrightarrow I\text{-}(M)_n^{\cdot}\ +\ I\cdot \qquad \text{Redissociation} \qquad (5b)$$

$$I\text{-}M\cdot\ +\ nM \longrightarrow I\text{-}(M)_{n+1}^{\cdot} \qquad \text{Reinitiation} \qquad (6a)$$

$$\text{cycle}$$

$$I\text{-}(M)_n^{\cdot}\ +\ nM \longrightarrow I\text{-}(M)_{2n}^{\cdot} \qquad \text{Reinitiation} \qquad (6b)$$

$$I\text{-}(M)_n^{\cdot}\ +\ I\cdot \longrightarrow I\text{-}(M)_n\text{-}I' \qquad \text{Retermination} \qquad (7a)$$

$$I\text{-}(M)_{2n}^{\cdot}\ +\ I\cdot \longrightarrow I\text{-}(M)_{2n}\text{-}I' \qquad \text{Retermination} \qquad (7b)$$

$$2\ I\text{-}(M)_n^{\cdot} \longrightarrow I\text{-}(M)_n\text{-}(M)_n\text{-}I \qquad \text{Dead Polymer} \qquad (8)$$

$$2\ I\text{-}(M)_n^{\cdot} \longrightarrow I\text{-}(M)_n\text{-}H\ + $$
$$I\text{-}(M)_{n-1}CH\text{=}CHX \qquad\qquad \text{Dead Polymer} \qquad (9)$$

which is formed on termination is itself highly strained due to steric hindrance between the three groups on the initiator fragment and the groups
situated at the end of the growing vinyl chain. Shown below are the M-I'
bonds which would be formed using either MMA or styrene.

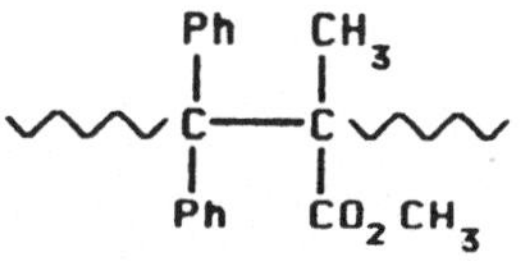

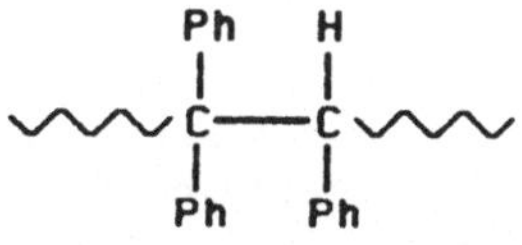

In Scheme 1 these bonds are denoted as M-I' bonds to distinguish them from
the stable, unhindered I-M bonds shown below which are produced by initiation with the same diarylsilylketyl radicals.

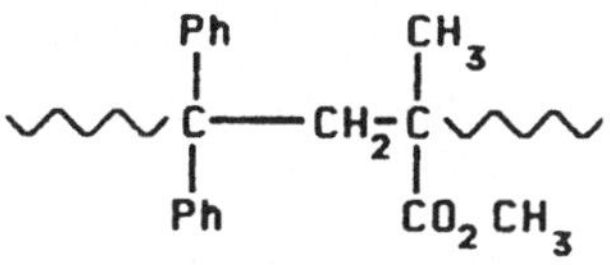

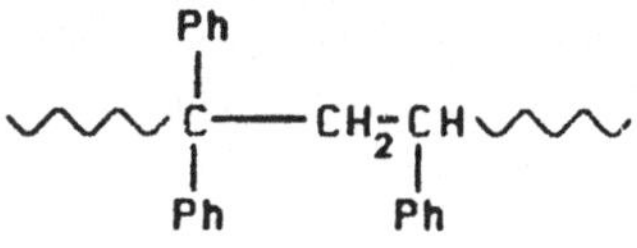

Although the steric strain of the M-I' bonds is less than that present in the bis(silylbenzopinacolate) initiators themselves, these bonds are also thermally dissociable. Thus, redissociation (steps 5a, 5b), followed by addition of monomer molecules (steps 6a, 6b) and retermination (steps 7a, 7b) constitute a cycle whereby the polymer can continue to grow by chain extension. Since the thermal threshold for dissociation of the M-I' bonds is considerably higher than for the I-M bonds, it should be possible to isolate polymers having structures I-$(M)_n$-I' and I-$(M)_{2n}$-I'. Indeed, by carrying out the polymerization of styrene using a macroinitiator containing bis(silylbenzopinacolate) groups at 55-60°C for long reaction times such that the I-I initiator groups are completely reacted after going through many half-lives, a block polymer containing styrene can be isolated. This polymer shows the ability to initiate the polymerization of styrene and other vinyl monomers by heating it at higher temperatures. Similarly, Bledzki and Braun[19] have observed that MMA oligomers produced using the related initiator, 1,2-dicyano-1,1,2,2-tetraphenylethane, can be used as macroinitiators for the polymerization of other vinyl monomers.

There are, as shown in steps 8-9, other processes which would be expected to occur as polymerization proceeds and the monomer and diarylsilylketyl radical concentrations decrease which result in permanent termination of the cycle. Conventional radical-radical coupling (step 8) and disproportionation (step 9) of the growing chains may occur which would result in "dead" polymer. In Figure 2, the departure of the curve from linearity is indicative of the complexity of the polymerization and underscores its less than "living" character. Given the propensity of such polymerizations to undergo slow but permanent termination, perhaps they are more aptly termed "long lived" rather than "living".

The success of using the PDMS macroinitiator approach for the synthesis of polydimethylsiloxane-vinyl monomer block polymers is dependent on two additional considerations. The first of these involves the control of free radical chain transfer processes. If substantial chain transfer to polymer occurs during the free radical polymerization, the block polymers which are formed may be branched rather than linear. Similarly, chain transfer to monomer or to solvents would result in the formation of homopolymers. We have sought, therefore, to minimize these possible sources of chain transfer by selection of those particular monomers and solvents which have low chain transfer constants and by careful purification of all the reagents which were used in the polymerization. A second factor which must be considered is the effect of the mode of termination on the microstructure of the final block polymer. Scheme 2 displays these effects schematically. If the primary mode of termination is coupling, multisequence $-(AB)_n-$ block polymers will be formed. On the other hand, if termination occurs exclusively by disproportionation, A-B-A triblock polymers will result. Many monomers, however, terminate by a mixture of termination modes so that in these cases a combination of tri- and multisequence block polymers will be produced. Thus, Scheme 2 predicts that the macroinitiator approach should lead to the formation of block polymers having architectures composed of at least three alternating vinyl monomer and PDMS blocks. At the same time, no diblock polymers should be formed. Although tri and higher block polymers having alternating hard and soft blocks are reported to have much better mechanical properties than the analogous diblock polymers, optimal mechanical properties are obtained for those polymers having the multisequence $-(AB)_n-$ microstructure.[20] Accordingly, our initial investigations have been directed toward the preparation of polymers having this microstructure. Since styrene undergoes termination predominantly by coupling, it has a low chain transfer constant, and because polydimethylsiloxane-b-polystyrene polymers have already been prepared by anionic techniques and have been thoroughly characterized, we have elected to concentrate our efforts on the synthesis of these materials.

Scheme 2.

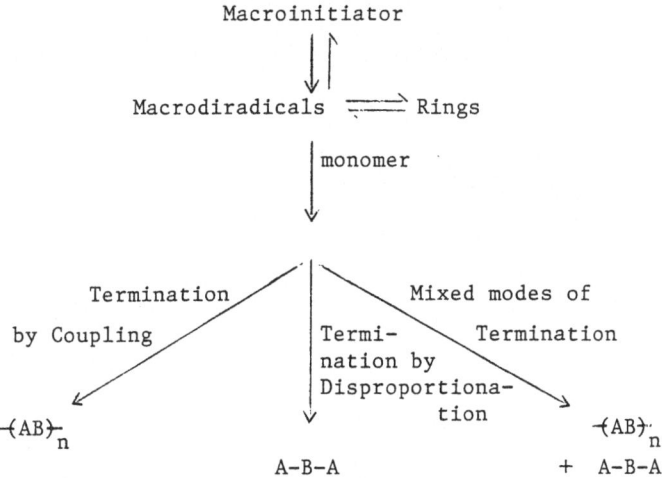

where: A represents the vinyl monomer block and B the PDMS block

The Characterization of Polydimethylsiloxane-b-Vinyl Block Monomer Polymers

In Figure 4 is shown a series of GPC curves performed on the same polydimethylsiloxane-b-polystyrene polymer using three different detectors. Besides the usual refractive index detector which assays the overall composition of the block polymer, a UV detector was used and set at 254 nm to detect aromatic groups in the polystyrene blocks and an IR detector set at 9.2μm which is the absorption band due to the Si-O-Si groups present in the PDMS blocks. As the figure shows, the response of the polymer to these three detectors is identical, indicating that the polymer is compositionally homogeneous throughout the entire molecular weight range. Further, the curves are strictly monomodal, suggesting the absence of homopolymers.

Table 2 shows data on five different polydimethylsiloxane-b-polystyrene block polymers in which different PDMS block lengths were used. In every case, high molecular weight polymers were obtained. Using the molecular weight data and the PDMS block lengths together with the percent composition of PDMS derived from the ^{1}H NMR spectra, one can calculate the number of sequences present in the block polymers. Here, one sequence consists of the combination of a PDMS and a polystyrene block. It is clear that the free radical synthesis described here yields multisequence block polymers with a large number of sequences. It should be mentioned that the sequence calculations are based on block polymer molecular weights determined using polystyrene standards. Shown also in Table 2 are the molecular weights for several of the polymers determined by osmotic pressure techniques. Since the molecular weights determined by the latter method are closer to the true values, the number of sequences reported here are probably considerably lower than the actual values.

Polydimethylsiloxane-b-polystyrene polymers prepared by the new free radical synthesis display other characteristics typical of block polymers which distinguish them from physical blends of the two homopolymers. For example, solutions of the polymers show intense iridescence due to diffraction phenomena ascribed to the presence of two discrete uniformly dispersed microphases. These polymers are soluble in solvents such as nitropropane,

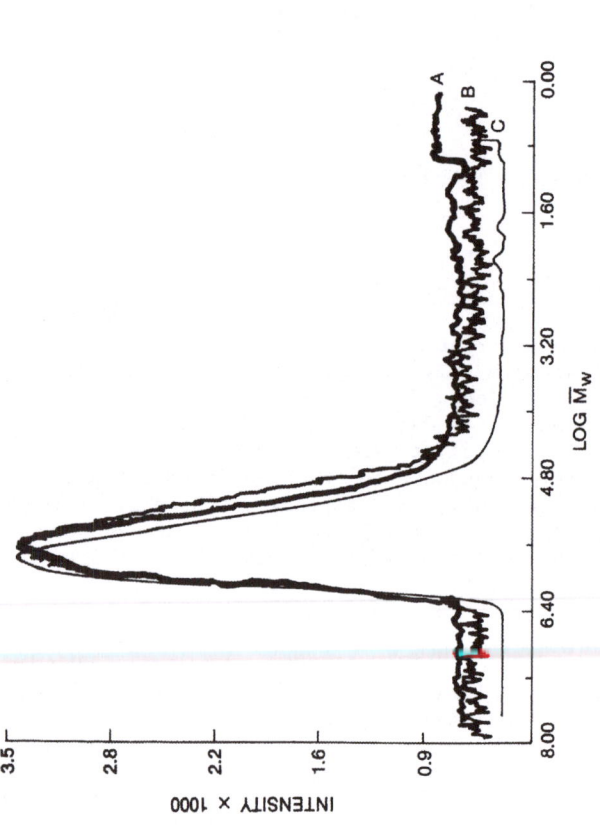

Figure 4. GPC curves for PDMS-b-polystyrene block polymer having a PDMS block length of 2634 g/mol and a PDMS content of 56%. Curve A; refractive index detector. Curve B; infrared detector at 9.2 μm. Curve C; uv detector at 254 nm. $\overline{M}_n$ = 243,000 g/mol. $\overline{M}_w$ = 513,700 g/mol.

Table 2. Composition of Polydimethylsiloxane–b–Polystyrene Block Polymers

Polymer No.	% Comp. PDMS/PS[a]	Mol. Weight[b] of Block Copolymers			PDMS Block $\overline{M}_n \left(\overline{DP}_n \right)$[d]	No. Sequences[e]
		$\overline{M}_n$	$\overline{M}_w$	$\overline{M}_w/\overline{M}_n$		
I	45/55	85,400	236,000	2.76	1390(18.8)	27.5
II	41/59	165,700 (218,000)[c]	571,600	3.45	2634(35.6)	25.8
III	42/58	99,400 (153,000)[c]	315,300	3.17	2634(35.6)	15.8
IV	40/60	286,200	492,800	1.72	4520(61.0)	25.3
V	57/43	162,100	281,100	1.73	8392(113.4)	11.0

a) Determined by ^{1}H–NMR; mole %

b) GPC data in $CHCl_3$ using polystyrene standards

c) Determined in toluene by membrane osmometry

d) Determined by ^{29}Si–NMR

e) The number of sequences was calculated using the expression:

N.S. = (fraction of PDMS)($\overline{M}_n$ of block polymer)/($\overline{M}_n$ of PDMS block)

Table 3. Physical Properties of Solvent Cast Films of
Polydimethylsiloxane-b-Polystyrene Block Polymers

Polymer No.	% Comp. PDMS/PS	Casting Solvent	Ten.Str. MPa(psi)	Elongation(%)	T_g
I	45/55	CH_2Cl_2	2.97(431)	45	98
		C_6H_{12}	3.00(435)	131	98
II	41/59	CH_2Cl_2	8.55(1240)	63	96
		C_6H_{12}	6.35(921)	209	ND[*]
III	42/58	CH_2Cl_2	13.65(1979)	344	93
		C_6H_{12}	15.20(2204)	315	ND
IV	40/60	CH_2Cl_2	12.03(1752)	88	ND
		C_6H_{12}	4.23(613)	142	ND
V	57/43	CH_2Cl_2	11.16(1618)	294	ND
		C_6H_{12}	8.86(1285)	503	ND

* ND = Not determined.

1,2-dichloroethane and cyclohexanone which are solvents for polystyrene but
not for PDMS. The stress-strain properties of solvent cast films of the
polymers are a function of the solvent used to prepare the film. Table 3
gives the mechanical properties of the same five polymers shown in Table 2
cast from both cyclohexane and CH_2Cl_2. Depending on the lengths of the
respective block lengths and their composition in the polymer, one can obtain polymers with good mechanical properties. For example, polymer III
gives films with tensile strengths of approximately 15 MPa and elongations
greater than 300%. The block polymers also display two discrete glass
transitions. One transition is observed near that reported for polystyrene
at 98°C and another is observed at −116°C which corresponds closely to the
T_g of −123°C observed for high molecular weight PDMS.[21]

CONCLUSIONS

A new free radical synthesis of polydimethylsiloxane-b-vinyl monomer
block polymers has been developed. The synthesis consists, first, of the
stoichiometric hydrosilylation of a bis(dialkylvinylsilyl pinacolate)
initiator with an α,ω-hydrogen functional PDMS oligomer in the presence of a
platinum containing catalyst to give a high molecular weight PDMS macroinitiator. Under normal circumstances, the macroinitiator is not isolated
but is carried forward into the polymerization step. A vinyl monomer is
added to the macroinitiator and the temperature raised to initiate the polymerization. The method gives high yields of block polymers essentially free
of homopolymers. In addition, this synthesis is adaptable to the preparation of a wide range of materials with properties which can be controlled by
the selection of the PDMS block lengths, the type of vinyl monomer and its
block length and their relative proportions in the block polymer. The results of some of these studies will be published elsewhere along with further details of block polymer syntheses.[22] In addition, considerable

future work is planned to prepare block polymers with interesting new architectures using this new synthetic method.

REFERENCES

1. J. W. Dean, J. Polym. Sci. Polym. Lett. Ed., 8, 677 (1970).
2. J. C. Saam, A. H. Ward and F. W. Gordon Fearson, Polym. Preprints, 13(1) 524 (1972).
3. P. C. Juliano, U.S. Patent 3,663,650, May 11, 1972 (to General Electric); Chem. Abstr. 77, 127417m (1972).
4. P. Bajaj, S. K. Varshney and A. Misra, J. Polym. Sci. Polym. Chem. Ed., 18, 295, (1980).
5. P. Chaumont, G. Beinert, J. Herz and P. Rempp, Europ. Polym. J., 15, 459 (1979).
6. C. H. Bamford and X. Han, Polymer, 22, 1299 (1981).
7. C. H. Bamford and S. U. Mullik, Polymer, 17, 98 (1976).
8. G. Smets and A. E. Woodward, J. Polym. Sci., 14, 126 (1954).
9. A. E. Woodward and G. Smets, J. Polym. Sci., 17, 51 (1955).
10. W. Heitz, C. Oppenheimer, P. S. Anand and X.-U. Qiu, Makromol. Chem. Suppl., 6, 46 (1984).
11. I. Piirma and L.-P. H. Chou, J. Appl. Polym. Sci., 24, 2051 (1979).
12. A. V. Toblosky and A. Rembaum, J. Appl. Polym. Sci., 8, 307 (1964).
13. N. Z. Erdy, C. F. Ferraro and A. V. Tobolsky, J. Polym. Sci., 8, 763 (1970).
14. E. Zaganiaris and A. V. Tobolsky, J. Appl. Polym. Sci., 11, 1997 (1970).
15. B. D. Karstedt, U.S. Patents 3,715,334, Feb. 6, 1973; 3,775,452, Nov. 27, 1973, and 3,814,730, Jun. 4, 1974 (to General Electric).
16. R. Calas, N. Duffaut, C. Biran, M. P. Bourgeois, F. Pisciotti and M. J. Dunogues, C. R. Acad. Sc. Paris, t. 267, 322 (1968).
17. M. Ziebarth and W. P. Neumann, Liebigs Ann. Chem., 1765 (1978).
18. T. Otsu and M. Yoshida, Makromol. Chem. Rapid Comm., 3, 127 (1982); T. Otsu, M. Yoshida and T. Tazaki, ibid., p. 133.
19. A. Bledski, D. Braun and K. Titzschkau, Makromol. Chem., 184, 745 (1983).
20. A. Noshay and J. E. McGrath, Block Copolymers, Academic Press, Inc., New York, 1977, p. 26.
21. J. Bandrup and E. H. Immergut, Polymer Handbook, Interscience, New York, 1966, pp. 111-183.
22. J. V. Crivello, D. A. Conlon and J. L. Lee, J. Polym. Sci. Polym. Chem. Ed., 24(6), 1197 (1986); ibid., p. 1251.

AN OVERVIEW OF THE CHEMICAL MODIFICATION OF NATURAL RUBBER

C.S.L. Baker and D. Barnard

Tun Abdul Razak Laboratory
MRPRA
Hertford, England

INTRODUCTION

Natural rubber (NR) is produced by the tree, Heyea brasiliensis, with a virtually perfect cis-1,4- polyisoprenic structure[1] and has excellent physical properties for its role as a general purpose elastomer. The main purpose of chemical modification would be to introduce functional pendent groups to provide new and improved methods of crosslinking or the grafting on of polymeric side chains. Enhanced resistance to aging, better bonding to glass, metals and textiles, and modified abrasion and friction characteristics could be other targets. The upgrading of physical properties such as oil resistance or air permeability could enable NR as a renewable resource to replace some oil-based speciality synthetic elastomers.

Such modifications of synthetic rubbers can usually be most easily achieved by the copolymerization of a second or third monomer carrying the appropriate functional group. Although the biochemical polymerization process leading to NR is well understood,[2] it has not yet been found possible to induce the enzymes concerned to incorporate functionalized 'monomer'. Therefore modification of NR has to be achieved by direct chemical reactions on the rubber itself.

NR is essentially a simple olefin and therefore, in theory, amenable to the very many known chemical reactions of such species. The trialkyl-ethylenic double bond is electron rich via inductive and hyper-conjugative effects although somewhat sterically hindered. The better known chemical modification of diene rubbers, including NR, have been reviewed[3] although the literature contains many other examples of more exotic reactions. Most reactions pertinent to a trialkylethylene have been tried on NR and, in general, they work with variable degrees of efficiency. However, usually only solution-phase chemistry has been employed.

GENERAL CONSIDERATIONS

In the work described in this paper emphasis has been placed on both the technological and economic viability of any method of chemical modi-fication. NR dissolves only slowly and partially in even the best solvents and the resulting solution is viscous at low concentrations of rubber. Any reaction which occurs only in solution need not be considered further.

Reaction Criteria

Certain criteria can be set for the conditions and extent of modification.

a) The modifying reagent should react with high efficiency when dispersed in latex (field or concentrate) or during conventional dry rubber mixing or curing operations. The heterogeneous nature of latex and surface layers of adsorbed protein and surfactant on the rubber particles can obviously give rise to difficulties as can diffusion control in reactions in solid rubber.

b) The reaction should preferably not require catalysts. Unpurified NR contains non-rubbers which can poison catalysts based on metals and interfere with free-radical systems.

c) There should be no unwanted changes induced in the NR molecule. Chain degradation, crosslinking and isomerization commonly occur with free-radical reactions and cyclization is caused by acids.

It is not surprising, therefore, that such restrictions cut the many candidate olefin reactions to a very few possibilities.

Viable levels of modification can also be defined.

1 mole % will scarcely affect physical properties but is adequate to provide functional sites for crosslinking, grafting, bonding, aging protection etc.

5 - 25 mole % could improve physical properties

25 - 100 mole % will give gross physical changes leading, essentially, to new materials.

Reagents which can carry pendent functional groups are certain to be expensive. It can be calculated that one costing 10 times as much as NR and having a molecular weight of 200 will add 30% to the price of the rubber if used at a level of 1 mole %. Reagents used for macro modification at levels of 10 mole % or higher must be very cheap to be viable. Bulk chemicals such as chlorine, hydrochloric acid or hydrogen peroxide have been used in the past and more will be said about the latter reagent in this paper.

THE 'ENE' REACTION

Many of the reagents that react with NR have been investigated with the above criteria in mind. The conclusion reached is that the most successful will operate by what has become known as the thermal 'ene' addition. This involves the 1,3-cycloaddition of an unsaturated group X=Y to an olefinic double bond in which there is a transfer of an allylic hydrogen and a shift of the double bond in probably a concerted reaction. The general reaction is illustrated in Fig.1.

The advantages of the ene reaction can be summarised as follows:-

a) it is purely a thermal reaction and does not rely on catalysts.

b) it can be highly efficient with the minimum of side reactions.

c) it is especially suited to the electron-rich double bond to NR.

d) it is relatively easy to synthesize ene reagents carrying
various functional groups as substituents.

The first ene reagent explored in detail was a nitroso-arene.   It
was found that with an electron-releasing p-substituent such as NH_2, NHR,
NHAr or OH on the arene ring reaction with NR in latex or in dry rubber at
$100-140^{\circ}$C occurred to give the expected pendent hydroxylamine.   In the
presence of reducing agents such as dithiocarbamates the latter was
efficiently transformed into a pendent sec-amino group.[4]   Technological
applications of this reaction are discussed in a later section.

The azo ene addition has proved the most rewarding area in that both
in latex and dry rubber almost quantitative efficiencies can be achieved.
The reaction of azodicarboxylates with olefins was first discussed by
Alder et al.[5]   Rabjohn[6] synthesised various examples and established
that their reaction with NR is essentially quantitative.   Bis-azo-
dicarboxylates were utilised by Flory and co-workers as near quantitative
crosslinking agents in their examination of the relationship between crosslink density and physical properties.[7]   It has been confirmed that their
reaction with olefins is insensitive to the presence of radical initiators
or scavengers[8] and that the transfer of primary allylic hydrogen to nitrogen
is preferred to that of secondary or tertiary hydrogens when more than one
type is available.[9]   This is in accord with our structural analysis of the
product of addition of azo compounds to NR depicted in Fig. 2.

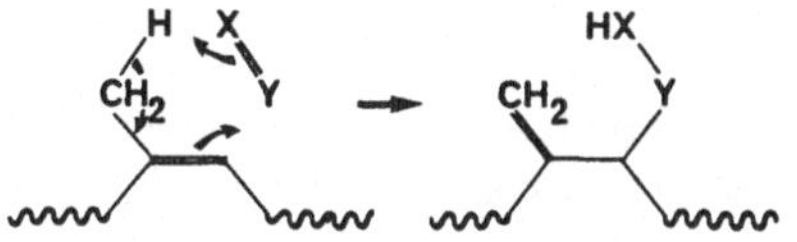

Fig. 1. The general ene reaction
X=Y can be N=O, N=N, C=S, C=O, C=C

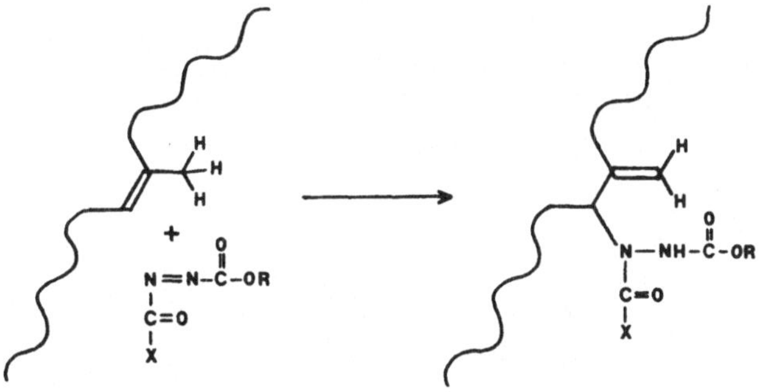

Fig. 2. The reaction of activated azo compounds
with polyisoprene

177

Another advantage of the azo ene addition is that it is relatively simple to obtain several decades spread in reaction rate by alterations to the structure of the azo compound.[10]   Practical considerations demand that reactivity is high enough to allow the addition to go to completion in a few minutes at dry rubber processing temperatures (or in several hours at more modest temperatures for latex) without being so high that non-homogeneous reaction occurs.   If an instantaneous surface reaction is required as in a dipping or spraying technique then an extremely high reactivity as in cyclic azo compounds is necessary.

Applications are again dealt with later.   The main limitation of the azo system is its reactivity towards certain functional substituents.   It is a mild dehydrogenating reagent and thiol, certain amine and hydroxyl groups cannot be attached to NR by the azo route.   Other functionalisation reactions have therefore been surveyed.

Aryl sulfonyl azides

Substituted aryl sulfonyl azides have been used to functionalize polymers with a variety of groups via their decomposition into nitrenes at elevated temperatures.[11]   While examining the utility of this method for NR we found that on mixing such azides into dry rubber they reacted via an efficient cycloaddition process at 100-120°C before decomposition to nitrenes could occur.   This again illustrates the reactivity of the electron-rich double bond of NR.   The triazoline initially formed decomposed to leave the aryl sulfonyl groups bound to the rubber via an imine bond. (Fig. 3).   This is very resistant to hydrolysis in the rubber environment and therefore serves as a stable functionalization.   As aryl sulfonyl azides can be easily synthesized carrying many types of substituent groups this new reaction looks to have great promise and is to be further explored.

TECHNOLOGICAL APPLICATIONS OF MODIFICATION

The physical effects of modification are well understood.   Most pendent groups are polar with respect to NR and because of this and the

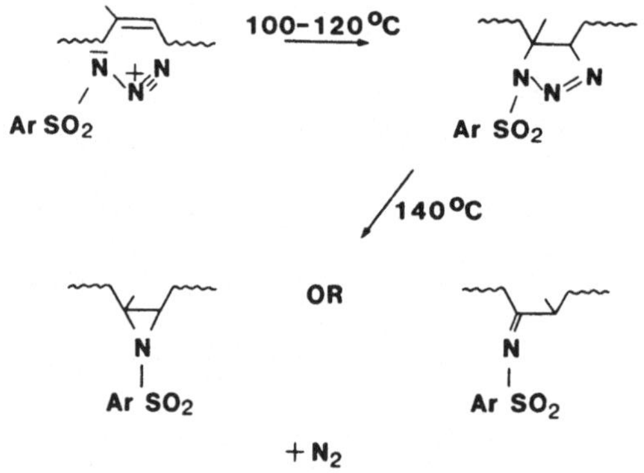

Fig. 3. The reaction of aryl sulfonyl azides with polyisoprene

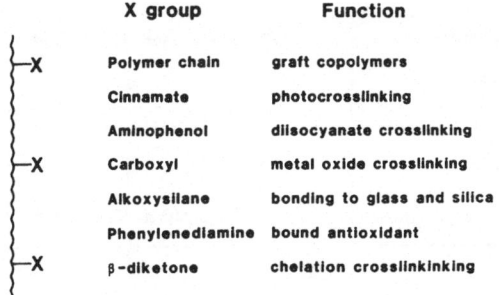

X group	Function
Polymer chain	graft copolymers
Cinnamate	photocrosslinking
Aminophenol	diisocyanate crosslinking
Carboxyl	metal oxide crosslinking
Alkoxysilane	bonding to glass and silica
Phenylenediamine	bound antioxidant
β-diketone	chelation crosslinkinking

Fig. 4. Some application of chemical modification
       of NR

progressive increase in Tg due to the restrictive effects of large groups
on segmental motion the oil resistance and air permeability of NR is
improved.   Significant changes require, however, modification in excess
of 10 mole % and, as noted earlier, this will normally be too costly.

     The chemical effects of modification around the 1 mole % level are
more useful and Fig 4 depicts specific areas where applications have been
found.

     These applications can be amplified one by one.

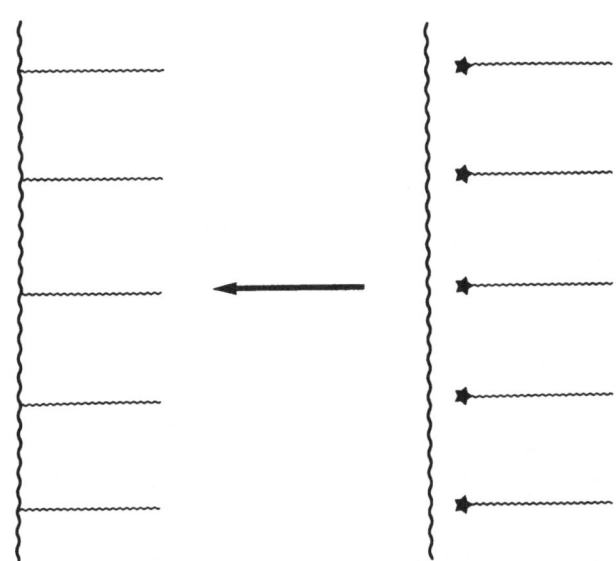

Fig. 5. The principle of reactive polymer grafting

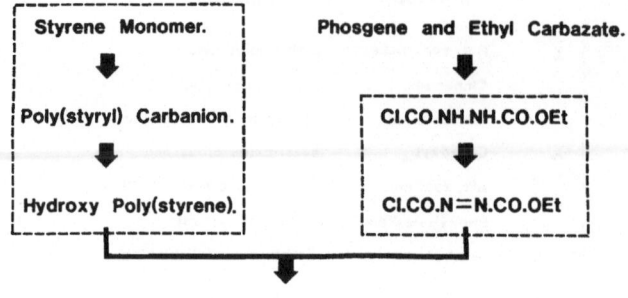

Fig. 6. Route to azo-tipped polymer

Grafting

The ability to carry out the controlled grafting of a range of polymers on to NR has been one of the more important outcomes of this work. The princple adopted was the synthesis of the polymer of desired molecular weight terminated by a reactive azo group which would lead to grafting by addition to NR. (Fig. 5)

The chemical route to an azo-tipped polymer is illustrated in Fig. 6 taking polystyrene as an example. Anionic polymerization is preferred as the molecular weight and molecular weight distribution can be accurately controlled. Termination of the living chains with ethylene oxide gives the hydroxy polymer which reacts with a separately prepared azo acid chloride to give the required functionalization in high yield (75-80%).

Against all expectations considering compatibility difficulties, simple dry mixing of azo polystyrene in an internal mixer at 100-120°C gave almost quantitative grafting (based on functionality) as assessed by gel permeation analysis.[12] This was for polyisoprene or purified NR; commercial grades of NR gave a few percent lower efficiency due to the reaction of the azo function with non-rubbers.

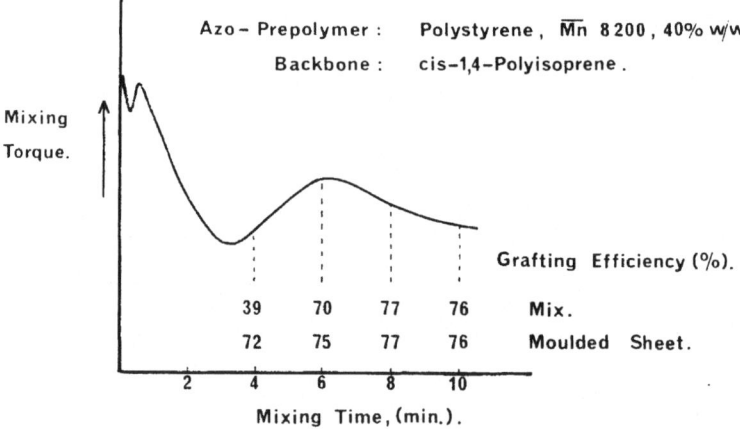

Fig. 7. A plot of torque vs mixing time with grafting efficiencies at different times

A typical torque/mixing time plot is given in Fig. 7. Initially the torque drops, then rises as grafting proceeds to fall again with continued degradation of the rubber. Grafting efficiency based on total polystyrene reaches a maximum after 8 minutes. If the mixing is stopped before this, the heating involved in making a moulded sheet is sufficient to raise the grafting to the maximum again.

The product of this grafting method is a comb graft with a rubber backbone and hard side chains. Such structures are known to possess the properties of a thermoplastic rubber[13] and this was found to be true here. For a polystyrene graft it was found that at optimum values of polystyrene content (40% w/w) and molecular weight ($\overline{M}_n$ 8,200) the product could be injection moulded and had a high tensile strength (24 MPa) without vulcanization.(Fig. 8)

This method of grafting has several advantages

a) The MW of the polymer to be grafted can be accurately determined

b) The number and therefore spacing of the grafted chains can be controlled

c) The method is theoretically applicable to all other polymers that can be reactively tipped, and indeed to mixtures of two or more such polymers, so that a very wide range of grafts is possible.

While all the examples of azo-tipped polymers prepared were found to graft to NR if a common solvent existed this was not the case for dry-mix grafting. A minimum compatibility of the polymer with NR at the mix temperature is required to allow the azo group to 'find' a double bond.

The solubility parameter (SP) of a polymer is a concept based on cohesive energy calculated from group molar attraction data and the difference in SP between two polymers is a measure of their compatibility.[14] Experimental results suggest that dry-mix grafting will occur only if the SP difference between the polymer and NR is not greater than 1.3 units (Table 1).

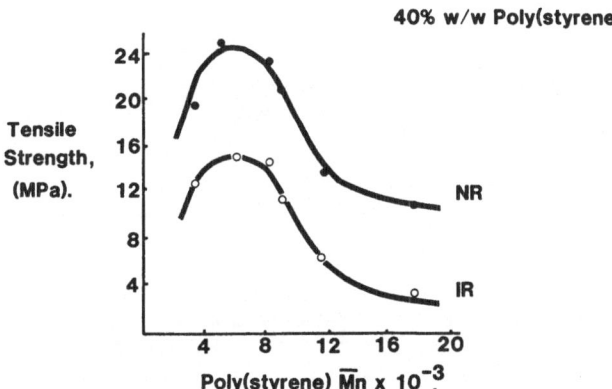

Fig. 8. Tensile strength as a function of polystyrene molecular weight for graft copolymers with NR and Cariflex IR 305.

Table 1. Solubility Parameter Difference versus Grafting

Polymer	SP Difference $(cal^{\frac{1}{2}} cm^{-3/2})$	Dry-Mix Grafting
polyethylene	0.1	yes
polypropylene	0.6	yes
poly(dimethylsiloxane)	0.8	yes
poly(α–methylstyrene)	0.8	yes
polystyrene	1.0	yes
poly(methylmethacrylate)	1.15	yes
poly(caprolactone)	1.3	no

Another restriction on dry-mix grafting is that the Tg of the polymer (or Tm for a crystalline material) must not be too high or it cannot be fluxed at a temperature at which the azo group and NR are stable. Equally, Tg should not be too low or physical properties will be very poor at room temperature.   It is thus possible to erect a feasibility diagram as in Fig. 9 to predict which polymers will dry-mix graft to NR and which might give the most useful properties as a thermoplastic rubber.   Both should be optimised in the bottom right-hand corner of the rectangle created.   The upper limit for Tm of about 180°C is raised by 30° for Tg as the latter is significantly depressed at the (usually) low molecular weight used for grafting.   Polypropylene appears to be well placed in the diagram.   Unfortunately the great difficulty of polymerizing and functionalizing this polymer at a useful degree of polymerization has so far prevented a good thermoplastic rubber being obtained.

It is indeed fair to say that although several of the grafted polymers do confer thermoplastic properties on NR none of the systems explored match the commercial block copolymers in cost-effectiveness.   However there is considerable interest in such grafts as adhesives or as blending aids.

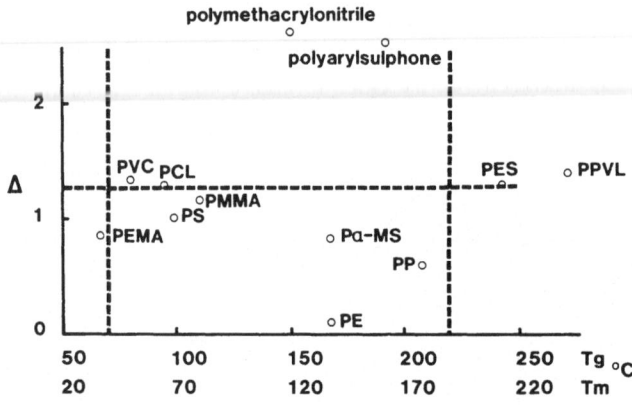

Fig. 9. A dry-mix grafting feasibility diagram obtained from the solubility parameter difference Δ between polymer and NR and Tg or Tm of the polymer where PEMA = poly(ethyl-methacrylate), PVC.= poly(vinylchloride), PCL = poly (caprolactone), PS = polystyrene, PMMA = poly(methyl-methacrylate), Pα-MS = poly (α-methylstyrene), PE = polyethylene, PP = polypropylene, PES = poly (ethylenesulphide), PPVL = poly(pivalolactone).

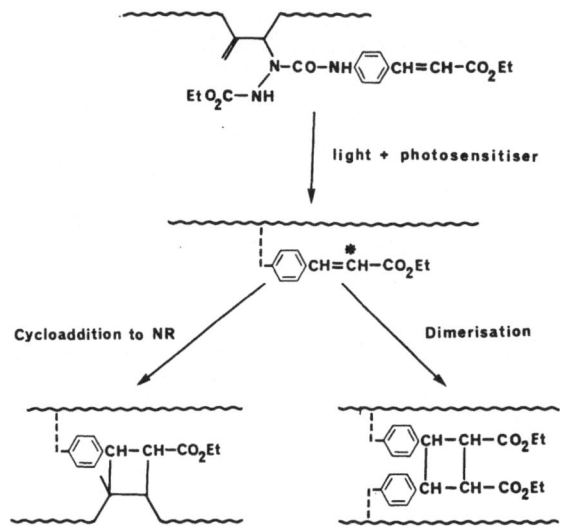

Fig. 10. Photocrosslinking via pendent cinnamate
groups

Photocrosslinking

A carbamoylazoformate carrying an N-cinnamate substituent has been
synthesized and found to add efficiently to NR in latex concentrate.   Thin,
dipped films were crosslinked on exposure to UV light or to ordinary light
if a photosensitizer were also present.   The mechanism of cinnamate photocrosslinking via dimerization of activated groups is illustrated in Fig.10.
Photoresists, as for printing rollers, were prepared by this method but the
main interest centered on the possibility of continuous vulcanization of thin
articles such as surgeons' gloves.   One advantage would be that the curing
agent is bound to the rubber and that the possible hazards of extractable
material would be avoided.   In practice the method worked but did not quite
achieve the crosslink densities necessary.   Either more efficient photo-
crosslinkers than cinnamate are required or else diffusion difficulties in
solid rubber are hindering the interaction of the activated centers.

Diisocyanate Crosslinking

The addition of nitrosophenols to NR to give pendent aminophenol groups[19]
has formed the basis of the now commercial NOVOR crosslinking system.   The
principle of this is shown in Fig.11.   The nitrosophenol in its tautomeric
oxime form reacts with an equivalent of a diisocyanate to give a stable,
non-toxic di-urethane.   When this is dispersed in NR and heated to vulcan-
izing temperature it dissociates into its component parts.   The resulting
aminophenol groups are converted into stable, urea crosslinks by reaction
with the diisocyanate (Fig.11).   The NOVOR vulcanizates are excellent for
demanding, high severity applications, combining the thermal stability of
a peroxide vulcanizate with the good tensile and fatigue properties usually
associated with a conventional sulfur curing system.

Alkoxysilane Modification

Azo reagents carrying an alkoxy silane substituent[16] have three
distinct areas of application.   Added to NR latex they react and are also

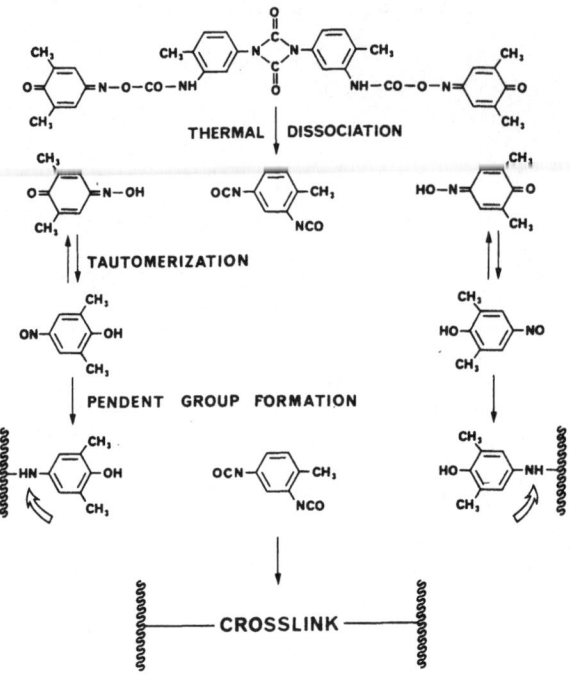

Fig. 11. The chemistry of the NOVOR crosslinking system

hydrolysed to give siloxane crosslinks (Fig. 12). They therefore offer a new route to prevulcanized latex.

Mixed into dry rubber they greatly enhance the reinforcing properties of silica fillers, presumably by binding the rubber to silica via the silane groups. The third attribute is to give excellent bonding of NR to glass in sheet or fibre form.

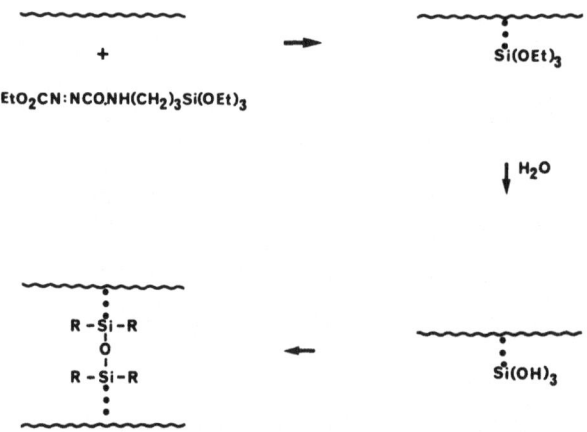

Fig. 12. Vulcanization of NR latex with azo alkoxysilanes

Bound Antioxidant

The reaction of nitroso-arylamines such as C-nitrosodiphenylamine with NR in latex or in dry mixing leads to pendent p-phenylenediamine groups in high yield.[17] The structure of these groups is typically that of an N-sec-alkyl, N-aryl-p-phenylenediamine which would be both a good antioxidant and antiozonant. NR modified in this way was protected against oxidation at least as efficiently as by the same molar quantity of the analogous phenylenediamine in free solution in the rubber, this protection being maintained after leaching with water and organic solvents. However, little if any protection against ozone was conferred, presumably because of the inability of the bound phenylenediamine to diffuse to the surface.

Commercial interest in bound antioxidants has not been large but with the advent of, for example, tires designed for long life and high mileages there could be an increased need for them in the future.

Chelation Crosslinking

The concept of using pendent chelating groups to form crosslinks via metal complexes is an attractive one. The nature of the group, the metal and the ligands already carried by the metal should enable the crosslink strength to be varied within wide limits (Fig. 13). From this spectrum it should be possible to obtain vulcanizates with interesting properties ranging from thermoplastic with heat-labile crosslinks to thermally stable examples. Aging, by the complexing of trace metals, and bonding to metals might also be significantly affected.

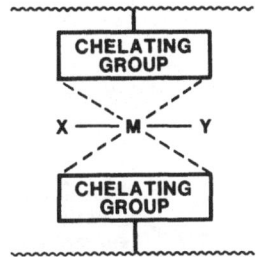

Fig. 13. Chelation crosslinking

MACRO-LEVEL MODIFICATION

As mentioned earlier, where high levels of modification are involved the reagents used must be cheap. Chlorination and hydrochlorination have

historically been the only commercial processes employed.   A modern and
more exciting example is that of epoxidation.   The treatment of latex with
peracids (Fig. 14) or, more economically, hydrogen peroxide/formic acid[18]
converts any desired percentage of the double bonds to oxirane groups.

   All data, including NMR, show, rather surprisingly, that epoxidation
is homogeneous throughout the rubber and randomly spaced.   Although the
oxirane oxygen profoundly alters some properties it has very little
effect on the stress crystallization of NR - the key to its strength.
Epoxidation increases T_g by approximately 1°C for each mole % and there is
a significant effect on oil resistance and air permeability.   ENR 50
(50 mole % epoxidized) is similar in respect of these properties to
medium nitrile or butyl rubber and has useful affinities for materials
such as PVC.   ENR 25 on the other hand gives both excellent wet grip[19] and
low rolling resistance when used in a tire tread compound (Fig. 15).
It is also as efficiently reinforced by silica as by carbon black.
ENR 10 is a useful engineering rubber with controlled damping characteristics.   Production is presently on a one tonne batch scale with plans
for a rapid scale up.

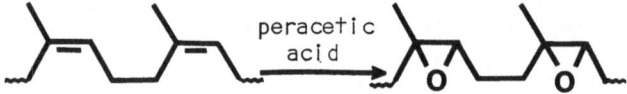

Fig. 14. Epoxidation of NR

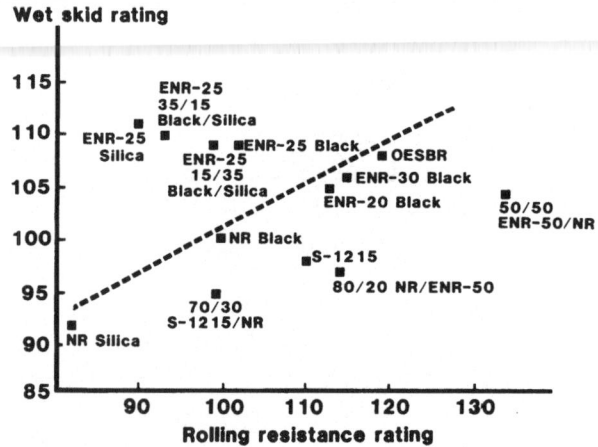

Fig. 15. Morton/Krol plot showing the high wet grip
         and low rolling resistance achieved by
         ENR-25 tire tread compounds.   S-1215 is a high
         vinyl SBR from Shell. (Diagram taken from
         Rubber Chem. Technol., 58, 67 (1985) Fig.13)

ABSTRACT

The objectives of the chemical modification of natural rubber include the enhancement of chemical and physical properties. It is possible to predict both the conditions (reaction in latex or in dry rubber) and the scale of such modification (usually 1 mole %) and, from these criteria, to define possible modifying reagents. The thermal ene reaction is shown to be particularly valuable. A new cycloaddition reaction of aryl sulfonyl azides to natural rubber is also promising. The main applications of chemical modification are discussed individually and include new methods for grafting, crosslinking, bonding and aging protection. The epoxidation of natural rubber as a cheap route to macro-modification is shown to give a new rubber, ENR, with interesting properties such as oil resistance and the ability to give tire-tread compounds with high wet grip and low rolling resistance.

REFERENCES

1. H.Y. Chen, Microstructures of polyisoprenes by high resolution NMR, J. Polym. Sci., B4, 891 (1966).
2. B.L. Archer, B.G. Audley, and F.J. Bealing, Biosynthesis of rubber in Hevea Braziliensis, Plastics and Rubber International, 7, 109 (1982).
3. D.N. Schulz, S.R. Turner, and M.A. Golub, Recent advances in the chemical modification of unsaturated polymers, Rubber Chem. Technol., 55(3), 809 (1982).
4. G.T. Knight and B. Pepper, A rationalisation of nitrosoarene-olefin reactions, Tetrahedron, 27, 6201 (1971).
5. K. Alder, F. Pascher, and A. Schmitz, Substitution reactions. I. Addition of maleic anhydride and azodicarboxylic esters to singly unsaturated hydrocarbons, Ber. dt. chem. ges., 76, 27 (1943).
6. N. Rabjohn, The synthesis and reactions of disazodicarboxylates, J. Am. Chem. Soc., 70, 1181 (1948).
7. P.J. Flory, N. Rabjohn, and M.C. Schaffer, Dependence of tensile strength of vulcanized rubber on degree of crosslinking J. Polym. Sci., 4, 225 (1949).
8. H.M.R. Hoffman, The ene reaction, Angew. Chem. Int. Edn., 8(8), 556 (1969).
9. W.A. Thaler and B. Franzus, The reaction of ethyl azodicarboxylate with mono-olefins, J. Org. Chem., 29, 2226 (1964).
10. D. Barnard, K. Dawes, and P.G. Mente, Chemical Modification of NR: Past, Present and Future, Proc. Internat. Rubber Conf. Kuala Lumpur 1975, 215.
11. A.A.R. Sayigh, B.W. Tucker and H. Ulrich, U.S. Patent 3,652,599, filed Mar. 20, 1970, issued Mar. 28, 1972 (to the Upjohn Company).
12. D.S. Campbell, D.E. Loeber, and A.J. Tinker, Efficient grafting of polystyrene on to polydienes by dry mixing, Polymer, 20, 393 (1979).
13. J.C. Falk, R.J. Schlott, D.F. Hoeg and J.F. Pendleton, New thermoplastic elastomers, Styrene grafts on lithiated polydienes and their hydrogenated counterparts, Rubber Chem. Technol., 46, 1044 (1973).
14. S. Krause, Polymer compatibility, J. Macromol. Sci. Rev., 7, 251 (1972).
15. C.S.L. Baker, Latest developments in the urethane crosslinking

of NR, Kautschuk Gummi Kunstoffe, $\underline{36}$, 677 (1983).

16. K. Dawes and R.J. Rowley, Chemical modification of NR – a new silane coupling agent, Plastics and Rubber : Materials and Applications, $\underline{3}$, 23 (1978).

17. M.E. Cain, G.T. Knight, P.M. Lewis and B.J. Saville, Development of network-bound antioxidants for improved ageing of NR, J. Rubber Res. Inst. Malaya, $\underline{22}$, 289 (1969).

18. I.R. Gelling, Modification of NR latex with peracetic acid, Rubber Chem. Technol., $\underline{58}$, 86 (1985).

19. C.S.L. Baker, I.R. Gelling and R. Newell, Epoxidized natural rubber, Rubber Chem. Technol., $\underline{58}$, 67 (1985).

STABILISATION OF RUBBERS IN AGGRESSIVE ENVIRONMENTS

Gerald Scott, Khirud B. Chakraborty and S. Mehdi Tavakoli

Department of Molecular Sciences

Aston University, Aston Triangle, Birmingham, B4 7ET

ABSTRACT

The synthesis of thiol-adduct antioxidants by a mechano-chemical procedure in elastomers is described. The yields of polymer-bound antioxidants so formed increase with increasing antioxidant loading and for most unsaturated polymers an optimal yield is observed at 20-30% loading. Bound antioxidant concentrates made by this procedure have been used as additives in a range of elastomers and it has been demonstrated that an adduct concentrate (masterbatch) in one polymer (e.g. EPDM) can be used in a range of other polymers without sacrificing efficiency.

INTRODUCTION

Diarylamines have long been the preferred antioxidants to protect rubbers from the effects of thermal oxidation.[1] However, it was recognised many years ago that the simplest member of the series, diphenylamine, is not very effective in an air oven test at high temperatures (100°C and above), and subsequent work in industrial laboratories has led to the development of higher molecular mass 4,4'- alkylated derivatives (see Table 1) which are not only less volatile but are also more soluble in most rubbers. They can therefore be incorporated at relatively high concentrations without "blooming" to the surface of the rubber. The diaryl-p-phenylene diamines (see Table 1) are also used for the same purpose and the N-alkyl-N'-phenyl-p-phenylene diamines which have the additional advantage of being able to protect rubbers against fatigue and ozone, are particularly important general purpose antioxidants.

Table 1. Arylamine Antidegradants for Rubbers

Arylamine		Trade Names	Antioxidant role
	R = t-C$_8$H$_{17}$	Octamine, Nonox OD	Thermal
	R = α-cumyl	Naugard 445	Thermal
	R = R' = β-naphthyl	Age Rite White	Thermal
	R = Ph, R' = i-Pr	Santoflex IP, Nonox ZA 4010NA, Flexzone C	Thermal, fatigue Ozone.
	R = R' = alkylphenyl	Wingstay 100	Thermal, fatigue

Hot fluid extraction

Recently deficiencies have been observed in the performance of the arylamines in rubbers subjected to surface contact with extractive fluids, particularly at high temperatures.[2] Two types of test are widely used to simulate these conditions. The first is an alternating exposure to hot fluid (generally at $120-150°C$) followed by exposure in an air oven (generally at the same temperature and for the same time). Most rubbers would fail after one exposure to such a cycle, and even oligomeric arylamines such as Flectol H (I), which has been found to confer excel-

lent heat ageing resistance on most rubbers, is relatively ineffective in the above test.

A second test which is even more discriminating is the so-called "contamination" test, in which the rubber is soaked in the fluid for several hours at room temperature and is then subjected to hot air ageing at high temperature. This test simulates the environment of many engine components (e.g. seals, gaskets, hoses, etc.). The main rubbers commonly used in such applications are nitrile rubbers which are not appreciably swelled by hydrocarbon oils and ethylene-propylene (EP) and their terpolymers (EPDM) which are resistant to the phosphate ester fluids (e.g. Skydrol) used in aircraft hydraulic systems.[2] The following discussion is concerned with the performance of antioxidants under these very aggressive conditions.

POLYMER-BOUND ANTIOXIDANTS

Research at the Malaysian Rubber Producers Research Association almost 20 years ago showed that antioxidants could be chemically attached to rubber during processing and fabrication.[3] Since that time, the development of a versatile chemically bound antioxidant system has increasingly been seen to be a potential solution to the problem of antioxidant loss from polymers under aggressive conditions. Although the nitroso antioxidants investigated by MRPRA were not developed commercially, they pointed the way forward. The potential advantages of polymer-bound antioxidants are that they cannot migrate to the rubber surface and hence cannot be lost by volatilisation or solvent leaching. A second important advantage, whose significance is not always appreciated, is that because adduct antioxidants are dispersed along the polymer chain, they are "infinitely soluble" and should therefore be more effective on a weight basis than conventional additives. A potential disadvantage which will be discussed later is that because they have restricted spatial mobility, their ability to scavenge radicals may be reduced.

Approaches to Polymer-Bound Antioxidants

Many scientific publications and patents have been directed in recent years to the problem of chemically attaching antioxidants and stabilisers to polymers.[4] These can be categorised as follows:

(1) Co-polymers of conventional vinyl monomers with vinyl antioxidants. This procedure has been successfully used by The Goodyear Tire & Rubber Company to introduce the antioxidant monomer (II) into a variety of polymers.[5] A grade of nitrile rubber containing this

191

bound antioxidant (Chemigum HR 665) is commercially available. Many
studies have been reported directed towards the incorporation of UV
stabilisers into plastics by the same procedure[4,6], but no commercial
UV stable polymers have so far been reported using this procedure.

$$\langle\bigcirc\rangle\text{-NH-}\langle\bigcirc\rangle\text{-NHCOC=CH}_2 \qquad \text{II}$$
$$\overset{CH_3}{|}$$

(2) Reaction of conventional antioxidants with functionalised polymers.
Many unsaturated rubbers can be made reactive toward conventional
antioxidants by chemical modification. Ways in which this can be
carried out have been recently reviewed[4,7] and will not be discussed
here.

(3) Reaction of reactive antioxidants with conventional polymers.
This method is potentially the most versatile, since it can be
carried out in a variety of ways and adds relatively little to the
cost of the final polymer. The nitroso-ene reaction in natural
rubber already referred to is one example of such a modification.[3]
However, in order to avoid the necessity to gear the modification
process to the vulcanisation reaction, we have concentrated in our
own work on antioxidant adduct formation to the rubber double bonds
either by vinyl grafting or by the Kharasch thiol addition reaction.[4,7]

Modification of rubbers with thiol antioxidants by the Kharasch reaction

The free radical addition of thiols to olefins provides a potentially
versatile method for antioxidant adduct formation in rubbers.[4,7] This is
shown for nitrile-butadiene rubber in Scheme 1, but it is applicable in
principle to any rubber which has a reactive double bond. Three main
methods have been used to carry out the reaction.

(1) Adduct formation in polymer latices using a redox initator.[4]

(2) Adduct formation in a formulated polymer using UV initiation.[4]

(3) Mechanochemical adduct formation in polymers in a high shearing
mixer.[7]

The first two methods have inherent limitations. The first is not
applicable to polymers not manufactured in latex form, and the second is
difficult to apply and is hence uneconomic in thick sections. The third
modification has proved to be highly versatile and capable of giving
bound antioxidant concentrates in a wide variety of unsaturated polymers.

MECHANOCHEMICAL FORMATION OF BOUND ANTIOXIDANTS

Mechano-oxidation is used to masticate polymers prior to incorpor-
ating compounding ingredients.[8] It involves the scission of the polymer
chain under shear, and is normally carried out on an open mill or in an
internal mixer in the presence of air which reacts rapidly with the
macroalkyl radicals (see Scheme 2).

It was shown many years ago by Watson and his co-workers[9] that the
initially formed macroalkyl radicals could be used to initiate free
radical reactions in the absence of oxygen, and indeed, they successfully produced block copolymers containing rubbery and glassy segments
(see Scheme 2). This classical use of mechanochemistry as a synthetic
procedure has been strangely neglected. We have been able to show that
high levels of thiol adduct formation can be made to occur with MADA at

192

Scheme 1. Reaction of Antioxidant thiols with nitrile-butadiene rubber (NBR)

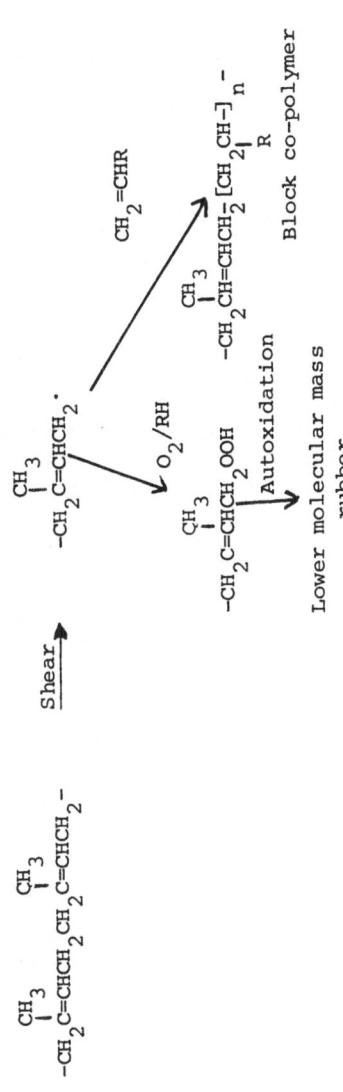

Typical examples of ASH

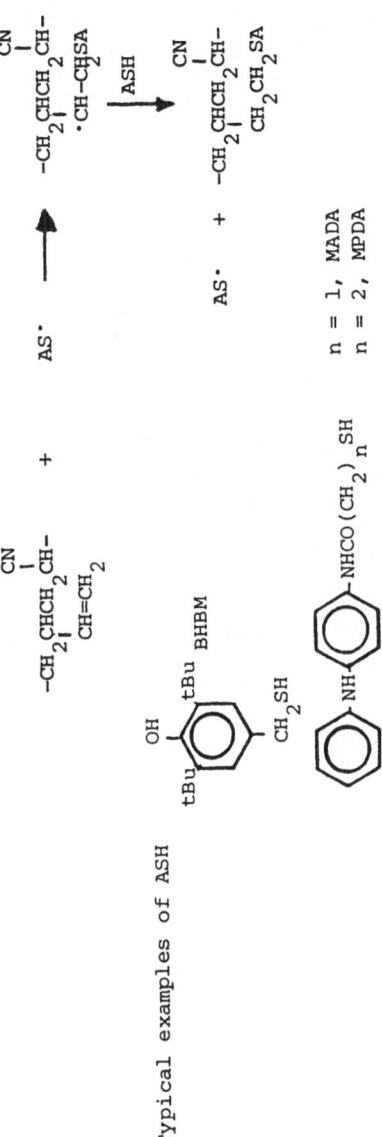

Scheme 2. Mechanomodification of Rubber during processing

antioxidant loadings in rubbers of up to 50%. This reaction has to be carried out essentially in the absence of oxygen, for the reasons explained above. However, there is evidence that the small amount of oxygen dissolved in the polymer leads to the formation of a small amount of hydroperoxide during the mechanochemical procedure, and this in turn leads to radical formation and completion of the adduct reaction.[10] About 50% of the antioxidant reacts during the first few minutes of processing when the applied torque is high, and the process is completed by hydroperoxide initiation.

The effect of antioxidant loading on the yield of adduct is unexpected, since it increases to a maximum for most polymers at about 20% loading and then decreases steadily as the loading increases further. Table 2 shows typical yields of MADA-B* in various rubbers at 20% loading in the polymer.[11,12]

Table 2. Yields of MADA-B in Rubbers at 20% Loading

Rubber	Temperature, $^\circ$C	Yield, %
NR	70	44
Cis-BR	70	78
SBR	70	75
Cis-IR	100	75
EPDM	70	78

Only natural rubber has failed to give yields greater than 50%, and the fact that good yields can be readily achieved with cis-IR[11] suggests that naturally occurring non-rubber constituents in NR interfere with the free radical chain reaction shown in Scheme 1. This is confirmed by the fact that prior extraction of raw rubber considerably improves the extent of binding.[11]

The restriction on the level of binding in natural rubber has not proved to be important in practice, since it has been found that the vulcanisation process brings about a further increase in binding[13] so that the level of adduct formation in the vulcanisate when the antioxidant concentration is reduced to 1% is not very different from that in other rubbers (see Table 3).

Table 3. Mechanochemical binding of MADA to Rubbers, before and after vulcanisation. (Reaction at 70°C).

Rubber	Extent of binding, % of theoretical	
	Before vulcanisation 10g/100g	After vulcanisation 1g/100g
NR	50	68
SBR	58	74

Masterbatch concentrates of antioxidant thiols as additives for rubbers

The masterbatch concentrate procedure has proved to be very important practically because it is a highly economical way of introducing bound antioxidants into polymers. Surprisingly, concentrated masterbatches of bound antioxidants are no less effective when reduced in concentration to normal antioxidant concentrations than are bound antioxidants made by modifying the whole of the rubber.[14] This is illustrated in Table 4 for nitrile-butadiene rubber using typical antioxidant thiols. It can be seen that even after exhaustive extraction the bound antioxidants are more

Table 4. Effect on NBR Stability of Interpenetrating Networks containing different Bound Antioxidant distributions (antioxidant concentration, 1%, temperature of test 150°C)

Antioxidant	Time to 1% oxygen absorption, hours						
	1*	5*	10*	20*	30*	40*	50*
BHBM-B (unextracted)	25	–	27	30	34	38	38
BHBM-B (extracted)	19	–	17	14	19	27	27
MADA-B (unextracted)	–	14	32	36	28	–	–
MADA-B (extracted)	–	11	21	25	23	–	–
Percentage of chains containing antioxidant	100	20	10	5	3	2.5	2

*Concentration of antioxidant in the masterbatch

effective when incorporated as concentrated masterbatches than when the whole rubber is initially treated. Thus, although only a small fraction of the chains contain antioxidant, the latter is capable of scavenging efficiently throughout the whole polymer.

Another practically important finding is that masterbatches of bound antioxidants in one rubber can generally be equally effective in another polymer as a masterbatch from that polymer. Table 5 shows that a 20% concentrate of MADA-B in EPDM is as effective at 2% in NBR in a compression relaxation test as is the best synergistic combination of conventional antioxidants at 4%. It is also somewhat superior to a commercial co-polymerised antioxidant in this test.[12]

Table 5. Retention of Sealing force (%) of Silica-filled NBR Vulcanisates* in a Compression Relaxation test at 150°C

Antioxidant	0 §	24 §	72 §
MADA-B (2% from 20% EPDM Masterbatch)	100	66	49
Chemigum HR 665	100	55	41
Naugard 445 (2%) + Wingstay 100 (2%)	100	68	49

§ Ageing time, hours

*Formulation: NBR (Krynac 825),100; ZnO,50; Stearic acid, 0.5; HVA-2,2; Maglite D,10; Ultrasil VN3,10; Durosil (SiO_2),40; Silane SI-69,4; Vulcanol OT,10; TE8,1; Sulphur, 1.6; TMTD,1.5; CBS,1; Cure-rite, 1.8; OTOS, 2.5.

A particularly important application of the EPDM masterbatch procedure is in EPDM itself. This rubber is widely used in the aerospace industry where resistance to phosphate ester fluids is a primary requirement. Table 6 compares the behaviour of MADA-B with the oligomeric antioxidant, Flectol H(I), which has been used in this application because of its relative resistance to extraction.[2]

Modulus change is a good indicator of oxidative degradation in this rubber and it can be seen that MADA-B at 1% is considerably superior to the conventional antioxidant at 2%, even after prior extraction.

Table 6.   Change in M_{100} (%) of EPDM Vulcanisates in a
                  Contamination test* with Skydrol

Antioxidant	Gum vulcanisates				Black vulcanisates			
	$1^{\S}$	$2^{\S}$	$3^{\S}$	$4^{\S}$	$1^{\S}$	$2^{\S}$	$3^{\S}$	$4^{\S}$
Control	−7	+9	+20	+35	+18	+50	+75	F
Flectol H (2%)	−6	+8	+14	+21	+16	+31	+52	+71
MADA-B (1%, unextracted)	−3	+1	+ 4	+ 8	+10	+20	+35	+43
MADA-B (1%, extracted)	−13	−6	− 1	+ 4	+12	+24	+38	+47

*6 hour immersion in fluid at room temperature, followed by ageing in an air oven at 120°C.

$\S$Time of ageing, weeks.

ACKNOWLEDGEMENTS

We are grateful to Robinson Bros. Ltd., for permission to publish previously unpublished results.

REFERENCES

1. Scott G, Atmospheric Oxidation and Antioxidants, Elsevier, London and New York, 1965, Chapters 4 and 9.
2. Thomas D. K, Developments in Polymer Stabilisation-1, Ed. G. Scott App. Sci. Pub., London, 1979, p.137.
3. Cain M. E, Knight G. T, Lewis P. M. and Saville B, Rubb. J., 50 (1968) 204.
4. Scott G, Developments in Polymer Stabilisation-4, Ed. G. Scott, App. Sci. Pub., London, 1981, p.181.
5. Horvath J. W, Grimm D. C. and Stevick J. A, Rubb. Chem. Tech, 48 (1975) 106.
6. Vogl O, Albertsson A. C. and Janovic Z, ACS Symposium Series, 280 (1985) 197.
7. Scott G, ACS Symposium Series, 280 (1985) 173.
8. Ref. 1, Chapter 10.
9. Ceresa R. J. and Watson W. F, J. Appl. Polym. Sci., 1 (1959) 101.
10. Ajiboye O. and Scott G, Polym. Deg. and Stab., 4 (1982) 415.
11. Scott G. and Suharto R, Europ. Polym. J., in press.
12. Chakraborty K. B. and Scott G, unpublished work.
13. Scott G. and Tavakoli S. M, Polym. Deg. and Stab., 4 (1982) 343.
14. Ajiboye O. and Scott G, Polym. Deg. and Stab., 4 (1982) 397.
15. Scott G. and Tavakoli S. M, unpublished work.

POLYMERS FROM HYDROGENATED POLYDIENES

PREPARED WITH NEODYMIUM CATALYSTS

H. L. Hsieh and H. C. Yeh

Phillips Petroleum Company
Research & Development
Bartlesville, Oklahoma 74004

ABSTRACT

The hydrogenation of various diene homopolymers and copolymers prepared with lanthanide coordination catalysts was investigated. The lanthanide-initiated polydienes which have been studied include: cis-1,4 polybutadiene, cis-1,4 poly-isoprene, random and block cis-1,4 copolymers of butadiene and isoprene, cis-1,4 poly(2,3-dimethyl butadiene), and trans-1,4-poly(2,4-hexadiene). The hydrogenated products were characterized by ^{13}C-NMR and infrared spectroscopy, melting temperature, glass transition temperature (Rheovibron), thermogravimetric analysis, density, intrinsic viscosity and gel permeation chromatography. The physical and mechanical properties of these hydrogenated polymers as well as the relationships between polymer structures and polymer properties are discussed.

INTRODUCTION

Chemical modifications of unsaturated polymers, particularly the hydrogenation of polydienes, have received a great deal of attention for many years.[1-8] Hydrogenation of unsaturated polymers can provide an alternate route to a variety of unique and useful elastomers and thermoplastics with specific structures and properties. Many polymers with novel monomer sequence distributions and compositions which are difficult or impossible to prepare directly through conventional polymerization methods can be easily obtained by the hydrogenation of polymers. Another important feature of the hydrogenation of polydienes is that the hydrogenated products normally have improved resistance to oxidative and thermal degradation as compared to the parent unsaturated polymers.

It has been shown that the hydrogenation of 1,4-polybutadiene (PB) can lead to a thermoplastic material with a structure identical to a linear polyethylene. Perfect alternating copolymers of ethylene with propylene can be achieved by hydrogenating 1,4-polyisoprene (PI). Head-to-head:tail-to-tail polypropylene can be obtained by hydrogenating 1,4-poly(2,3-dimethyl butadiene)(PDMB), 1,4-poly(2,4-hexadiene)(PHXD) or head-to-tail 1,4-poly(3-methyl-1,3-pentadiene).

In the past few years, the hydrogenation of a variety of diene polymers and copolymers made with anionic initiators and d-orbital transition metal catalysts has been studied extensively. It is of interest to investigate the hydrogenated polymers derived from the stereospecific polymerizations of dienes using 4f-orbital lanthanide catalysts. The lanthanide coordination catalysts are known to be highly stereospecific for producing high-cis polybutadiene, high-cis polyisoprene, and ideally random and block (di- and tri-block) copolymers of butadiene and isoprene with high-cis content in both monomer moieties, as well as other novel stereoregular polydienes from a variety of conjugated 1,3-diene monomers.[9-10]

The reagents frequently used for the hydrogenation[11] of polydienes are (a) insoluble metal and noble metal catalysts, (b) homogeneous organometallic catalysts, and (c) diimide generated in situ. It has been known that complete hydrogenation of polydienes could be accomplished stereospecifically[12] (i.e., cis-addition onto the olefin) by repeated treatment with diimide generated from p-toluenesulfonylhydrazide (TSH). Diimide appears to be a good hydrogenation reagent for hydrogenating unsaturated units in polymers because of the simple apparatus and technique required for diimide reactions as well as the ease of purification of the hydrogenated products from reaction residues.[3,4] We therefore used diimide as the reducing agent to study the hydrogenations of lanthanide-initiated polydienes.

In this paper we wish to report the hydrogenation of cis-1,4-polybutadiene, cis-1,4-polyisoprene, cis-1,4-poly(2,3-dimethylbutadiene), and trans-1,4-poly(2,4-hexadiene) which were prepared with lanthanide polymerization catalysts. The physical and mechanical properties of these hydrogenated polymers are discussed.

EXPERIMENTAL

The high 1,4-polydienes were prepared with lanthanide polymerization catalysts as described before.[9,10] The polydienes were hydrogenated with diimide generated in situ from p-toluenesulfonylhydrazide (TSH) by thermal decomposition in refluxing xylene solvent.[3] The hydrogenation reactions were carried out in the presence of 5.0 parts per one hundred parts of rubber (phr) Irganox 1010 antioxidant to prevent side reaction and chain degradation during the hydrogenation processes. Generally, polymer samples (5 g) with Irganox 1010 (0.25 g) were dissolved in 700 ml xylene at 50°C under N_2 with stirring. The mixture was then refluxed under N_2 with an initial 2:1 mole ratio of TSH to unsaturated units of the parent polydiene. The hydrogenation was run for a period of 24 h with the repeated additions (3 times) of a one-fold excess of TSH (i.e., total five-fold excess of TSH). The warm mixture was then neutralized with a base and poured slowly into methanol following the work-up procedures of Harwood et al.[4]

The microstructures of the parent polydienes and the degrees of hydrogenation of the hydrogenated products were determined either by IR spectroscopy or by ^{1}H- and ^{13}C-NMR. Inherent viscosity (I.V.) was determined in toluene at 25°C for polydienes, and in 1,2,4-trichlorobenzene (TCB) at 130°C for hydrogenated polymers. The MW and MWD of parent polymers were obtained by GPC analyses using a Waters Associates chromatograph model 200 or 202. GPC chromatograms for the hydrogenated and some parent polymers were obtained at 130°C using a Waters 150-C GPC equipped with spherical polystyrene gel columns (Shodex GPC A-80 M) and an infrared detector. A 0.125 wt. % polymer solution in TCB solvent containing BHT antioxidant was used with a flow rate of 1.0 ml/min. Molecular weights for the hydrogenated polymers were calculated from the calibration standard polyethylene or polypropylene samples. The glass transition temperatures were derived from the dynamic-mechanical spectrum at 11 Hz with a Rheovibron viscoelastometer. A Perkin-Elmer DSC-2C differential scanning calorimeter (DSC) was used for the determination of the melting temperature. Samples were compression molded at 160°C and scans were run at 20°C/min. The thermal degradation studies were carried out by thermogravimetric analysis (TGA) and differential thermogravimetric analysis (DTGA) using a Perkin-Elmer TGS-2 thermogravimeter. Thermogravimetric analyses were done using a programmed temperature increase of 10°C/min. Density was measured at room temperature by a standard density gradient column. The samples were compression molded at 160°C and tested for mechanical properties. Stress-strain data were obtained in uniaxial tension on an Instron using dumbbell specimens of 0.05 cm thickness, 0.32 cm width, and 2.54 cm effective length. Aluminum tabs were attached to the ends of the specimens to prevent slippage from the Instron clamps.

RESULTS AND DISCUSSION

It is well known that the properties of hydrogenated polymers depend primarily on the structure of the parent polymers. The residual unsaturation units and their distribution along the polymer chain would also affect the polymer properties. In general, hydrogenation of linear unbranched polymers, such as cis-1,4-polybutadiene leads to thermoplastics and hydrogenation of linear short chain branched polymers, such as cis-1,4-polyisoprene, 1,4-poly(2,3-dimethyl butadiene), and 1,4-poly(2,4-hexadiene), etc., containing methyl branching, gives amorphous elastomers.

The use of TSH as a hydrogenation agent has merits in its efficiency and simplicity to hydrogenate unsaturated units in polymers.[4] The percentage of unsaturation and complete hydrogenation can be controlled by the concentration of TSH and hydrogenation conditions (hydrogenation time, solvent, and temperature).

The relative activity in the hydrogenation of 1,4-polydienes with diimide, followed by evaluating the percent hydrogenation, showed the following order:

cis-PB > cis-PI >> trans-2,4-PHXD > cis-PDMB > trans-PDMB

Repeated hydrogenations and long reaction times appeared to be necessary for the hydrogenation of PHXD and PDMB to reduce the unsaturation to a negligible level.

Hydrogenation of cis-1,4 polybutadiene (cis-PB)

The polybutadiene polymers obtained with lanthanide catalysts have a high cis-1,4-content which can be varied from less than 80% to as high as 99%.[10] The polymer vinyl contents (i.e., 1,2-addition) are always less than 0.8%. That is the total 1,4-structure remains higher than 99% and is not affected by the polymerization parameters.

Cis-1,4 PB can be structurally regarded as an alternating copolymer of ethylene and acetylene. Hydrogenation of the double bonds progressively increases the ethylene units and finally leads to a pure polyethylene (PE). The hydrogenated products, H(cis-PB), resemble a linear high density PE in both physical and mechanical properties. Generally, the stress-strain curves of hydrogenated cis-PB showed a yield point at 10-60% elongation, a high initial modulus around 10-17 MPa, higher tensile strength (17-35 MPa), low elastic elongation and high permanent set as compared to unhydrogenated cis-PB. The hydrogenated products also showed higher thermal stabilities than those of the parent polymers (Table 1).

No glass transition temperature (Tg) between -100 to 100°C was observed for completely hydrogenated cis-1,4 PB. The melting temperature is around 125 to 129°C which is inversely related to the molecular weight (Table 2). The density of H(cis-PB) is in the range of 0.937 to 0.940 and is also inversely affected by the MW. The density is directly related to the crystallinity which was calculated from heat of fusion, ΔH_f, i.e., the area under the DSC melting curve.

GPC chromatograms (Figures 1 and 2) for the parent and hydrogenated PB show that the higher MW polymer (I.V. = 9.6) experienced slight chain degradation under repeated and vigorous hydrogenation conditions. The medium or lower MW polymer (I.V. = 1.6) did not show MW changes. ^{13}C-NMR spectra of H(cis-PB) showed no incorporation of interfering fragments of TSH into the polymer chain.

Table 1. Properties of cis-1,4 Polybutadiene and Hydrogenated Product

	Parent polymers	Hydrogenated polymers
Hydrogenation (%)	0	>99
Appearance	clear elastomer	translucent thermoplastic
I.V. in Toluene @ 25°C (dL/g)	1.9 – 3.7	N.D.
Intrinsic Viscosity in TCB at 130°C (dL/g)	N.D.	2.6 – 4.4
T_g (°C)	-99[a]	none
T_m (°C)	-8[b]	129 – 130[c]
ΔH_f	N.D.	40 – 48
Crystallinity (%)	N.D.	57 – 70[d]
Density (g/cm³)	N.D.	0.937 – 0.940
Thermogravimetric Analysis (TGA) 10% wt. Loss Temp. (°C)	420	435
Differential TGA (DTGA) Max. Degradation Temp. (°C)	450	470
Tensile Strength at Breake (MPa)	2.5 – 7.0[f]	19 – 26
Elongation at Breake (%)	400 – 900[f]	100 – 400

[a] Determined by Rheovibron, temperature of E" maximum at 11 Hz.
[b] Determined by Rheovibron, temperature of break in E'.
[c] Determined by DSC.
[d] The percentage of crystallinity calculated from ΔH_f; with 100% crystallinity = 70 cal/g.
[e] Uniaxial stress-strain at room temperature; 0.17 sec⁻¹ strain rate.
[f] The parent polymers were cured with 0.75 phr of Di-cup 40 C.

N.D. = Not determined.

Table 2. The Effect of MW on the Properties of Completely Hydrogenated cis-1,4 Polybutadiene

Sample No.	1	2	3	4
Inherent Viscosity in Toluene at 25°C (before hydrogenation)	9.6	3.7	2.3	1.6
Density (g/cm^3)	0.937	0.939	0.939	0.940
Tm (°C)	125	128	126	129
ΔH_f (cal/g)	ca. 26	43	42	47.6
Crystallinity (%)	37	61	60	68
TGA				
10% wt. Loss Temp (°C)	–	435	–	428
Max. degradation temp. (°C)	–	472	–	468
Tensile Strength at Break[a] (MPa)	35.0	19.0	24.0	17.0
Elongation at Break (%)	380	150	450	720
100% modulus (MPa)	16.8	17.5	12.6	10.8
300% modulus (MPa)	27.0	–	16.7	11.0
Yield strength (MPa)	none	17.5	12.6	12.0
Elongation at Yield (%)	–	60	50	10

[a]Uniaxial stress-strain at RT; 0.017 sec^{-1} strain rate.

Table 3. Properties of cis-1,4 Polyisoprene and Hydrogenated Products

Polymers	Parent Polymers	Hydrogenated
Appearance elastomer	clear elastomer	clear
I.V. in Toluene at 25°C	3.0 – 4.5	ca. 1.0 – 3.5
Tg[a] (°C)	-59	-57
Tm (°C)	none	none
Density (g/cm^3)	0.901	<0.865
TGA 10% wt. Loss Temp. (°C)	345	405
DTGA Maximum Degradation Temp. (°C)	375	445
Tensile Strength at Break[b] (MPa)	5.0 – 7.0[c]	0.5 – 0.6
Elongation at Break[b] (%)	1500 – 2000[c]	100 – 190

[a]Determined by Rheovibron, temperature of E″ maximum at 11 Hz.
[b]Uniaxial stress-strain at RT; 0.17 sec^{-1} strain rate.
[c]The parent polymers were cured with 0.75 phr of Di-cup 40 C.

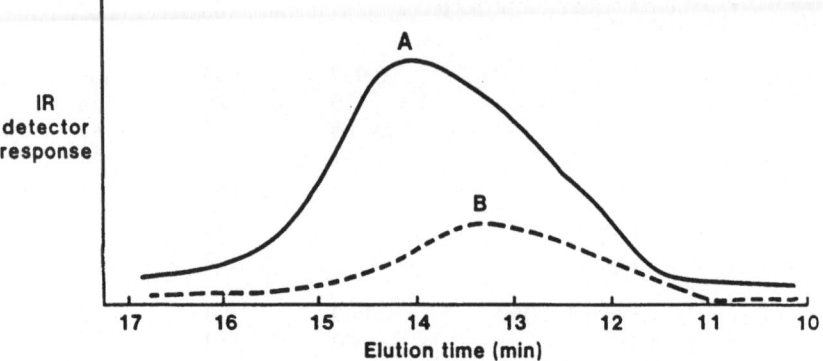

Fig. 1. GPC curves for hydrogenated cis-PB (A)
and its parent polymer (B).
cis-PB: $\bar{M}_w$, 625,000; $\bar{M}_n$, 145,000
Hydrogenated cis-PB: $\bar{M}_w$, 310,000; $\bar{M}_n$, 85,000

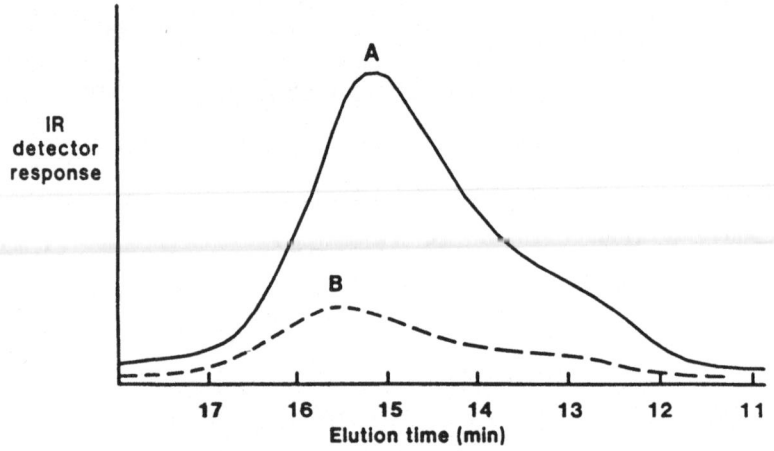

Fig. 2. GPC curves for lower mw hydrogenated cis-PB (A)
and its parent polymer (B).
cis-PB: $\bar{M}_w$, 120,000; $\bar{M}_n$, 30,700
Hydrogenated cis-PB: $\bar{M}_w$, 149,000; $\bar{M}_n$, 31,200

Hydrogenation of cis-1,4 polyisoprene (cis-PI)

Polyisoprene polymers with cis-1,4 content greater than 96% were obtained with lanthanide catalysts.[10] ^{1}H-NMR spectra of the hydrogenated samples showed the residual unsaturation levels to be below 1%, indicating complete hydrogenation. ^{13}C-NMR analyses showed that the hydrogenated cis-PI with 3,4-addition less than 3% exhibits four signals corresponding to 4 types of carbon atoms from a 1-1 alternating ethylene and propylene copolymer.[3,13]

The hydrogenated polymer was a completely amorphous elastomer with a glass transition temperature at -57°C which is close to that of the parent polymer (Tg = -59°C) in Table 3. The hydrogenated cis-PI polymers have a relatively low tensile strength (less than 0.6 MPa), low modulus and low permanent set (i.e., excellent recovery from deformation). The physical and mechanical properties of cis-PI and their hydrogenated products are listed in Table 3. The mechanical properties of H(cis-PI) appear to resemble those of the uncured rubber and those of nearly equimolar EP commercial copolymers (EPsyn® EPM rubber). GPC chromatograms for cis-PI and H(cis-PI) showed no significant chain scission for high MW (I.V. 7.0) hydrogenated polymers (Figure 3).

Hydrogenation of Cis-Copolymers of Butadiene-Isoprene

Butadiene and isoprene were copolymerized with lanthanide catalysts to random and block copolymers via the methods of initial monomer charge and the sequential monomer addition, respectively.[10] The lanthanide catalyst systems offer several unique features in the copolymerization of B and I monomers, such as:

(1)   Ideally random copolymers with monomer reactivity ratios of B and I equal to unity,

(2)   High cis-1,4 microstructures in both monomer units which are independent of copolymer composition, and

(3)   Di- and tri-block copolymers under certain favorable conditions.

The copolymers obtained normally have a cis-1,4 content above 97% in both butadiene and isoprene moieties. In general, the vinyl contents in butadiene units are less than 1% and the 3,4-addition of isoprene units is usually less than 3%.

The hydrogenation of cis-1,4 copolymers of B and I would lead to polyolefins with composition and sequence distribution consisting of ethylene (E) blocks and alternating ethylene/propylene (E/P) blocks. These novel polyolefins are difficult or almost impossible to obtain directly by simple polymerization of E and P monomers using any existing polymerization catalysts. Since structural variations in these polyolefins, such as composition and monomer sequence distribution, would significantly affect the polyolefin properties, the hydrogenated cis-1,4 B/I copolymers with uniformly random distribution of E and E/P units may serve as model polymers to study structure-property relationships and be useful as polymers with unique properties.

A. Hydrogenated Cis-1,4-B/I, H(B/I), Random Copolymers

The properties of H(B/I) random copolymers can be varied from those of semi-crystalline plastics (i.e., linear LDPE with methyl branches) to those of essentially amorphous elastomers. A gradual transition from translucent thermoplastics to transparent rubbers was observed as the molar ratio of E (or HB) to E/P (or HI) decreased.

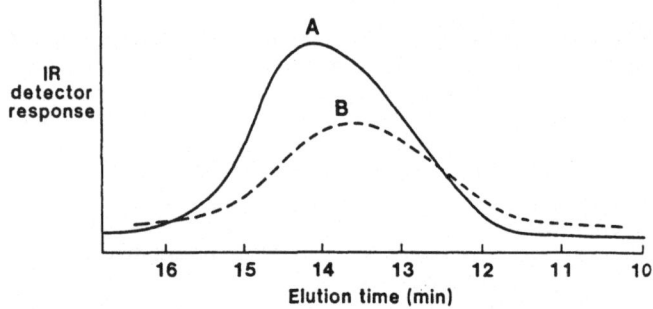

Fig. 3. GPC curves for hydrogenated cis-PI (A)
and its parent polymer (B).
cis-PI: $\bar{M}_w$, 397,000; $\bar{M}_n$, 133,000
Hydrogenated cis-PI: $\bar{M}_w$, 296,000; $\bar{M}_n$, 115,000

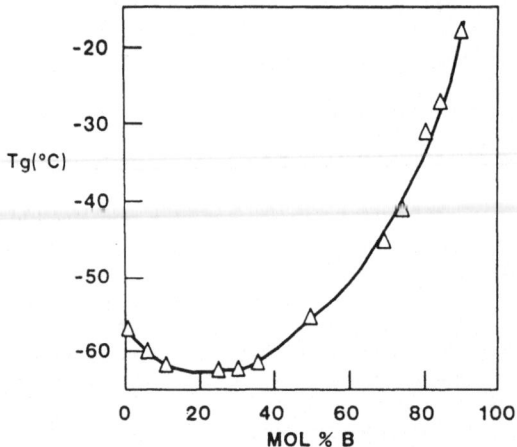

Fig. 4. The effect of butadiene content on the Tg of
hydrogenated cis-B/I random copolymers

The stress-strain properties of H(B/I) random polymers with low isoprene content (less than 6% isoprene) shows yielding behavior similar to H(cis-PB) polymers but elongation to break is significantly better than that of HB polymers (Table 4). The increase in the isoprene content (e.g., from 10 to 30%) gradually lowers the stress to break and the initial modulus, but the elongation to break is again increased. The H(B/I) polymers with higher HB contents showed strain hardening (or haze) which indicated the occurrence of strain-induced crystallization. No strain-induced crystallization was observed when the isoprene content was greater than 50 mole %. The stress-strain properties of H(B/I) polymers with higher isoprene contents (above 65%) behaved as uncured rubbers resembling hydrogenated cis-polyisoprene (Table 5).

The melting behavior of H(B/I) random copolymers decreased gradually from 130°C for H(cis-PB) to 113°C for H(B/I = 95/5) and 80°C for H(B/I = 70/30), and eventually to a completely amorphous material with no Tm (Table 6). The glass transition temperature changed from -18°C (B/I = 90/10) to -45°C (B/I = 70/30) and reached a minimum at -62°C with butadiene contents in the range of 20 to 30 mole % (Figure 4). The density and the degree of crystallinity (calculated from ΔHf of DSC melting curve) also decreased as the molar ratios of B to I decreased. The dependence of density on the H(B/I) composition is shown in Figure 5. It appears that H(B/I) random copolymers with B contents between 98 to 70 mole percent have properties resembling linear low density polyethylene (LLDPE) or very low density PE, and that H(B/I) copolymers with B contents less than 50 mole % behave like EPM or EPDM elastomers.

The effects of MW on the properties of hydrogenated cis-1,4 B/I random copolymers containing various compositions (B/I = 90/10 to B/I = 50/50) are listed in Tables 7 through 10. In general, for the polymers with the same composition, the density increased as the MW decreased. However, the tensile strength increased directly with MW.

GPC chromatograms showed that the hydrogenated B/I copolymer (50/50) has about the same MW distribution position as the parent polymer and the shape of the chromatogram remained unimodel and broadened, indicating no chain scission (Figure 6). GPC data of H(B/I) and parent polymers with various compositions are listed in Table 11.

Thermal stability as determined by TGA and DTGA showed that the maximum rate of degradation of the hydrogenated polymers occurred around 450-460°C, and that the polymers with higher ethylene contents appeared to be more stable thermally than those with lower ethylene contents.

13C-NMR spectra of hydrogenated B/I = 50/50 polymer showed that the polymer was completely hydrogenated. The triad distribution indicated the E and E/P units are randomly distributed along the polymer chain.

B. Hydrogenated Cis-1,4 B-I-B Block Copolymers

A series of high-cis B-I-B triblock copolymers were prepared with lanthanide catalysts by incremental monomer addition. The hydrogenated products lead to E-(E/P)-E triblock copolymers with hard (semi-crystalline) ethylene block at both chain ends and soft (amorphous; rubbery) alternating E/P block in the center segment.

Rheovibron studies showed that the E-(E/P)-E triblock copolymers all had the same glass transition temperature of -57°C independent of polymer composition and block length (Table 12). The transition temperature at -57°C is the Tg of the alternating E/P rubber block. The constant Tg and the opaque appearance of the E-(E/P)-E triblock copolymers indicated that the rubber phase of the E/P blocks and the plastic phase of the E blocks are incompatible in the solid state.

Table 4. Hydrogenated cis-1,4 B/I Random Copolymers
(Low isoprene content, same MW copolymers)

Sample No.	1	2	3
B/I (molar ratio)	98/2	96/4	94/6
I.V. in Toluene at 25°C (before hydrogenation)	1.0	1.0	1.0
$\bar{M}_W$ x 10-3	85	82	85
$\bar{M}_n$ x 10-3	45	43	44
Tm (°C)	115	114	112
ΔH_f (cal/g)	34.1	32.2	ca. 27
Tensile Strength at Break[a] (MPa)	26.0	22.4	22.7
Elongation at Break (%)	750	700	730
Yield Strength (MPa)	10.7	10.1	10.1
Elongation at Yield (%)	60	60	60

[a]Uniaxial stress-strain at room temp.; 0.017 sec-1 strain rate.

Table 5. Hydrogenated cis-1,4 I/B Random Copolymers
(Isoprene content greater than 65 mole %; polymers of similar MW)

Sample No.	1	2	3
I/B (molar ratio)	65/35	70/30	75/25
IV in toluene at 25°C (before hydrogenation)	1.9	2.2	2.2
$\bar{M}_W$ x 10-3 (before hydrogenation)	360	410	420
$\bar{M}_n$ x 10-3 " "	91	94	100
Density (g/cm3)	0.875	0.872	0.874
Tm (°C)	none	none	none
Tg (°C)	-61	-62	-62
Tensile Strength at Break[a] (MPa)	0.4	0.3	0.3
Elongation at Break (%)	260	230	180

[a]Uniaxial stress-strain at room temp.; 0.017 sec-1 strain rate.

Table 6. Hydrogenated cis-1,4 B/I Random Copolymer

Sample No.	1	2	3	4	5	6
B/I (molar ratio)	90/10	85/15	80/20	75/25	70/30	50/50
E/P (molar ratio)	95/5	92.5/7.5	90/10	87.5/12.5	85/15	75/25
$[\eta]$ in TCB at 140°C	2.5	2.5	2.2	1.6	2.1	2.1
$\overline{M}_w \times 10^{-3}$	260	285	245	151	242	370
$\overline{M}_n \times 10^{-3}$	61	60	60	49	53	102
Density (g/cm^3)	0.926	0.915	0.910	0.900	0.896	0.865
Tm (°C)	107	101	96	90	~80	none
ΔHf (cal/g)	18.1	12.1	7.4	3.3	2.0	–
Tg (°C)	-18	-27	-31	-41	-45	-55
TGA						
10% wt. Loss Temp. (°C)	420	415	414	414	412	–
Max. Degradation Temp. (°C)	460	458	457	457	450	–
Tensile Strength at Break[a] (MPa)	30	28	27	24	19	7.5 [b]
Elongation at Break (%)	670	690	750	900	920	2000
100% Modulus (MPa)	7.9	6.4	5.0	4.0	3.1	1.0
300% Modulus (MPa)	10.5	9.1	7.3	5.2	4.6	1.2

[a]Uniaxial stress-strain at room temp.; 0.17 sec^{-1} strain rate.
[b]Room temperature, 0.017 sec^{-1} strain rate.

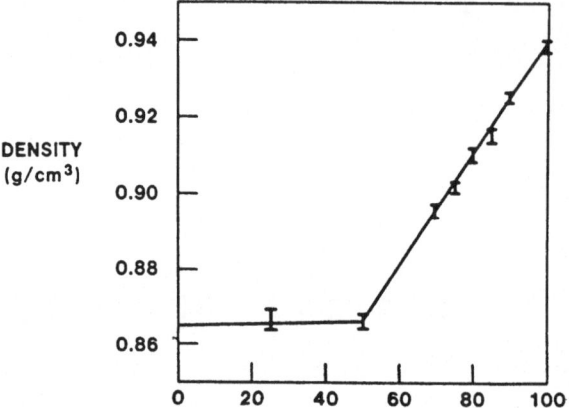

Fig. 5. The effect of butadiene content on the density of hydrogenated cis-B/I random copolymers

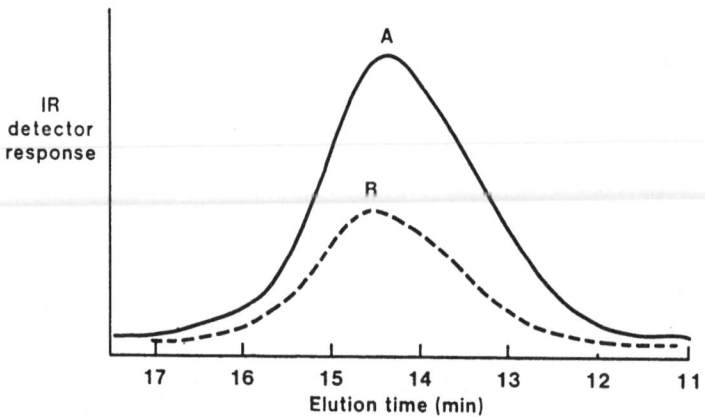

Fig. 6 GPC curves for hydrogenated cis-1,4-B/I (50/50) random copolymer (A) and its parent copolymer (B). cis-1,4-B/I; $\bar{M}_w$, 200,000; $\bar{M}_n$, 81,000
Hydrogenated cis-1,4 B/I: $\bar{M}_w$, 250.000; $\bar{M}_n$, 93,600

Table 7. Effect of Molecular Weight on Properties of Hydrogenated cis-1,4 B/I Random Copolymers

	(B/I molar ratio = 90/10)				
Sample No.	1	2	3	4	5
I.V. in Toluene at 25°C (before hydrogenation)	7.9	6.4	2.1	1.9	1.2
[η] in TCB at 140°C	–	–	2.5	–	–
$\overline{M}_w$ x 10^{-3}	543	585	260	182	123
$\overline{M}_n$ x 10^{-3}	153	119	61	71	47
HI	3.5	4.9	4.3	2.6	2.6
Density (g/cm^3)	0.918	0.919	0.926	0.921	0.923
Tm (°C)	104	–	107	105	105
ΔH_f (cal/g)	–	–	18.1	14.3	17
Tensile Strength at Break[a] (MPa)	30.2	25	29	33.1	27.6
Elongation at Break (%)	510	470	670	720	780
100% Modulus (MPa)	8.1	8.5	7.9	7.8	7.1
300% Modulus (MPa)	13.7	14.0	10.5	10.7	9.3

[a]Uniaxial stress-strain at room temp.; 0.017 sec^{-1} strain rate.

Table 8. Effect of Molecular Weight on Properties of Hydrogenated cis-1,4 B/I Random Copolymers

	(B/I molar ratio = 80/20)			
Sample No.	1	2	3	4
I.V. in Toluene at 25°C (before hydrogenation)	7.0	5.6	2.6	1.4
$\overline{M}_w$ x 10^{-3} (before hydrogenation)	–	–	415	142
$\overline{M}_n$ x 10^{-3} " "	–	–	120	56
Density (g/cm^3)	0.900	0.903	0.905	0.908
Tensile Strength at Break[a] (MPa)	35.7	26.0	27.0	21.0
Elongation at Break (%)	800	700	720	750
100% Modulus (MPa)	5.3	4.8	5.3	4.8
300% Modulus (MPa)	8.1	7.5	8.1	6.6

[a]Uniaxial stress-strain at room temp.; 0.017 sec^{-1} strain rate.

Table 9. Effect of Molecular Weight on Properties of Hydrogenated cis-1,4 B/I Random Copolymer

(B/I molar ratio = 70/30)

Sample No.	1	2	3	4
I.V. in Toluene at 25°C (before hydrogenation)	6.6	1.9	1.7	1.2
$\overline{M}_w$ x 10-3 (before hydrogenation)	–	250	210	100
$\overline{M}_n$ x 10-3 " "	–	60	75	40
Density (g/cm^3)	0.893	0.895	0.894	0.895
Tm (°C)	–	75	77	75
ΔH$_f$ (cal/g)	–	0.8	1.2	0.5
Tensile Strength at Break[a] (MPa)	24.5	29.0	28.8	17.6
Elongation at Break (%)	610	1200	1200	1350
100% Modulus (MPa)	4.4	3.0	3.0	2.7
300% Modulus (MPa)	8.5	4.2	4.3	3.5

[a]Uniaxial stress-strain at room temp.; 0.017 sec^{-1} strain rate.

Table 10. Effect of Molecular Weight on Properties of Hydrogenated cis–1,4 B/I Random Copolymers

(B/I molar ratio = 50/50)

Sample No.	1	2	3	4
I.V. in Toluene at 25°C (before hydrogenation)	4.9	2.1	1.4	0.7
$\overline{M}_w$ x 10-3	–	370	186	68
$\overline{M}_n$ x 10-3	–	102	58	29
Density (g/cm^3)	0.868	0.866	0.865	–
Tensile Strength at Break[a] (MPa)	9.2	7.5	5.0	0.15
Elongation at Break (%)	1100	2000	2200	380
100% Modulus (MPa)	1.4	1.0	1.1	0.1
300% Modulus (MPa)	2.0	1.2	1.2	0.15

[a]Uniaxial stress-strain at room temp.; 0.017 sec^{-1} strain rate.

Table 11. Comparison of Molecular Weights of Hydrogenated B/I Random Polymers and Their Parent Polymers

Sample	Parent Polymer [a]		Completely Hydrogenated Product [b]	
B/I (molar ratio)	$\overline{M}_w \times 10^{-3}/\overline{M}_n \times 10^{-3}$	I.V. in Toluene at 25°C (dL/g)	$\overline{M}_w \times 10^{-3}/\overline{M}_n \times 10^{-3}$	$[\eta]$ TCB, at 140°C (dL/g)
90/10	282/78	2.1	260/61	2.5
85/15	310/79	2.2	285/60	2.5
80/20	305/76	2.0	245/60	2.2
75/25	270/84	1.8	155/50	1.7
70/30	260/65	1.8	245/55	2.1
50/50	375/102	2.1	250/94	2.4

aGPC of parent polymers was obtained in THF at room temperature.
bGPC of hydrogenated polymers was measured in TCB at 140°C.

The melting temperatures of the triblock copolymers remain in the vicinity of 128 ± 2°C regardless of compositional variation and chain length. This result again indicates the incompatibility between E and E/P blocks in the solid state. The heat of fusion, ΔHf, and the density of triblock copolymers were found to increase with the ethylene block content. However, the normalized values of ΔHf (normalized to E block content) of the triblock copolymers were all around 35 ± 2 cal/g.

The stress-strain properties of the triblock copolymers showed improved elongation at break compared to those of hydrogenated homopolymers of PB and PI. The tensile strengths and moduli increased with increasing ethylene content.

C. Hydrogenated Cis-1,4 B-I Block Copolymers

Diblock copolymers of E-(E/P), where E and (E/P) are block segments of ethylene and alternating ethylene/propylene, respectively, were obtained from the hydrogenation of cis-1,4 B-I diblock copolymers.

The opaque appearance and Tg at -57°C as well as the fixed melting temperature around 127°C (Table 13) suggested that the hard ethylene block and the soft E/P block of E-(E/P) diblock copolymers are incompatible in the solid state.

The mechanical properties of diblock E-(E/P) were similar to those of the triblock E-(E/P)-E copolymers. The tensile strengths at break and moduli are strongly influenced by the increase in the proportion of the hard ethylene block. The elongation at break was around 400-600% and was not affected by compositional variations.

The effect of MW on the properties of hydrogenated B-I diblock copolymers (B:I = 80:20) is shown in Table 14. The melting temperature and the density slightly decreased with the increase in MW. An increase in the MW led to the increase of tensile strength at break and modulus.

Hydrogenation of trans-1,4-poly(2,4-hexadiene) (PHXD)

Poly(2,4-hexadiene) with a trans-1,4 content greater than 98% was obtained with lanthanide catalysts.[9] The completely hydrogenated product will be a head-to-head polypropylene.

Trans-1,4 PHXD with 1,2-disubstituted trans double bonds and sterically hindered methyl branching adjacent to the double bonds appeared to be more difficult to hydrogenate than cis-PB and cis-PI. Under mild hydrogenation conditions,^{1}H-NMR showed that the hydrogenated polymer contained 20% unsaturated double bonds. The 80 percent hydrogenated material exhibited rubbery behavior with a glass transition temperature lower than that of the parent polymer (Table 15).

The completely hydrogenated product was obtained by repeated addition of TSH (5 times with a 1/1 mole ratio of TSH to unsaturated double) under reflux. GPC chromatograms (Figure 7) showed that the hydrogenated product has a significantly smaller size (i.e., lower MW) than the parent polymer. These data indicate that appreciable chain degradation occurred during the hydrogenation reaction.

The thermal stability of the hydrogenated polymers was greatly improved as compared to that of the parent polymer. The thermal stability also increased with the increase in the degree of hydrogenation for the partially hydrogenated polymers.

212

Table 12. Hydrogenated cis-1,4 B-I-B Triblock Copolymers

Sample No.	1	2	3	4
B-I-B (molar ratio)	42-16-42	35-30-35	27-46-27	19-62-19
I.V. in Toluene at 25°C (before hydrogenation)	4.8	4.4	4.7	5.1
$\bar{M}_w$ x 10-3 (before hydrogenation)	1000	950	1030	1300
$\bar{M}_n$ x 10-3 " "	110	130	120	180
Density (g/cm^3)	0.924	0.918	0.906	0.894
Tm (°C)	127	130	129	126
ΔH_f (cal/g)	25.3	22.1	18.7	11.3
Tg (°C)	-57	-57	-57	-57
Tensile Strength at Break[a] (MPa)	17.2	25	17	4.2
Elongation at Break (%)	300	520	620	350
100% Modulus (MPa)	12.8	13.8	10.7	3.0
300% Modulus (MPa)	17.2	18.0	13.8	4.0

[a]Uniaxial stress-strain at room temp.; 0.017 sec^{-1} strain rate.

Table 13. Hydrogenated cis-1,4 B-I Diblock Copolymers

Sample No.	1	2	3	4	5
B/I (molar ratio)	87/13	70/30	55/45	37/63	15/85
I.V. in Toluene at 25°C (before hydrogenation)	5.0	5.1	4.5	5.0	5.5
$\bar{M}_w$ x 10-3 (before hydrogenation)	940	1030	1150	1180	1250
$\bar{M}_n$ x 10-3 " "	130	150	160	160	190
Density (g/cm^3)	0.921	0.902	0.899	0.887	<0.870
Tm (°C)	129	127	127	127	126
ΔH_f (cal/g)	27	24	18	13	6.5
Tg (°C)	-57	-56	-56	-56	-57
Tensile Strength at Break[a] (MPa)	19	ca. 17	16	5.8	1.3
Elongation at Break (%)	430	260	580	600	400
100% Modulus (MPa)	12.5	11.0	7.0	3.1	1.2
300% Modulus (MPa)	15.5	–	8.9	4.3	1.2

[a]Uniaxial stress-strain at room temperature; 0.017 sec^{-1} strain rate.

Table 14. Effect of Molecular Weight on Properties of Hydrogenated cis-1,4 B-I Block Copolymer

(B:I molar ratio = 80:20)

Sample No.	1	2	3	4
I.V. in toluene at 25°C (before hydrogenation)	3.4	1.6	1.1	0.7
$\bar{M}_w \times 10^{-3}$ (before hydrogenation)	572	160	105	51
$\bar{M}_n \times 10^{-3}$ " "	190	77	53	29
Tm (°C)	124	125	126	127
ΔH_f (cal/g)	18	20	23	26
Density (g/cm^3)	0.918	0.923	0.926	–
Tensile Strength at Break[a] (MPa)	26.5	20.0	9.0	5.0
Elongation at Break (%)	530	660	300	50
100% Modulus (MPa)	9.4	8.8	7.8	–
300% Modulus (MPa)	13.0	10.3	8.0	–

[a]Uniaxial stress-strain at room temperature; 0.017 sec^{-1} strain rate, sample 1 showed no tensile yield point. Samples 2,3, and 4 have yield strength at 10–20% elongation.

Table 15. Properties of trans-1,4-poly(2,4-Hexadiene)

	Parent Polymer	Hydrogenated Polymers	
Appearance	opaque elastomer	clear elastomer	clear elastomer
Hydrogenation (%)	0	80	>98
I.V. in Toluene at 25°C	0.32	0.1	–
$\bar{M}_w \times 10^{-3}$	52	3.7	–
$\bar{M}_n \times 10^{-3}$	16	2.5	–
Tg[a] (°C)	–11	–17	ca. –25
Tm[b] (°C)	90; 76 (shoulder)	none	none
Crystallinity	low	none	none
TGA 10% wt. loss temp (°C)	295	–	ca. 305
DTGA Maximum degradation temp. (°C)	345	–	435

[a]Determined by Rheovibron, temperature of E" maximum.
[b]Determined by DSC, heat rate at 20°C/min.

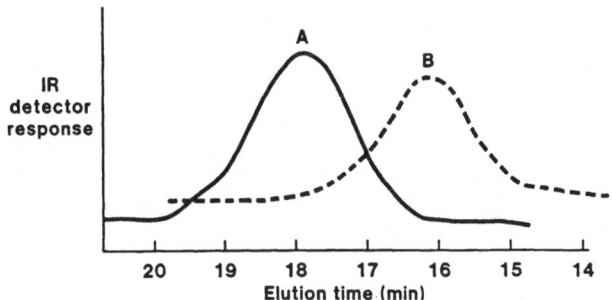

Fig. 7. GPC curves for trans-1,2-poly(2,4 hexadiene) (B)
and its hydrogenated product (A).
Trans-1,4-PHXD: $\bar{M}w$, 52,300; $\bar{M}n$, 15,700
Hydrogenated trans-1,4-PHXD: $\bar{M}w$, 3,700; $\bar{M}n$, 2,500

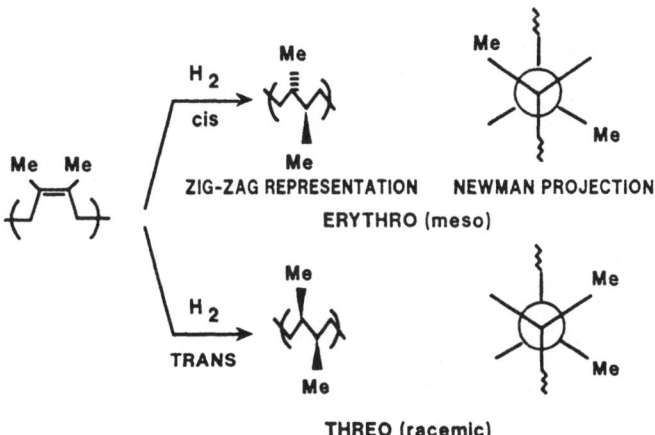

Fig. 8. Hydrogenation of cis-1,4-poly(2,3-dimethylbutadiene)

Hydrogenation of cis-1,4-poly(2,3-dimethylbutadiene) (PDMB)

Polymers obtained with lanthanide catalysts had mostly cis-1,4 structure.[9] The cis-1,4 PDMB with tetra-substituted double bonds in the polymer chain was found to be more difficult to hydrogenate than those with di- or tri-substituted double bonds. The partially or completely hydrogenated products of PDMB would lead to the possibility of erythro (meso) and threo (racemic) stereoisomers (Figure 8). The hydrogenation using TSH reagent has been known to be a highly stereospecific cis-addition.[12]   One would expect that the hydrogenated polymers derived from cis-PDMB with TSH would have the erythro configuration.   ^{13}C-NMR showed that the saturated units of the hydrogenated products were predominantly in the meso configuration which is consistent with the cis-hydrogenation mechanism of a cis-1,4 PDMB (Table 16).   A small amount of racemic units was found in some of the partially hydrogenated polymers presumably due to the hydrogenation of trans-1,4 units. The trans-1,4 units could have been formed by isomerization during the repeated hydrogenation steps.   ^{13}C-NMR also indicated that the cis-1,4 units were easier to hydrogenate than the trans-1,4 units. The ^{13}C-NMR signals showed that the cis/trans ratio for the parent polymers was higher than the meso/racemic ratio for the hydrogenated cis-1,4 PDMBs.

Molecular weights of partially hydrogenated PDMBs were lower than those of the parent polymers (Figure 9). This result suggests that chain degradation might occur during the prolonged hydrogenation process. The methyl branching of PDMB might contribute to the formation of alkyl radicals or 3° carbon radicals and lead to chain scission during hydrogenation.   In comparison, no significant chain degradation was observed during hydrogenation of cis-PB and cis-PI.

The parent cis-PDMB and lightly hydrogenated (less than 5%) polymers were crystalline plastics with Tg's above room temperature. The hydrogenated polymers containing 25% or more hydrogenated units were amorphous elastomers with properties resembling atactic head-to-tail polypropylene.   The Tg progressively decreased from 33 to -17°C as the degree of hydrogenation increased from 0 to 60%. These data suggest that the hydrogenated PDMB chain is more flexible than that of the parent polymer. Plots of Tg vs. degree of hydrogenation and the extrapolation of the curves (assuming the variation of Tg is a linear function of the percentage of hydrogenation) showed that the Tg of completely hydrogenated cis-PDMB is around -30°C. Thermogravimetric analyses showed that the thermal stability increased as the degree of hydrogenation was increased (Table 17).

CONCLUSIONS

Hydrogenated polymers derived from lanthanide-initiated 1,4-polydienes were investigated.   GPC molecular weight measurements showed that hydrogenated cis-1,4-polybutadiene, cis-1,4-polyisoprene and cis-1,4-copolymers of butadiene and isoprene experience little or no chain scission during hydrogenation with diimide. In contrast, hydrogenations of trans-1,4-poly(2,4-hexadiene) and cis-1,4-poly(2,3-dimethylbutadiene) showed significant chain degradation (Table 18) under similar hydrogenation conditions.

The mechanical and physical properties of the hydrogenated products varied with the type of 1,4-polydiene and its chain length (MW). Hydrogenated cis-PB is a plastic material resembling linear PE or medium HDPE. Hydrogenated cis-PI is a completely amorphous material that behaves like uncured rubber. The properties of hydrogenated cis-B/I copolymers are strongly influenced by compositional variations and monomer sequence distribution. For the hydrogenated random cis-B/I copolymers, the randomness of ethylene and alternating ethylene/propylene block distribution requires higher ethylene contents to achieve good elastic and mechanical properties. The hydrogenated trans-1,4-poly(2,4-hexadiene) and cis-

Table 16. Structure[a] of cis-1,4 PDMB and Partially Hydrogenated Products

	Parent Polymer	Hydrogenated Polymers	
% Hydrogenation	0	25	56
Unsaturated Units			
% cis-1,4	>98	75	41
% trans-1,4	<1	0	3
Saturated Units			
Meso	0	25	53
Racemic	0	0	3

[a]Determined by ^{13}C-NMR.

Table 17. Properties of cis-1,4-poly(2,3-Dimethylbutadiene)

	Parent Polymer	Hydrogenated Polymers			
Hydrogenation (%)	0	2 - 5	25	~40	55 - 60
Appearance	highly crystalline opaque plastics	crystalline opaque plastics	clear rubber	clear rubber	clear rubber
I.V. in TCB at 140°C	0.63	0.4	0.4	–	0.05
Tm[a] (°C)	195 (sharp)	180 (broad)	none	none	none
ΔH_f (cal/g)	15	6.5	–	–	–
Tg[b] (°C)	33	25	-3	-8	-17
Solubility in CHCl$_3$ at 25°C	insol.	insol.	sol.	sol.	sol.
TGA 10% wt. loss temp.(°C)	297	300	325	330	340
DTGA Maximum degradation temp.(°C)	352	355	373	382	395

[a]Determined by DSC.
[b]Determined by Rheovibron, temperature of E" maximum.

Table 18. Comparison of Molecular Weights of Hydrogenated Polydienes and Their Parent Polymers

Sample	Parent Polymer			Hydrogenated Polymer		
	$\overline{M}_w \times 10^{-3}$	$\overline{M}_n \times 10^{-3}$	Intrinsic Viscosity in TCB at 140°C	$\overline{M}_w \times 10^{-3}$	$\overline{M}_n \times 10^{-3}$	Intrinsic Viscosity in TCB at 140°C
PB	120	30.7	~1.6	149	31.2	2.4
PI	397	133	–	296	115	–
B/I random (50/50)	200	81	–	250	93.6	–
PDMB	63.5	9.4	0.36	1.9	0.9	0.17
PDMB	113	39	0.63	4.3	2.1	0.05
PHXD	52.3	15.7	0.32	3.7	2.5	–

Both GPC chromatograms of parent and hydrogenated polymers were run in TCB at 140°C or 145°C.

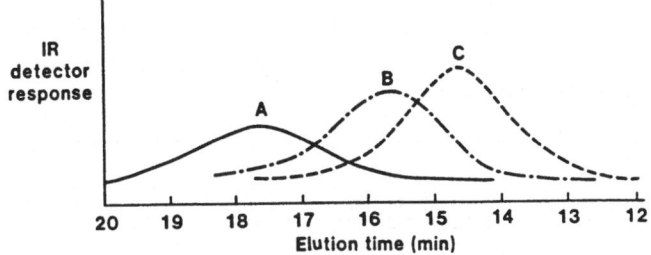

Fig. 9. GPC curves for cis-1,4-poly(2,3-dimethylbutadiene) and its hydrogenated products: A - 60% hydrogenation, B - 5% hydrogenation, C - parent polymer.

polymer A: $\bar{M}_w$, 4,300; $\bar{M}_n$, 2,100
polymer B: $\bar{M}_w$, 31,600; $\bar{M}_n$, 13,700
polymer C: $\bar{M}_w$, 113,000; $\bar{M}_n$, 39,000

Table 19. Thermal Properties of Hydrogenated cis-Polydienes and Their Parent Polymers

	Parent Polymer		Hydrogenated Polymers	
	10% wt. Loss temp. (°C)	$T_{max.}$ Degradation	10% wt. Loss temp (°C)	$T_{max.}$ Degradation
Polybutadiene	425	450	435	470
Polyisoprene	345	375	405	445
Random B/I (90/10)	390	443	420	460
Random B/I (70/30)	394	435	413	453
Poly(2,4-hexadiene)	295	345	305	435
Poly(2,3-dimethyl-butadiene)	297	352	340[a]	395[a]

----
[a]60% hydrogenation.

1,4-poly(2,3-dimethylbutadiene) are amorphous materials with properties resembling atactic-polypropylene. The thermal properties of the hydrogenated 1,4-polydienes are better than those of their parent polymers (Table 19).

REFERENCES

1. C. W. Moberly, in Encyclopedia of Polymer Science and Technology, Interscience Publ., N.Y. 1964, Vol 7, p. 557.

2. R. V. Jones, C. W. Moberly, and W. B. Reynolds, Ind. Eng. Chem., 45, 1117 (1953).

3. L. A. Mango and R. W. Lenz, Makromol. Chem., 163, 13 (1973).

4. H. J. Harwood, D. B. Russell, J. A. Verthe, and J. Zymonas, Makromol. Chem., 163, 1 (1973).

5. Y. Mahajer, G. L. Wilkes, I. C. Wang, and J. E. McGrath, Polymer Prep. Am. Chem. Soc., 22 (2), 138 (1981).

6. J. C. Falk and R. J. Schlott, Angew. Makromol. Chem., 21, 17 (1972).

7. Y. Mohajer, G. L. Wilkes, I. C. Wang, and J. E. McGrath, Polymer, 23, 1523 (1982).

8. W. P. Gergen, Kautsch. Gummi Kunstst., 37, 284 (1982).

9. H. C. Yeh, and H. L. Hsieh, Polymer Prepr. Am. Chem. Soc., 25 (2), 52 (1984).

10. H. L. Hsieh and H. C. Yeh, Rubber Chem. Technol., 58 (1), 117 (1985).

11. H. Rachapudy, G. G. Smith, V. R. Raju, and W. W. Graessley, J. Polym. Sci., Polym. Phys. Ed., 17, 1211 (1979).

12. M. Moller and H.-J. Cantow, Macromolecules, 17, 733 (1984).

13. Y. Tanaka, H. Sato, and A. Ogura, J. Polym. Sci. Polym. Chem., 14, 73 (1976).

A NEW CROSSLINKING REACTION OF POLYACRYLIC

ELASTOMER BY ONIUM SALT CATALYST

T. Nakagawa, S. Yagishita, K. Hosoya, and M. Inagami

Research and Development Center
Nippon Zeon Co., Ltd.
1-2-1, Yako Kawasaki-ku, Kawasaki, 210 Japan

ABSTRACT

Polyacrylic elastomer is a chemically saturated polymer of acrylic acid esters. Since esters are rather inert, other groups such as epoxy, halogen and carboxyl must be incorporated into the polymer to introduce sites for curing. We have studied the effect of various ammonium and phosphonium salt catalysts on the rate of crosslinking of polyacrylic elastomers which contain carboxyl and epoxy groups as cure-sites. We find that the catalytic activity of onium salts correlates with nucleophilicity of their anions.

The onium salt catalyzed reaction of a polyacrylic elastomer containing epoxy groups as cure-sites was enhanced by the addition of a dicarboxylic acid as an external crosslinking agent. This enhancement of cure rate decreased by increasing the alkylene chain length of the dicarboxylic acid. This behavior can be explained by the dissociation constants, K_1 and K_2, of dicarboxylic acid.

INTRODUCTION

Polyacrylic elastomers contain chemically saturated backbones. Since these rubbers are not unsaturated, they cannot be vulcanized with conventional sulfur cure systems. Monomers with cure-sites must therefore be introduced into the polymer chain.

Various types of monomers have been used to introduce cure-sites into polyacrylic elastomers. Polyacrylic elastomers containing active chloride, epoxy, and carboxyl groups as active cure-sites have been developed for industrial applications. Polyacrylic elastomers which contain epoxy groups are vulcanized with ammonium salts and dithiocarbamate salts, whereas elastomers which contain active chloride are vulcanized with sulfur-metal soap, organic amines, etc.

Crosslinking of polyacrylic elastomers through carboxyl group is well established in rubber technology[1]. The crosslinking mechanism of polyacrylic elastomers which contain both carboxyl and epoxy groups as cure-sites has also been studied for the low molecular weight polymers[2].

In this paper, we describe the effect of various onium salt catalysts on the crosslinking reaction of polyacrylic elastomers which contain both carboxyl and epoxy groups as the active cure-sites. In addition, the effect of onium salt catalysts on the crosslinking reaction of polyacrylic elastomers containing epoxy groups in the presence of a dicarboxylic acid, as an external crosslinking agent, is also discussed.

Physical properties of the vulcanizates obtained by this curing system were investigated.

EXPERIMENTAL

Polymer Synthesis

Table 1. Polymerization of Ethyl acrylate (EA) with various Comonomers: Allyl glycidyl ether (AGE), Glycidyl methacrylate (GMA) and Methacrylic acid (MAA)

		EA/MAA/AGE Terpolymer	EA/GMA Copolymer	EA/AGE Copolymer
Ethyl acrylate	(grams)	977	975	970
Allyl glycidyl ether	(grams)	18	–	30
Glycidyl methacrylate	(grams)	–	25	–
Methacrylic acid	(grams)	5	–	–
Water	(grams)	2000	2000	2000
Sodium lauryl sulfate	(grams)	20	20	20
Sodium salt of polymerized naphthalene sulfonic acid	(grams)	5	5	5
Tetrasodium ethylenediamine-tetraacetate	(grams)	0.3	0.3	0.3
Sodium ferric salt of ethylenediamine-tetraacetic acid	(grams)	0.2	0.2	0.2
Sodium sulfate	(grams)	5	5	5
Sodium formaldehyde sulfoxylate	(grams)	4	4	4
Paramenthane hydroperoxide	(grams)	0.3	0.3	0.3
Reaction temperature	(°C)	10	10	10
Reaction time	(hours)	12	12	12
Conversion	(%)	95	97	93

The polyacrylic elastomers were prepared by emulsion polymerization. The monomers, water, emulsifiers (i.e., sodium lauryl sulfate and sodium salt of polymerized naphthalene sulfonic acid), tetrasodium ethylenediaminetetra-acetate, and sodium ferric salt of ethylenediaminetetraacetic acid were charged to the reactor vessel equipped with an agitator. The temperature of the reactor vessel was controlled at 10°C. The reactor was evacuated to remove air, closed and the emulsion agitated. Sodium sulfate, sodium formaldehyde sulfoxylate and paramenthane hydroperoxide were then charged to the reactor. The reaction time was 12 hours. The conversions to copolymers were 93 - 97% (Table 1).

After polymerization, the emulsion was removed from the reactor, and coagulated with a dilute solution of calcium chloride to isolate the elastomer. The elastomer obtained was washed thoroughly with water and air-dried.

Characterization of Polymers

Table 2. Polymer Characterization

Copolymer Composition	Polymer Mooney (ML 1+4,100°C)	Content of σ (ephr)[a]	Content of COOH (ephr)
EA/MAA/AGE	38	0.0053	0.0042
EA/GMA	40	0.012	-
EA/AGE	53	0.010	-

a) ephr: equivalent mole per hundred grams rubber

The carboxyl content of the polymer was determined by potentiometric titration with KOH in methyl ethyl ketone solution. The epoxy content of the polymer was determined by potentiometric titration with KOH in methyl ethyl ketone solution, using an excess of hydrogen chloride-acetone solution prior to the titration.

The Mooney viscosity (ML 1+4) was measured by JIS-K6300 procedure at 100°C. The results are shown in Table 2.

Compounding Formulations

Table 3. Compounding Recipes

Formulation	EA/MAA/AGE[a]	EA/GMA[b]	EA/AGE[c]
Copolymer	100	100	100
Stearic Acid	1.0	1.0	1.0
MAF Carbon Black	50	50	50
Onium Salt	0.005 (mole)	-	-
Octadecyltrimethylammonium Bromide	-	2.0	2.0
Dicarboxylic Acid	-	0.006 (mole)	0.005 (mole)

The compounding recipes are shown above in Table 3.
All parts are by weight except as indicated.
a) EA/MAA/AGE: Ethyl acrylate/Methacrylic acid/Allyl glycidyl ether terpolymer
b) EA/GMA: Ethyl acrylate/Glycidyl methacrylate copolymer
c) EA/AGE: Ethyl acrylate/Ally glycidyl ether copolymer

RESULTS AND DISCUSSION

It is well known that amines are efficient catalysts for the esterification of carboxylic acids with alkylene oxides[3-6]. Similarly, the catalytic effect of onium salts on the reaction of carboxylic acids and alkylene oxides has also been studied.[2,7,8]

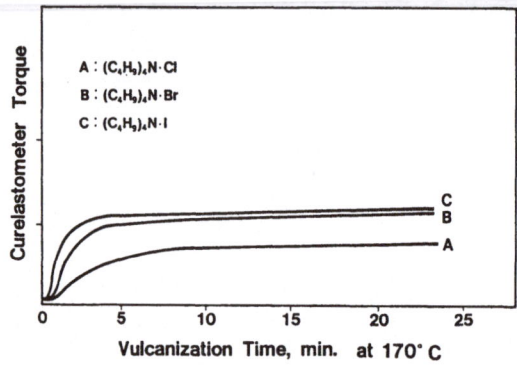

Fig.1. Effect of counter anions of quaternary salts on the crosslinking of EA/MAA/AGE terpolymer. Curing temperature: 170°C

The vulcanization curves at 170°C for the model terpolymer EA/MAA/AGE catalyzed by three different quaternary ammonium salts are shown in Figure 1. The cure rate constants, K_2, of various onium salts are given in Table 4, assuming that crosslink formation is a first order reaction after an induction period. High crosslinking rates were achieved when onium salts were used as catalysts.

Table 4. Cure Rate Constant, K_2, of Various Onium Salts for EA/MAA/AGE Terpolymer at 170°C

ONIUM SALT	Cure Rate Const., K_2
$(C_4H_9)_4N \cdot Cl$	0.21
$(C_4H_9)_4N \cdot Br$	0.47
$(C_4H_9)_4N \cdot I$	0.67
$(C_{12}H_{25})(CH_3)_3N \cdot Br$	0.31
$(C_{17}H_{35})(CH_3)_3N \cdot Br$	0.34
$(C_{18}H_{37})(CH_3)_3N \cdot Br$	0.33
$(C_4H_9)_4P \cdot Br$	0.48
$[\bigcirc N C_{16}H_{33}] Br \cdot H_2O$	0.36

The crosslinking rate was affected by the type of anion of the onium salt and increased in the following order:

$$I > Br > Cl$$

The structures of alkyl groups of the ammonium or phosphonium salts did not significantly affect the crosslinking rate.

Based on our results, the following reaction mechanism is proposed:

The onium salt dissociates in the polymer to form the corresponding ions, i.e., the active center. The degree and rate of this dissociation are affected by polarity of the polymer.

$$NR_4\ X \rightleftharpoons N^+R_4 + X^- \tag{1}$$

These ions open the epoxy ring and an alcoholate onium salt with high basicity is formed.

$$N^+R_4 + X^- + CH_2{-}CH\text{\wedge\wedge} \longrightarrow X{-}CH_2{-}CH\text{\wedge\wedge} \tag{2}$$
$$\underset{O}{\diagdown\!\diagup} \qquad\qquad \underset{O^-\ N^+R_4}{|}$$

This alcoholate onium salt reacts immediately with a carboxylic acid cure-site to form an active carboxylate onium salt, having loose ion-pair, and a free alcohol.

$$X{-}CH_2{-}CH\text{\wedge\wedge} + \text{\wedge\wedge}C \longrightarrow \text{\wedge\wedge}C\ N^+R_4 + X{-}CH_2'{-}CH\text{\wedge\wedge} \tag{3}$$

Reaction (2) and (3) represent the initiation reactions.

Dissociation of the carboxylate onium salt, in effect, enhances the nucleophilicity of carboxyl group which leads to rapid esterification of the epoxy moiety. And an ion pair is formed as shown in reaction (4). This ion pair reacts with carboxyl group attached to the polymer backbone, giving the carboxylate onium salt again.

Reaction (4) and (5) represent the propagation reactions and the termination of this reaction is due to disappearance of one or both cure-sites.

$$\text{\wedge\wedge}C\ N^+R_4 + CH_2{-}CH\text{\wedge\wedge} \longrightarrow \text{\wedge\wedge}C{-}O{-}CH_2{-}CH\text{\wedge\wedge} \tag{4}$$

$$\text{\wedge\wedge}C{-}O{-}CH_2{-}CH\text{\wedge\wedge} + \text{\wedge\wedge}C \longrightarrow \text{\wedge\wedge}C\ N^+R_4 + \text{\wedge\wedge}C{-}O{-}CH_2{-}CH\text{\wedge\wedge} \tag{5}$$

As shown in Figure 1, crosslinking reactivity varies with the onium salts used. The catalytic activity of amines depends on the types of substituent groups on the amines. On the other hand, the catalytic activity of quaternary ammonium salts depends primarily on the nature of the anion and is not affected appreciably by the size or structure of substituents on the nitrogen atom.

The catalytic activity shows good agreement with the nucleophilicity of the anion proposed by Swain or Pearson. The nucleophilic constants of the anions are shown in Table 5. In other words, the reaction rate is controlled by the rate of opening of epoxy ring in the initiation reaction and increase of the nucleophilicity of counter anion.

225

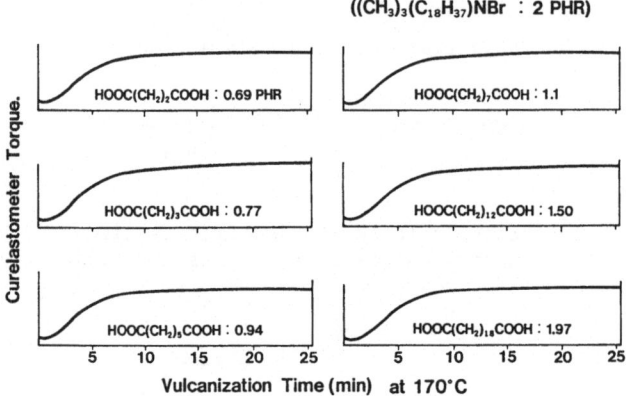

$((CH_3)_3(C_{18}H_{37})NBr : 2 PHR)$

Fig.2 . Effect of Various Dicarboxylic Acids
on Cure Rate of EA/GMA Copolymer

Table 5. Relationship Between Cure Rate Constant and
Relative Nucleophilic Reactivity Constant

Onium Salt	Cure Rate Constant K_2	Relative Nucleophilicity Constant of X	
		by Swain[9]	by Pearson[10]
$(C_4H_9)_4N.Cl$	0.21	3.04	4.37
$(C_4H_9)_4N.Br$	0.47	3.89	5.70
$(C_4H_9)_4N.I$	0.67	5.04	7.42

Similarly, the model EA/GMA copolymer containing only epoxy groups showed a high crosslinking rate with onium salt when a dicarboxylic acid was used as the external crosslinking agent as shown in Figure 2. The crosslinking reaction is affected by the length of the alkylene chain of dicarboxylic acid and tends to become slower when the alkylene chain length of dicarboxylic acid is longer.

Figure 3 shows the characteristics of 1st order cure rate constant K_2, which is calculated from vulcanization curves of Figure 2, with respect to the number of methylene carbons of dicarboxylic acid used as the crosslinking agent. The larger the number of methylene carbons of dicarboxylic acid, the lower the K_2.

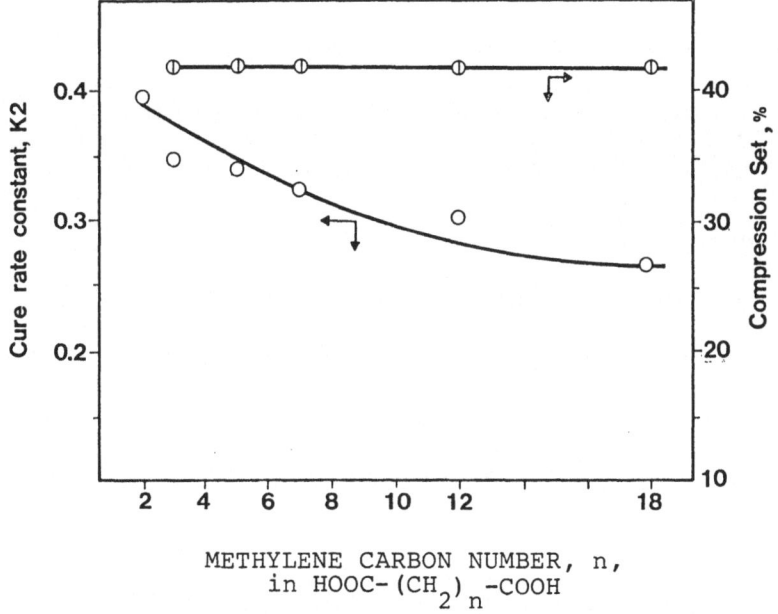

METHYLENE CARBON NUMBER, n, in HOOC-$(CH_2)_n$-COOH

Fig.3. Effect of Number of Methylene Carbons of Dicarboxylic Acid, HOOC-$(CH_2)_n$-COOH on Rate of Curing and Compression set of EA/GMA copolymer

This phenomenon correlates very well with the dissociation constants, K_1 and K_2 of the dicarboxylic acid. Increasing the length of alkylene chain decreases K_1 and K_2 constants as shown in Table 6 and retards the rate of crosslinking.

Table 6. Dissociation Constants K_1 and K_2
of $HOOC(CH_2)_n COOH$ $(X10^5)$

n	0	1	2	3	4	5	6	7	8
K_1^a	3.8×10^3	177.0	7.36	4.60	3.90	3.33	3.07	2.82	2.8
K_2^b	5.9	0.437	0.450	0.534	0.529	0.487	0.471	0.464	–

a) K_1: Dissociation Constant at First Stage

b) K_2: Dissociation Constant at Second Stage

Also, good compression set property was achieved without post curing as shown in Figure 3.

The crosslinking mechanism is considered to be basically the same as the onium salt crosslinking mechanism of polyacrylic elastomer with epoxy and carboxyl groups. However, there is a difference in that it is accomplished in two stages. In the first stage, only one of the carboxyl groups of the dicarboxylic acid reacts with the epoxy group in the polymer. Then, the crosslinking reaction occurs in the second stage when the second carboxyl group reacts. Thus the crosslinking reaction occurs slower in this case than for the EA/MAA/AGE terpolymer containing epoxy and carboxyl groups.

The effect on the crosslinking rate by trimethyloctadecylammonium bromide and octadecamethylene dicarboxylic acid was measured for an EA/GMA copolymer and EA/AGE copolymer as shown in Figure 4. The crosslinking rate is faster with GMA copolymer than with AGE copolymer.

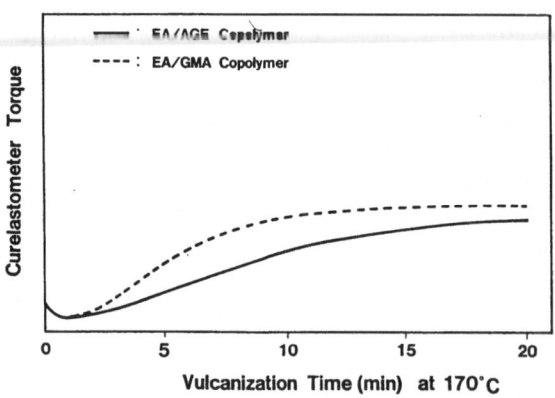

Fig.4. Comparison of Cure Rate Between EA/GMA copolymer and EA/AGE copolymer.
Epoxy Content of EA/AGE Copolymer: 0.010 ephr;
Epoxy Content of EA/GMA Copolymer: 0.012 ephr.

Since the nucleophilic activity is considered to be intensified by electron attraction ability of the carbonyl group of GMA in the polymer with GMA, the epoxy ring opens more easily than that of the polymer with AGE.

The effects of dicarboxylic acid and onium salt contents on scorch time and compression set of EA/GMA copolymer in this curing system were evaluated as shown in Figure 5 and 6. With the increase in the amount of $HO_2C(CH_2)_{18}CO_2H$, both scorch time and compression set increase. The increase of quaternary ammonium salt decreases both scorch time and compression set.

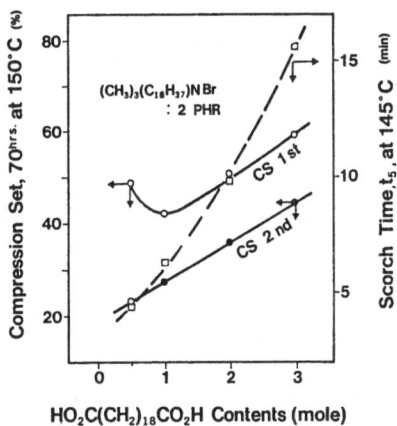

Fig.5. Effect of Dicarboxylic Acid Concentration on Scorch Time and Compression Set of EA/GMA Copolymer. $HO_2C(CH_2)_{18}CO_2H$ Contents show Mole Ratio of Carboxyl Groups against Mole Content of Epoxy Groups in the Copolymer.

Therefore, in order to apply this new curing system industrially, it will be necessary to find the optimum quantities of dicarboxylic acid and onium salt to be added. Our conclusion is that the best results could be obtained when quaternary ammonium bromide is 1.8 phr and dicarboxylic acid 1.6 phr.

The curing systems for acrylic rubber with epoxy groups as cure-sites generally use ammonium benzoate or a dithiocarbamate salt. We have carried out a limited amount of work for a comparison between these curing systems and the new curing system (Table 7). The curing curves of three types of curing systems are shown in Figure 7. The new curing system indicates a fast curing rate, in contrast to the very slow curing rate obtained with the ammonium benzoate (AB) system.

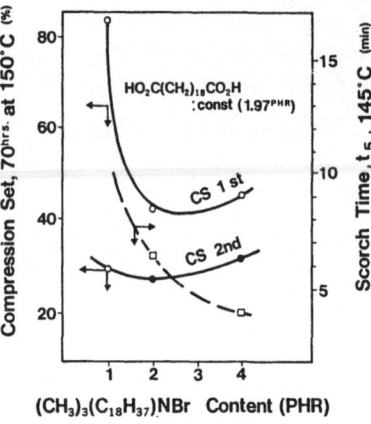

Fig.6. Effect of Onium Salt Contents on Scorch Time and
Compression Set of EA/GMA Copolymer at 1.97 phr of
$HOOC-(CH_2)_{18}-COOH$
CS 1st: First cured at 170°C x 20 min.,
CS 2nd: Post cured at 150°C x 16 hrs.

Table 7. Compounding Formulations for EA/GMA Copolymer

Compound Formulation	1	2	3
Polymer	100	100	100
Stearic Acid	1	1	-
AC Polyethylene	-	-	1
MAF Carbon Black	50	50	50
Ammonium Benzoate (AB)	1.3	-	-
Zinc Dimethyldithiocarbamate (PZ)	-	1.0	-
Ferric Dimethyldithiocarbamate (TTFe)	-	0.5	-
Octadecamethylene Dicarboxylic Acid	-	-	2.0
Octadecyltrimethylammonium Bromide	-	-	2.0

The physical properties of the vulcanizates obtained by these curing
systems were measured. The vulcanizate obtained by the new curing system
shows excellent compression set, without any need for post curing (2nd cure)
as shown in Table 8.

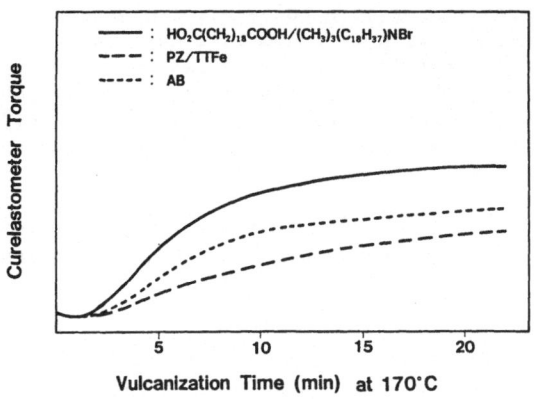

Fig.7. Curelastometer Curves of EA/GMA Copolymer
for Various Curing Systems

Table 8. Physical Properties of EA/GMA Copolymer Vulcanizates

Curing System	1 AB	2 PZ/TTFe	3 $HO_2C(CH_2)_{18}CO_2H$ + $(CH_3)_3(C_{18}H_{37})N \cdot Br$
Mooney Scorch (125°C, ML)			
Vmin	62.0	56.0	55.5
t_5 (min)	6.9	14.5	11.5
t_{35} (min)	13.4	41.4	44.3
Stress-Strain Properties (Cured 20 min. at 170°C)			
Tensile Strength(MPa)	11.0	10.8	10.9
Elongation (%)	340	360	270
Hardness (JIS)	66	66	70
Compression Set, 70 Hours at 150°C			
Cured 20 min. at 170°C, (%)	64	79	34
Post Cured 16 Hours at 150°C,(%)	42	58	20

CONCLUSIONS

1. In crosslinking reaction of polyacrylic elastomer which contains carbox-
    yl and epoxy groups as cure-sites by onium salt catalyst, the effective
    center is an active carboxylate onium salt. The catalytic activity
    showed good agreement with the nucleophilicity of the counter anion
    of onium salt and followed I > Br > Cl.

2. The rate of crosslinking of a polyacrylic elastomer containing epoxy groups as cure-sites with an onium salt catalyst is enhanced by the addition of a dicarboxylic acid as an external curing agent. However, the enhancement in the cure rate decreases as the number of methylene carbons in the alkylene chain of the dicarboxylic acid increases. This behavior can be explained by the dissociation constants, K_1 and K_2, of the dicarboxylic acid.

3. The vulcanizates cured by using onium salt curing system showed good physical properties and excellent compression set without post curing.

REFERENCES

1. H.P. Brown, Rubber Chem. Technol., 36, 931 (1963).
2. E. Giannetti and coworkers, Rubber Chem. Technol., 56, 21 (1983).
3. H. Kaiuchi and Y. Tanaka, J. Org. Chem., 31, 1559 (1966).
4. K. Hisauma and coworkers, Kogyo Kagaku Zasshi, 70, 169 (1967).
5. V.F. Shrets and A.V. Romashkin, Kinet. Catal., 13, 796 (1972).
6. Y. Enoki and coworkers, Nippon Kagaku Kaishi, 132 (1973).
7. R. Ueshima and H. Munakata, Nippon Kagaku Kaishi, 1496 (1973).
8. H. Kamatari, Nippon Kagaku Kaishi, 1505 (1977).
9. C.G. Swain and C.B. Scott, J. Am. Chem. Soc., 75, 141 (1953).
10. R.G. Pearson, H. Sobel and J. Songstad, J. Am. Chem. Soc., 90, 319 (1968).

FATIGUE RESISTANCE OF POLYBUTADIENES AND EFFECT OF MICROSTRUCTURE

Luciano Gargani and Mario Bruzzone

EniChem Elastomeri S.p.A.

Assago, Milan, Italy

SUMMARY

Rubber failure upon application of stress is delayed by: i) viscoelastic energy dissipation and ii) crystallization upon strain of the base polymer. A peculiar rubber failure, namely fatigue failure, is also affected by the same factors. The presence of carbon black or other reinforcing agents does not diminish the contribution of the base polymer on fatigue resistance.

A base polymer particularly interesting as a model for fatigue resistance tests is represented by polybutadiene. Polybutadiene microstructure can be changed at will by making use of existing catalyst systems. In particular, the high vinyl structures are able to show high viscoelastic dissipation and therefore to delay fatigue failure by the mechanism indicated under i). The high cis structures of polybutadiene are capable of crystallizing upon strain and therefore are able to delay fatigue failure by the mechanism indicated under ii).

In this work it is shown that the mechanism ii) is more effective than i) in improving fatigue resistance of polybutadiene. Also a small decrease of chain defects (in particular 1,2 units) in the vicinity of 100% cis content is able to show a substantial improvement in fatigue resistance, both in reinforced and in pure gum polybutadienes.

INTRODUCTION

Rubber failure by application of stress has been studied extensively owing to its overwhelming scientific and practical interest. In particular, two mechanisms have been put forward as relevant in delaying rubber failure. The first mechanism is based on viscoelastic energy dissipation, which can be increased by increasing the glass transition of the base polymer, or by other routes such as the use of additives or controlled network imperfection. The second mechanism is based on polymer crystallization upon strain. It is known that this mechanism can be effective also when the polymer glass transition is very low. .

Limiting our analysis to the effect of the base polymer on rubber failure, the polymer structure could be tailored as a dissipative one by increasing Tg up to the maximum value acceptable for an elastomer or toward a strain crystallizable structure by increasing the micro-structure regularity up to a point in which a rapid strain induced crystallization is achieved.

A base polymer particularly interesting for studying the effect of the aforementioned mechanisms on rubber failure is polybutadiene. In fact, polybutadiene microstructure can be changed in an extremely wide range, by making use of the host of catalyst systems developed by the ingenuity of chemists, just starting with the same monomer.

In particular, the high vinyl structures characterized by high Tg, are good models for measuring the improvement related to an increased viscoelastic energy dissipation, whereas the very high cis structures are good models for showing the improvement due to an increased strain crystallization. These two main structures of polybutadiene have been taken into consideration in this work in order to answer the question often raised by the polymer chemist as to which polybutadiene structure is better from a practical stand point. The "trans" structures of polybutadiene are not being considered here[1].

In this work, a specific rubber failure, namely, the fatigue to failure has been selected for comparing the aforementioned polybutadiene microstructures. In practical applications, fatigue to failure is a very complicated phenomenon in which mechanical, thermal and chemical effects play a role. However, the phenomenon can be simplified, at least in part, by avoiding overwhelming thermal and chemical effects. This is done, in practice, by performing a fatigue to failure test on sufficiently small specimens and by avoiding an excessive energy input per cycle, in order to achieve a substantially isothermal condition for all the specimens involved, independently of their different viscoela-stic energy dissipation. Another requirement of the fatigue to failure test is that the results must be based on a rigorous statistical analysis of a large number of specimens. In this work the number of tested specimens was not less than 48 for each polybutadiene microstruc-ture and for each crosslink density. The failure probability curves compare polybutadienes of different microstructure, selected at the same crosslink density.

EXPERIMENTAL

Materials

Polybutadiene samples of different microstructure and molecular weight, as shown in Table I.

TABLE I: Microstructure of polybutadienes.

Catalyst	1,2, %	cis-1,4, %	trans-1,4, %	$\overline{M}w \times 10^{-3}$	ML-4, 100°C
Li	10.7	42.7	46.6	293	49
Li	28	28.8	43.2	235	34
Li	52.5	16.8	28.5	201	42
Li	72.4	11.1	16.5	289	35
Ti	4.0	92.7	3.3	363	45
Ni	1.9	96.3	1.8	406	39
Co	1.3	97.3	1.4	376	43
Nd	0.8	98.3	0.9	457	33
Nd	0.8	98.3	0.9	577	43
Nd	0.8	98.3	0.9	657	55

Recipes

	Gum	Black
Polymer	100	100
Zinc Oxide	3	5
Stearic Acid	2	3
Antiox. 2246	1	1
Sulfur	—	variable
MBTS	variable	variable
DTM	variable	—
TMTD	0.1	—
N 330 carbon black	—	50

TMTD = tetramethylthiuram disulphide
DTM = 4,4'-dithiomorpholine
MBTS = dibenzothiazyl disulphide

Methods

Microstructure by 1-H NMR (300.00 MHz) and 13-C NMR (75.46 MHz) spectroscopy.

Molecular weights by Waters ALC-GPC model 244 B, with Shodex columns, using K and α values of 4.57×10^{-4} and 0.693, respectively.

Stress-strain measurements at different temperatures by Instron dynamometer, ring specimens, 4 mm thick, 44.6 mm internal and 52.6 mm external diameter, 0.005 s^{-1} extension rate.

Crosslink density by swelling in n-heptane, with the following values of rubber-heptane interaction parameters[2]:

Li-polybutadienes   = 0.37 + 0.52 Vr
High cis polybutadienes = 0.45 + 0.35 Vr

where Vr is the swelling ratio by volume.

Viscoelastic properties in tensile mode by Rheovibron DDV II at 110 Hz. Viscoelastic properties in shear mode by Rheometrics (DMS) at 1 Hz. Heating and cooling cycles at 1°C/min.

Fatigue life by Fatigue-to-Failure Tester (Monsanto) at room temperature and at least 48 specimens for each polymer and each crosslink density. Experiments were performed on several sets of specimens, both notched and un-notched, gum and black compounds, at several crosslink densities and strain amplitudes. Data reported here were obtained on un-notched specimens, 136% maximum strain amplitude in the case of black compounds; and notched specimens, 68% maximum strain amplitude in the case of gum compounds.

VISCOELASTIC DISSIPATION AND RUBBER FAILURE

Polymer failure takes place at stress levels much lower than expected on the basis of bond strength. Theoretical strength values[3] are actually orders of magnitude greater than experimentally observed.

To rationalize this inconsistency, a hypothesis of local distortion of stress field has been put forward and attributed to defects, which are always present[4]. These defects are assumed to cause a local stress concentration and, consequently, a premature failure. In the distorted stress field, some stress components are inversely proportional to the defect radius. This means that some stress components are very high in the neighbourhood of the crack tip, and approach infinity as the radius of the crack tip tends to zero, so that the rubber should hardly sustain any stress in the presence of very sharp flaw tips[5]. In conclusion, these models foresee a too strong or a too weak material, in contrast with experimental results.

The energy criterion introduced by Griffith[6] overcomes the difficulty by recognizing that crack growth requires work to be supplied by the elastic energy stored in the material upon deformation. This model indicates a possible mechanism for delaying the elastomer failure. If a large amount of deformation energy is converted into heat, less energy is available for the crack growth. Therefore, the fracture resistance of viscoelastic materials should be enhanced by increasing the viscous component. Experiments largely confirm this assumption, since fracture resistance and energy dissipation are interrelated in amorphous rubbers. Figure 1 displays the similar patterns of the two properties for a typical amorphous rubber, assuming elongation at break as a measure of the elastomer capability in sustaining a deformation, and loss modulus (Rheovibron) as a measure of the elastomer capability of dissipating energy. Unfortunately, a reinforcement mechanism based on viscoelastic dissipation is not appropriate for several rubber applications, in which both a high strength and a low hysteresis are simultaneously required.

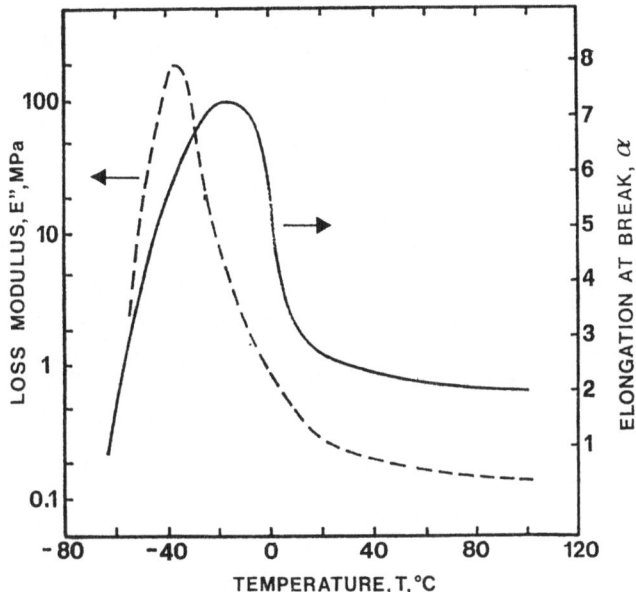

Fig. 1. Loss modulus and elongation at break as a function of temperature for a typical amorphous rubber: SBR cured to 1.8×10^{-4} mol cm^{-3} crosslink density. SBR used was an emulsion rubber containing 23.5 wt.% styrene. (Our data.)

STRAIN CRYSTALLIZATION AND RUBBER FAILURE

A different mechanism of reinforcement is available, based on the capability of some stereo-rubbers to undergo a strain induced crystallization. High cis-polyisoprene and high cis-polybutadiene are noteworthy examples of this behaviour[7]. Figure 2 shows the behaviour of a synthetic high cis-polyisoprene prepared by a titanium-based catalyst. Unlike Figure 1, this plot shows that strain crystallization is able to delay rubber fracture over an extended range of temperature, well above the glass transition temperature of the polymer. In this way, the polymer can display both a high deformation capability and a low hysteresis.

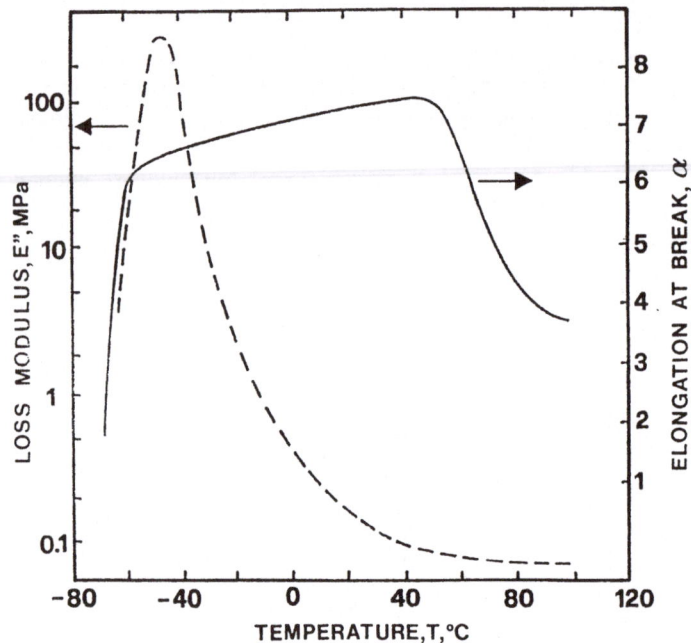

Fig. 2. Loss modulus and elongation at break as a function of temperature for a typical strain crystallizable rubber: polyisoprene
cured to 1.7×10^{-4} mol cm^{-3} crosslink density. (Our data.)

FATIGUE TO FAILURE

So far we have dealt with the resistance of elastomers to steady
fracture. When discussing that peculiar fracture resistance taking
place after repeated deformations, generally referred to as fatigue
resistance, which in principle encompasses mechanical, thermal and
chemical phenomena, a strict definition of the experimental conditions
is necessary in order to lessen the number of variables coming into
play. In our experiments, the specimen deformation amplitude and frequency have been kept sufficiently low as to allow a constant temperature
to be maintained during the fatigue experiment. In these conditions,
isothermal crack growth is the factor controlling the fatigue life of
the specimen; thermal and chemical degradation effects can be reasonably disregarded. A second important point is the treatment of the
fatigue experimental data. The statistical nature of fatigue failure
has been taken into account in order to compare different materials at
a suitable level of statistical significance. Therefore experimental
data were collected on a large number of specimens, and data were
processed according to a statistical method in order to describe the
materials behavior within tight confidence limits. From the point of

view of the properties investigated, we have focussed on the effects of hysteresis and of crystallization on the fatigue life of elastomers under the experimental conditions mentioned earlier.

Correlations between chemical structure and mechanical behaviour have been established by evaluating polybutadienes of different micro-structures as model elastomers with behaviour ranging from highly hysteretic to highly crystallizable. So, a second target has been reached, providing some guidelines for the synthesis of the most appropriate polymer structure for attaining an improved fatigue resistance starting from the same monomer, i.e., butadiene.

POLYBUTADIENES OF DIFFERENT MICROSTRUCTURES

Figure 3 summarizes the main routes towards improved fatigue resistance, together with the structural features in the case of polybutadiene. Of course, a choice has to be made between a high vinyl and a high cis structure, whereas an increase of molecular weight can be beneficial in both cases. Suitable Ziegler-Natta catalysts are available to obtain both atactic high vinyl polybutadiene and extremely high cis-polybutadiene.

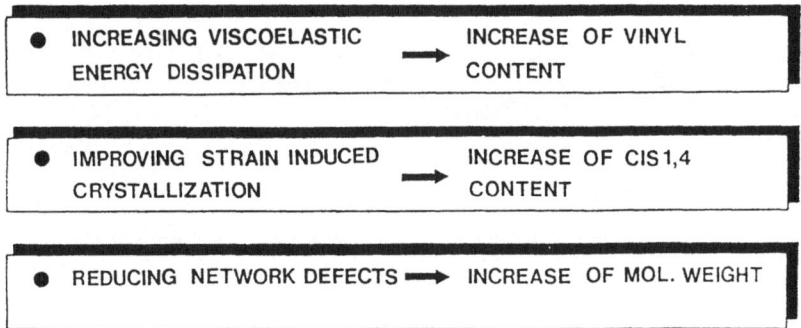

Fig. 3. Routes towards improved fatigue resistance in polybutadiene.

The influence of the vinyl content on the viscoelastic behaviour of polybutadienes is shown in Figure 4. Measurements of $\tan \delta$, the phase angle between stress and strain under sinusoidal deformation, have been performed by Dynamic Mechanical Spectrometry (Rheometrics). Looking at the shift along the temperature axis due to the different vinyl content, a maximum vinyl content of 72% has been chosen, since beyond this limit the polymer can hardly be regarded as a rubber.

Moreover, the shift along the temperature axis eventually entails an increase of dissipation at temperatures of practical interest for rubber applications, as shown in Figure 5 referring to cured samples. The crossover of the curves for the raw polymers shown in Figure 4, disappears after curing, so that the amount of dissipation increases at relatively high temperatures according to the vinyl content, as seen in Figure 5.

Polybutadienes with vinyl contents of 10.7 and 72.4% have been selected for further fatigue-to-failure analysis. Fatigue life measurements have been performed on a large number of specimens by means of the Fatigue-to-Failure Tester (Monsanto). Carbon-black-filled compounds with 50 phr N330 carbon black were cured with conventional curing system at 145°C. Curative levels have been chosen to obtain vulcanizates at several levels of 300% modulus. Fatigue tests were performed on un-notched specimens at room temperature and at several deformation amplitudes. Data discussed here were obtained at 136% strain amplitude.

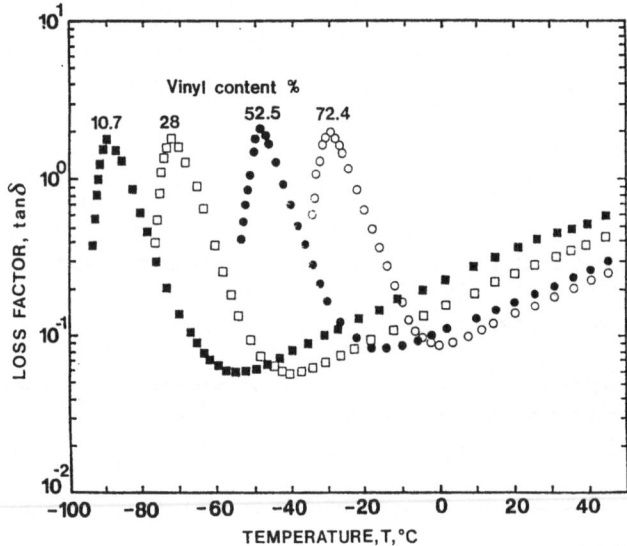

Fig. 4. Viscoelastic behavior of raw polybutadienes as a function of vinyl content.

Experimental data on the distribution of fatigue lives have been tentatively fitted with several probability distribution functions. According to the average correlation coefficients, the best fit was provided by the Weibull function, see Figure 6. The two parameters characterizing a given Weibull distribution (characteristic life, b, and shape factor, α) were determined by numerical regression so that both the probability density function and the cumulative distribution of probability of failure could be evaluated. Figure 7 shows that the maximum probability of failure shifts from 3 kcycles for 10.7% vinyl to 13.5 kcycles for 72.4% vinyl, a four-fold improvement.

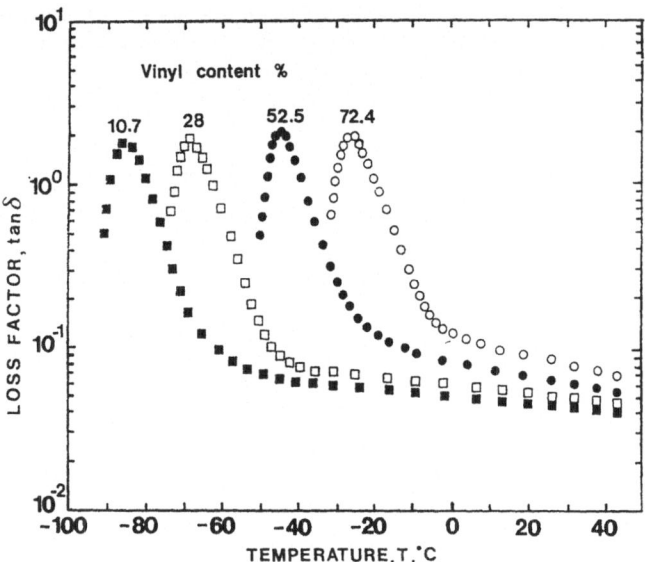

Fig. 5. Viscoelastic behavior of cured polybutadienes as a function of vinyl content.

FUNCTION	EQUATION	CORRELATION C.
DOUBLE EXP	$F(n) = 1 - \exp\left(-\exp\left(\alpha(n - x_0)\right)\right)$	0.91
NORMAL	$F(n) = \int_{-\infty}^{n} \frac{1}{\sigma\sqrt{2\pi}} \exp\left(-\frac{1}{2}\left(\frac{t-\mu}{\sigma}\right)^2\right) dt$	0.92
LOG NORMAL	$F(n) = \int_{-\infty}^{n} \frac{1}{\sigma\sqrt{2\pi}} t^{-1/2} \exp\left(-\frac{1}{2}\frac{(\ln t - \mu)^2}{\sigma}\right) dt$	0.92
EXPONENTIAL	$F(n) = \exp(-\lambda(n - c))$	0.95
WEIBULL	$F(n) = 1 - \exp\left(-\left(\frac{n}{b}\right)^{\alpha}\right)$	0.99

Fig. 6. Average fitting of several probability distribution functions to fatigue data.

The same data plotted in Figure 8 as cumulative functions allow either to compare the behavior of the two polymers at each level of failure probability or to evaluate the failure probability after a given number of cycles. In conclusion, it is inferred that an hysteresis increase in polybutadienes, obtained by increasing the vinyl content, is a viable route for improving fatigue resistance.

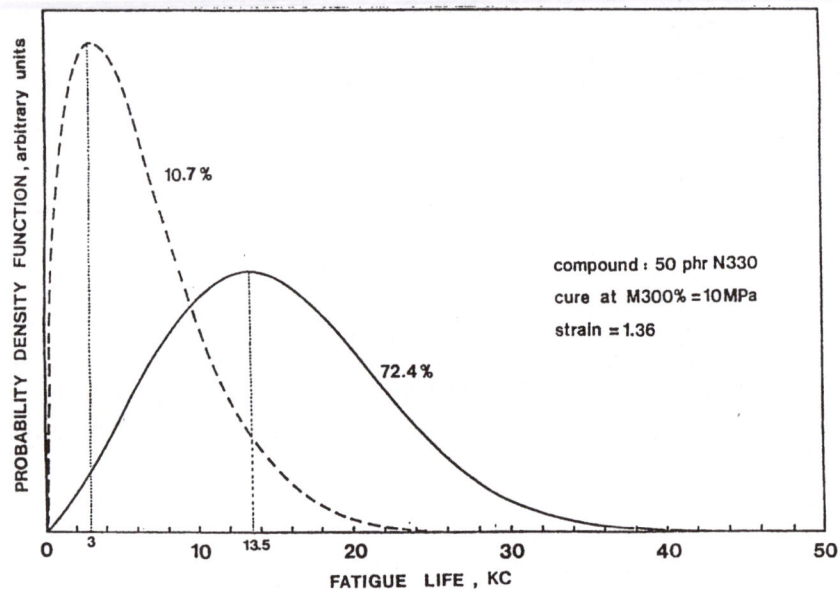

Fig. 7. Failure probability density as a function of number of cycles in polybutadienes with different vinyl content.

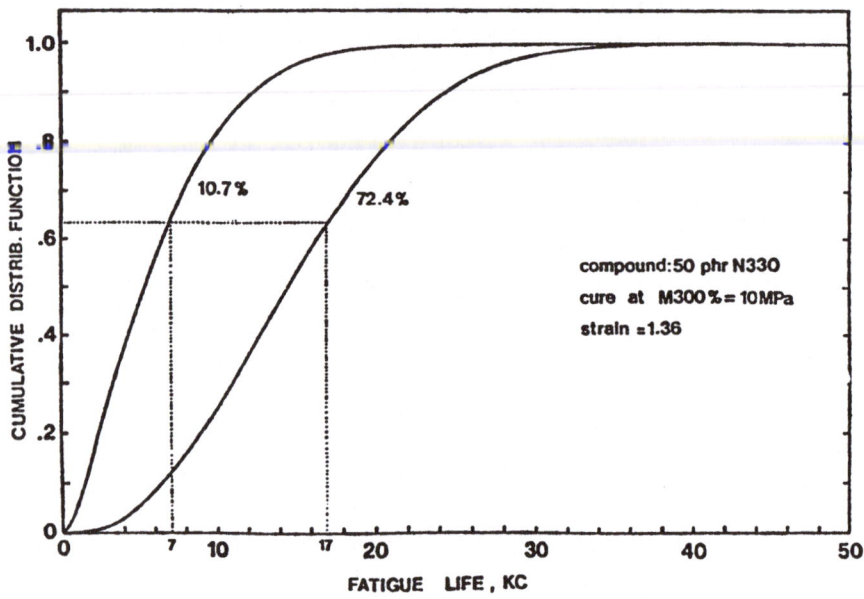

Fig. 8. Failure cumulative distribution as a function of number of cycles in polybutadienes with different vinyl content.

However, a far better fatigue resistance can be obtained by following the second route shown in Figure 3, i.e., strain induced crystallization. As previously mentioned, this mechanism is operative in high cis-tactic polybutadienes.

Figure 9 shows the fatigue behavior of several strain crystallizable polybutadienes with cis contents ranging from 92.7 to 98.3%.

All of them perform better than the amorphous polybutadiene with the highest vinyl content (dotted line), confirming that the strain crystallization mechanism is by far more efficient than the viscoelastic one in determining an enhancement of fatigue resistance. The Weibull mean life, plotted in Figure 10, gives a more impressive picture of the experimental data.

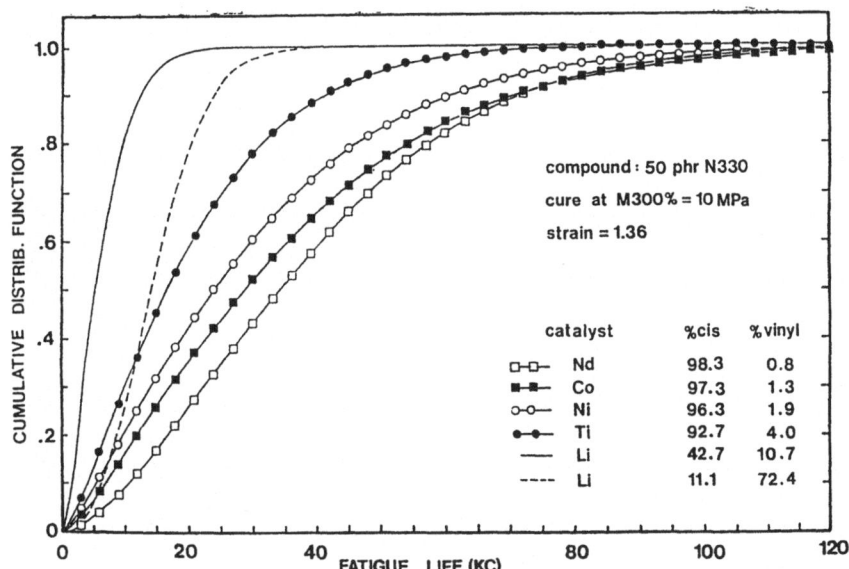

Fig. 9. Fatigue life of several polybutadienes with different microstructure.

In order to confirm that the behaviour shown in Figure 9 is controlled by the polybutadiene base polymer and that the interaction with carbon black does not effect the relative position of the curves shown in the plot, additional experiments on fatigue resistance of polybutadiene gum stocks were performed. Results shown in Figure 11 and the substantial agreement with Figure 9, confirm the predominant effect of the base polymer in positioning the different polybutadiene curves of the plot of Figure 9.

Another aspect that needs stressing is that the differences among high cis polybutadienes cannot be accounted for by their different hysteresis. In fact, even by assuming an influence of the residual vinyl content, if any, its influence on hysteresis should go exactly in the opposite direction.

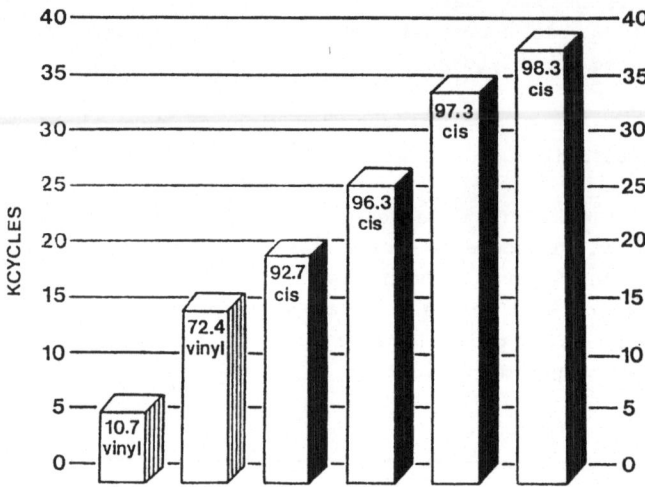

Fig. 10. Weibull mean life as a function of microstructure of polybutadiene.

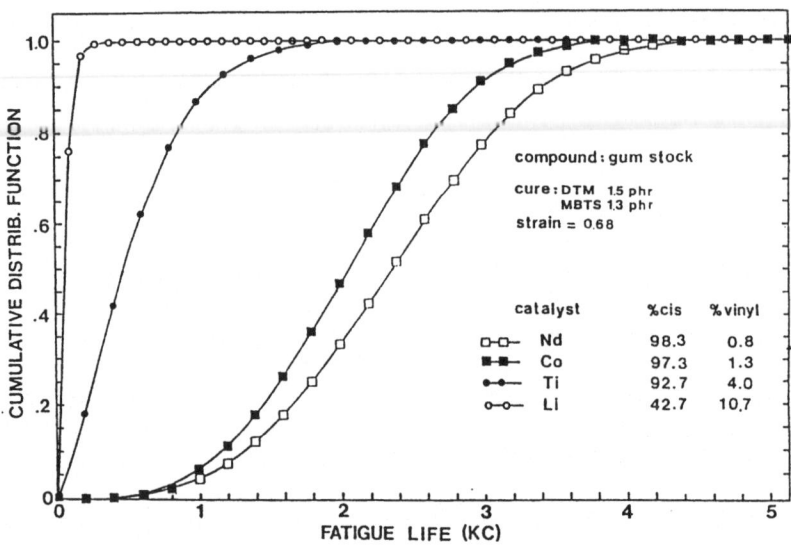

Fig. 11. Fatigue life of unfilled compounds of several polybutadienes
with different microstructure.

ANALYSIS OF STRAIN-INDUCED CRYSTALLIZATION

The surprising sensitivity of fatigue resistance to small variations of cis content in the proximity of 100% cis prompted us to
determine the sensitivity of strain crystallization to cis content.

Conventional isothermal stress-strain measurements were performed
at several temperatures, and the data obtained at each strain value
have been plotted against temperature to provide non-equilibrium stress-
temperature profiles, as shown in Figure 12. Each curve corresponds to
a given elongation α. The dotted curves, referring to low cis polybutadiene, show a linear decrease of stress with decreasing temperatures.
That means that viscoelastic transitions are negligible down to -60°C
under our experimental conditions. Viscoelastic effects should be even
less important for a high cis polybutadiene, which displays a lower
glass transition temperature with respect to low cis-polybutadiene.
Therefore, the sudden upturn shown by the curves of the high cis
polybutadiene can be explained only by the onset of strain-induced
crystallization.

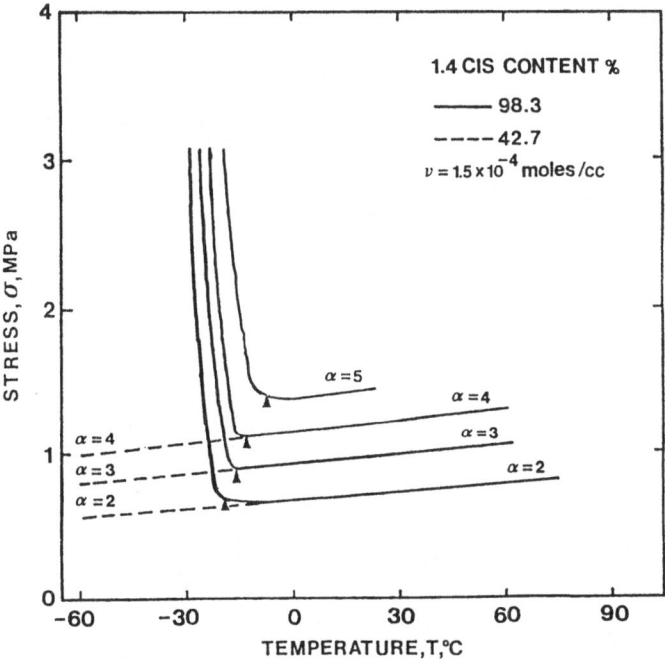

Fig. 12. Non-equilibrium stress-temperature profiles for an amorphous
(42.7% cis) and a crystallizable (98.3% cis) polybutadiene at
different elongation ratios, α. Gum vulcanizates were used.

The temperatures of incipient crystallization, indicated by triangles in Figure 12, are plotted in Figure 13 against the elongation ratio α for three different cis contents. The effectiveness of small increases in the cis content in enhancing crystallization is confirmed. By increasing the cis content, the crystallization is promoted at both higher temperature and lower elongation ratio.

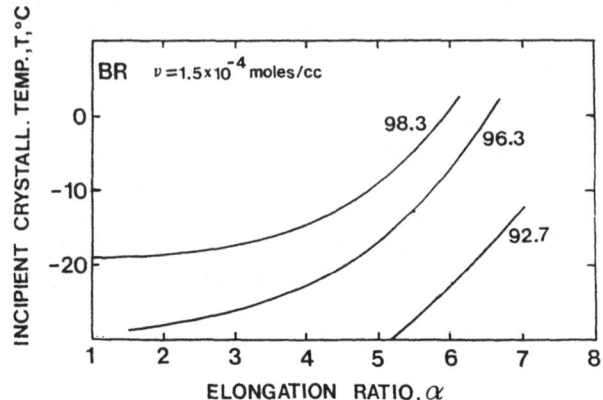

Fig. 13. Onset of strain-induced crystallization for polybutadienes of different cis contents.

In order to get an insight into the mechanism of strain induced crystallization and its consequences on mechanical properties, a more sensitive method for detecting crystallization has been developed using measurements of the elastic component of the complex modulus of specimens both in undeformed conditions (isotropic crystallization) and in strained conditions.

The influence of cis-tacticity on isotropic crystallization is shown in Figure 14, where the effects of cooling and heating cycles on the storage modulus are illustrated for three polybutadiene raw rubbers.

At a 42.7% cis content, the path is reversible and the transition zone from rubber-like to glass-like behaviour lies in the region around -90°C. At very high levels of cis-tacticity, 92.7% and 98.3%, a sharp increase of the modulus occurs well above the glass transition temperature, indicating the development of crystallinity. The melting point is indicated by the sharp return to the rubbery behaviour as the temperature increases. The region of melting has been further analyzed to get additional information on the crystallization mechanism, mainly in the strained state.

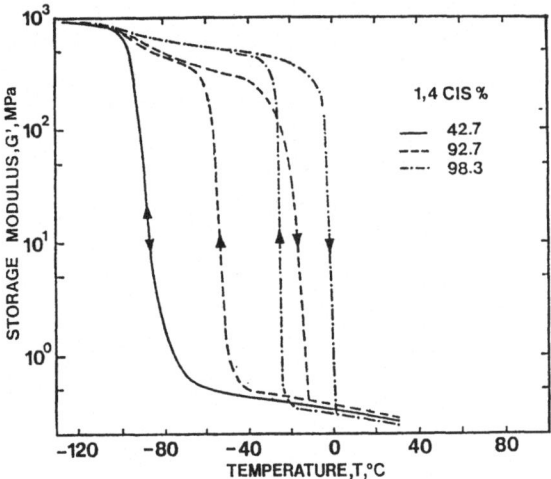

Fig. 14. Temperature induced crystallization and melting detected by Dynamic Mechanical Spectroscopy (Rheometrics).

In fact, beyond a given strain value the sharp melting temperature just mentioned splits into two, as shown in Figure 15.

The first melting point indicates the completion of the primary melting, involving the major amount of crystallinity as indicated by the substantial drop of modulus. This melting point is slightly dependent on strain and is probably related to chain-folded crystallites[8].

The second melting point, corresponding to a smaller decrease of modulus, probably indicates the melting of a few fibrillar crystals[8] and is highly sensitive to strain. The trend of the melting points as a function of the elongation ratio is summarized in Figure 16 for a polybutadiene with 98.3% cis content.

Also strain, as expected, has a strong effect on crystallization rate, which increases by an order of magnitude for each 100% of incremental elongation, as shown in Figure 17. The time required to raise the modulus to as much as three-fold the initial value (0.5 increment in logarithmic scale), is 60 min. at 1.46 elongation, and becomes 6 min. at 2.31 elongation and about 6 sec. at 3.64 elongation.

In Figure 18, in which an extrapolation of the experimental results is drawn, we show that crystallization induced by high elongation can develop fast enough to be useful for many practical applications.

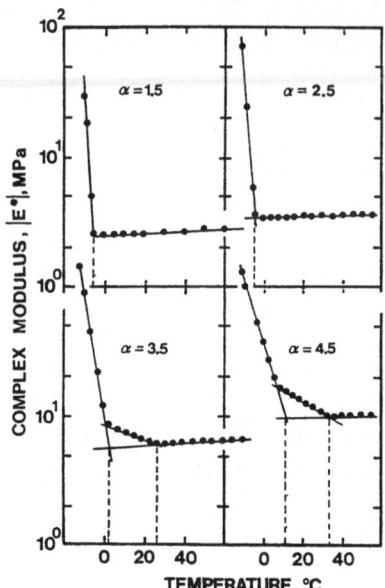

Fig. 15. Melting points of strained specimens of polybutadiene with 98.3% cis content; elongation ratios, α, as indicated.

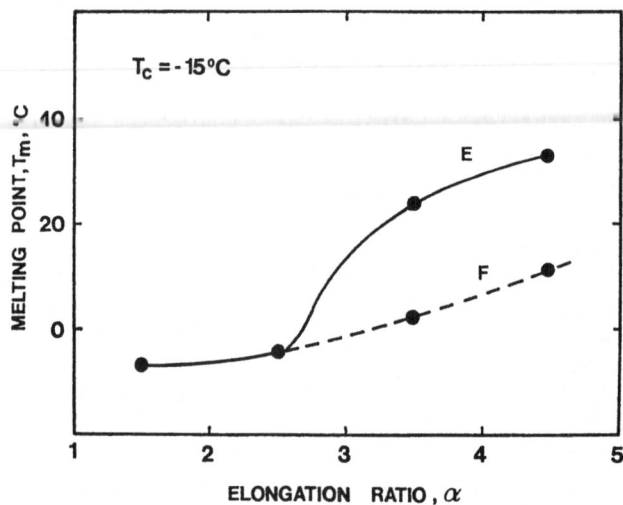

Fig. 16. Melting points of extended chain (E) and chain folded (F) crystals.

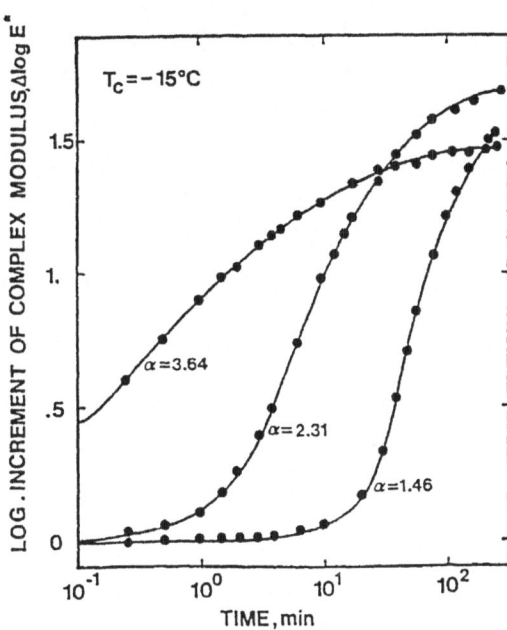

Fig. 17. Influence of elongation ratio, α, on crystallization rate of 98.3% cis polybutadiene.

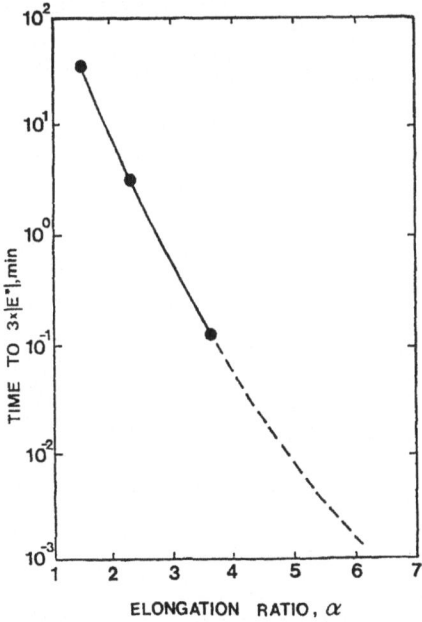

Fig. 18. Time required for crystallization to increase the modulus to three-fold the initial value, versus elongation ratio, α, for a 98.3% cis polybutadiene.

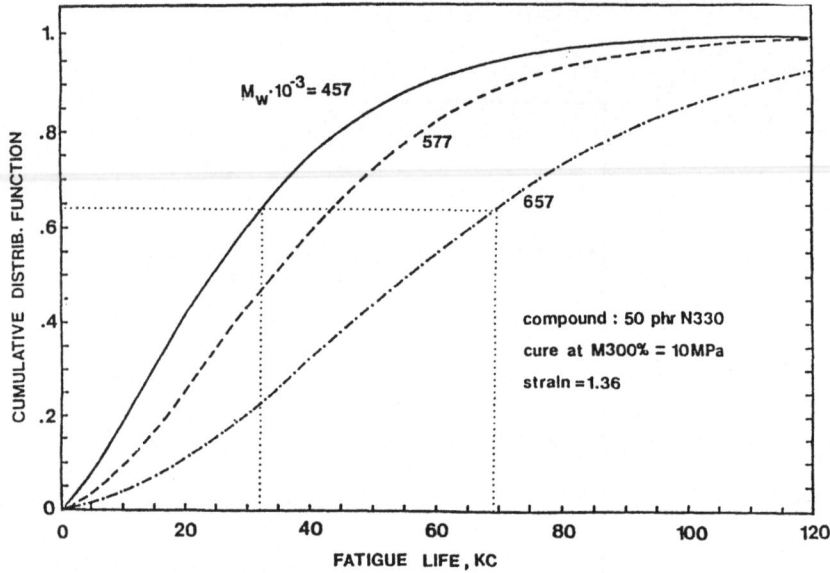

Fig. 19. Fatigue life of 98.3% cis polybutadiene as a function of molecular weight.

MOLECULAR WEIGHT AND FATIGUE RESISTANCE

As previously mentioned, a significant additional improvement in fatigue resistance can be achieved by increasing the polymer molecular weight. Figure 19 shows that a 44% increase of MW improves fatigue life by about 112%. A cis-tactic polymer of this MW can be processed by conventional equipment without difficulty.

CONCLUSION

An extremely high cis content (98.3%) and an enhanced molecular weight proved to be beneficial for fatigue resistance for polybutadiene rubbers.

REFERENCES

1. M. Bruzzone, A. Carbonaro, L. Gargani, Rubber Chem. and Technol., 51, 907 (1978).
   A. Carbonaro, L. Gargani, E. Sorta, M. Bruzzone, Proc. Intl. Rubber Conf., Venice, 1979, p. 312.

2. A.S.M. Burton, Handbook of Solubility Parameters, C.R.C. Press, Boca Raton, Florida, U.S.A. (1983).

3. F. Bueche, Rubber Chem. and Technol., 32, 1269 (1959).

4. C.E. Inglis, Trans. Inst. Nav. Arch., London, 55, 219 (1913).

5. A.N. Gent, Strength of Elastomers in "Science and Technology of Rubber", F.R. Eirich ed., Academic Press, 1978.

6. A.A. Griffith, Philos. Trans. Royal Soc., London, A 221, 163 (1920).
   A.A. Griffith, Proc. Intl. Congr. Appl. Mech., Delft, 55 (1924).

7. M. Bruzzone, Synthetic Hydrocarbon Rubbers in "Giulio Natta, Present Significance of His Scientific Contribution", Editrice di Chimica, Milano, Italy (1982), p. 63.

8. M. Cesari, G. Perego, A. Zazzetta, L. Gargani, Makromol. Chim., 181, 1143 (1980).

THE INFLUENCE OF CHEMICAL STRUCTURE ON THE STRENGTH OF RUBBER

A. N. Gent

Institute of Polymer Science
The University of Akron
Akron, Ohio 44325, U.S.A.

Abstract

The detailed chemical structure of the polymer molecule has surprisingly little effect upon many of the important physical properties of crosslinked elastomers. For example, the elastic modulus, extensibility, tensile strength and tear strength are all much the same for many common elastomers. However, the strength of elastomers under some conditions is strikingly different. Two particular modes of fracture are considered here: tearing, and abrasive wear. Certain elastomers crystallize rapidly on stretching and become self-reinforcing, so that their tear strength is greatly enhanced. Factors that govern the speed of strain-induced crystallization are reviewed. In abrasive wear, the macroradicals generated by molecular rupture are highly reactive and their reactions affect both the nature and the extent of wear. The wear processes that occur in various reinforced elastomers are described and compared.

1. Introduction

Many of the physical properties of crosslinked elastomers do not depend directly upon the local chemical structure of the molecule at all. Instead, they depend upon other quantities, for example, the number ν of molecular strands per unit volume, upon their contour length L and mass M_c and upon the local rate ϕ of Brownian motion of molecular sub-units, consisting of small portions, about 5 main-chain atoms in length, of a molecular strand. (The actual number of main-chain atoms per molecular sub-unit is denoted q below; it is a measure of chain stiffness and hence it is somewhat smaller for more flexible molecules and larger for stiffer ones.) Some of these quantities are only indirectly related to the local molecular structure. As a result, many physical properties are found to be quite similar for elastomers that have markedly different chemical structure. For example, the tensile modulus of elasticity E for a network made up of flexible molecular strands is predicted by the statistical theory of rubber elasticity to be given by (1):

$$E = 3\nu kT \tag{1}$$

where k is Boltzmann's constant, and T is the absolute temperature. Although the analysis leading to equation 1 takes into account the limited flexibility of elastomer molecules, this feature of their chemical structure does not appear in the final result. Thus, whether the individual molecules are relatively stiff or relatively flexible is unimportant in so far as the elastic modulus of the network is concerned.

Even the maximum extensibility of the network is only slightly dependent upon the molecular flexibility, over the range that might be expected for simple elastomeric polymers. It is principally determined by the molecular length L, and hence molecular weight M_c between points of molecular inter-linking (crosslinking) (1). It can be characterized by the ratio λ_m of the fully-stretched-out molecular length L to the average distance L_o between the ends of molecular strands in the unstretched state. The former quantity is given by

$$L = n\ell \tag{2}$$

where n is the number of molecular sub-units in a molecular strand and ℓ is the length of a sub-unit. The latter quantity is given by

$$L_o = n^{\frac{1}{2}}\ell \tag{3}$$

if it is assumed that the sub-units are connected together by freely-rotating joints. Thus,

$$\lambda_m = n^{\frac{1}{2}}, \tag{4}$$

where n is related to the number n_o of main-chain atoms per molecular strand by

$$n = n_o/q, \tag{5}$$

to the molecular weight M_c of a network strand by

$$n = M_c/qM_o, \tag{6}$$

where M_o is the molecular weight per main-chain atom, and to the number ν of network strands per unit volume by

$$n = A/\nu q M_o, \qquad (7)$$

where A denotes Avogadro's number. Equations 4-7 show that λ_m depends upon $q^{-\frac{1}{2}}$.

2. Tear Strength of Non-Crystallizing Elastomers

When elastomeric networks are torn apart under conditions of minimum strength, i.e., when no additional energy is expended in various dissipative processes (for example, viscous motion of molecular strands or detachment from filler particles), then the work $G_{c,o}$ of fracture per unit area torn through is given by (2)

$$G_{c,o} = \nu' n_o U \qquad (8)$$

where ν' is the number of strands crossing a randomly chosen plane of unit area (the fracture plane, for example) and U is the dissociation energy of a main-chain bond. The value of ν' is directly related to the number ν of strands per unit volume and the average distance L_o between their ends (2):

$$\nu' = (3/8)^{\frac{1}{2}} \nu L_o \qquad (9)$$

Thus, from equations 3, 7, 8 and 9:

$$G_{c,o} = (3/8)^{\frac{1}{2}} \rho A U q^{\frac{1}{2}} \ell M_c^{\frac{1}{2}} / M_o^{3/2} \qquad (10)$$

In terms of Young's modulus E, from equations 1, 6 and 7:

$$G_{c,o} = (9/8)^{\frac{1}{2}} (\rho A)^{3/2} (qkT)^{\frac{1}{2}} U \ell / M_o^{3/2} E^{\frac{1}{2}} \qquad (11)$$

Equation 1 indicates, and experiments confirm, that the modulus of elasticity E depends primarily upon the number of network strands and not upon their detailed structure. On the other hand, equations 10 and 11 show that the tear strength depends significantly upon the mass M_o per main-chain atom, as well as upon the number of strands and hence E. Experimental measurements of tear strength under threshold conditions, i.e., at high temperatures and low rates of tearing, are in good agreement with these theoretical predictions, as shown in Figures 1 and 2 (3). Values of the work of fracture G_c are found to increase in proportion to $M_c^{\frac{1}{2}}$, and to decrease in proportion to $E^{-\frac{1}{2}}$, for networks prepared by crosslinking to different degrees. And, for the same values of M_c or of E, substantial differences are found between different polymers, those with larger values of mass M_o per main-chain atom having lower tear strengths, as low as 1/5 of the tear strength of the simple hydrocarbon elastomers.

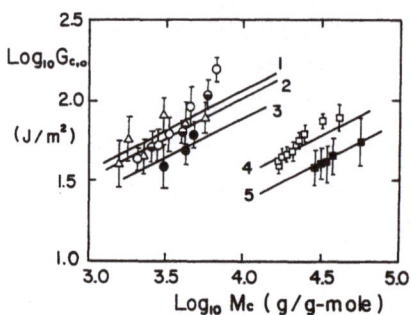

Fig. 1. Threshold tear strength $G_{c,o}$ vs. molecular weight M_c of network strands (3). 1, polybutadiene, M_o=13.5a.m.u., (Δ); 2, cis-polyisoprene, M_o=17a.m.u., (0); 3, trans-polyisoprene, M_o= 17 a.m.u., (●); 4, polydimethyl siloxane, M_o=37a.m.u., (□); 5, phosphonitrilic fluoroelastomer, M_o=185a.m.u., (■).

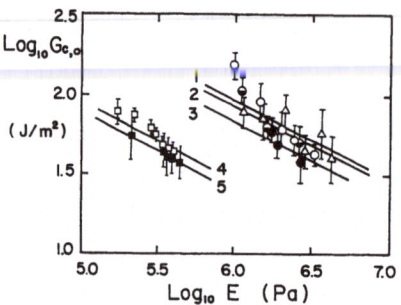

Fig. 2. Threshold tear strength $G_{c,o}$ vs. Young's modulus E (3). Symbols as in Figure 1.

Under normal conditions the tear strength is many times larger than the small threshold value $G_{c,o}$, about 50–100 J/m^2, because of energy expended in various dissipative processes. For simple viscoelastic materials the tear strength is governed by the local viscosity, i.e., by the rate ϕ of Brownian motion of molecular sub-units. In turn, ϕ is directly related to the temperature difference $T-T_g$, where T is the test temperature and T_g is the glass transition temperature of the elastomer (4):

$$\log_{10}(\phi_T/\phi_{T_g}) = 17.6(T-T_g)/(52+T-T_g) \qquad (12)$$

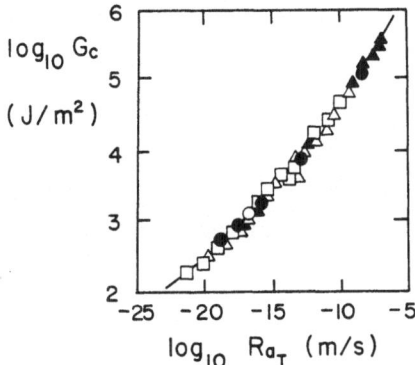

Fig. 3.   Tear strength G_c plotted against the effective rate of tearing at T_g (5). Results are shown for six butadiene-styrene (SBR) and butadiene-acrylonitrile (NBR) elastomers, with T_g ranging from -30°C to -80°C.

where ϕ_{T_g} denotes the rate of sub-unit motion at T_g, about 0.1 jumps/sec. Using a scaling factor $a_T = \phi_{T_g}/\phi_T$ for the rate of tearing, measurements of tear strength for several elastomers at many temperatures can be super-imposed to give a master curve for tear strength as a function of the effective rate of tearing at T_g, Figure 3. This demonstrates that the tear strength depends only upon $T-T_g$, and not upon the local chemical structure of the elastomer except insofar as that determines the value of T_g (5).

Far above T_g, under threshold conditions, the tear strength depends significantly upon the molecular structure as discussed earlier, and there is some evidence that the same relative differences are maintained under non-equilibrium conditions. But the primary variable for determining the tear strength is $T-T_g$.

The question now arises; which fracture processes, if any, are <u>strongly</u> affected by the local chemical structure? Two examples are considered below: tearing and crack growth, and abrasive wear. Under certain conditions these failure processes are found to depend upon particular features of the elastomer molecule and they are therefore distinctly different, even for closely-related chemical structures. Natural rubber can usefully be compared with <u>cis</u> 1, 4-polybutadiene in this respect, because, although their chemical structures are superficially similar, large differences are observed in their resistance to tearing and in the mechanism of wear.

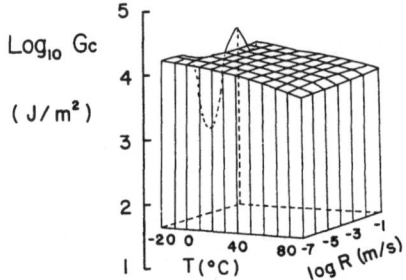

Fig. 4.   Tear strength $\underline{G}_c$ of natural rubber as a function of test temperature $\underline{T}$ and rate $\underline{R}$ of tearing (6).

3. Tearing and Crack Growth in Strain-Crystallizing Elastomers

Certain elastomers, notably natural rubber, crystallize on being stretched by several hundred per cent. They become much stiffer, and rather inelastic due to delays in crystallization and in melting on release. At a crack tip, rubber is highly stressed even when the overall strain is relatively small. The loss of energy associated with crystallization and, later, melting in this region leads to enhanced tear strength at low rates and high temperatures (6), and a much improved resistance to crack growth under repeated stressing (7), as shown in Figures 4, 5 and 6, in comparison with a non-crystallizing elastomer. Strain-induced crystallization is thus a specific, and highly desirable, feature of elastomers. The physical and chemical factors responsible for it are discussed below.

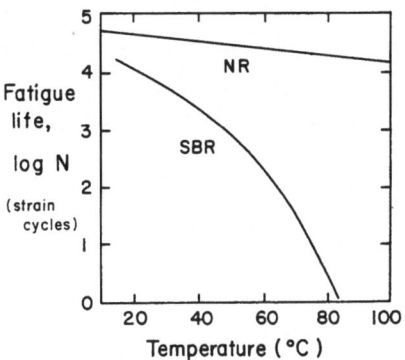

Fig. 5.   Fatigue life <u>N</u> of natural rubber (NR) and a butadiene-styrene
rubber (SBR) plotted against the test temperature <u>T</u> (7).

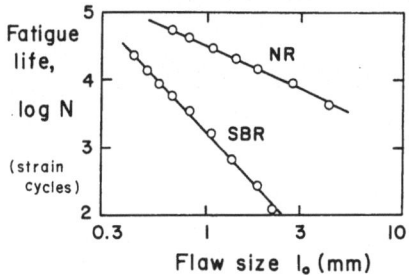

Fig. 6.   Fatigue life <u>N</u> of natural rubber (NR) and a butadiene-styrene
rubber (SBR) plotted against the depth l_o of an initial edge
crack (7).

4. Strain-Induced Crystallization

The phenomenon of rapid crystallization in the strained state and rapid melting on release can be attributed to three main causes:

(i) In the unstrained state the crystal melting temperature T_m lies below ambient temperature and the material is therefore non-crystalline.

(ii) On stretching, the melting temperature is raised markedly, to values well above ambient, so that crystallization is thermodynamically favored and the free energy change on fusion is large.

(iii) The glass transition temperature T_g is quite low, well below ambient. Molecular sub-units are consequently highly mobile at ambient temperature and are able to enter the crystalline state rapidly when the free energy change is favorable.

Many polymers have low values of T_g and therefore satisfy condition (iii). However, many of them have either relatively large latent heats of fusion h or, more commonly (8), relatively small latent entropies of fusion s, so that their melting temperatures T_m, given by

$$T_m = h/s, \tag{13}$$

lie above ambient temperature. They are therefore normally crystalline in the unstrained state. Some common examples are: polyethylene, polyethylene oxide, trans 1, 4-polybutadiene and trans 1, 4-polyisoprene (see Table 1). These materials do not satisfy condition (i).

Table 1: Melting temperatures T_m and latent heats of fusion h for some representative crystallizing polymers.[a]

	T_m (^{0}C)	h (kJ/kg)
Polyethylene	141	280
Polyethylene oxide	66	200
Trans 1, 4-polyisoprene	74	190
Trans 1, 4-polybutadiene	148	187
Cis 1, 4-polybutadiene	6	163
Trans 1, 4-polychloroprene	80	95
Cis 1, 4-polyisoprene	30	65

a Taken from reference 8 and "Physical Constants of Linear Homopolymers", by O. G. Lewis, Springer-Verlag, New York 1968.

Of the remaining elastomeric materials, some will meet condition (ii) more successfully than others. The reasons for this can be readily deduced from Flory's approximate theoretical treatment for the melting temperature T_{m_λ} of crystallites in a molecular network held at a stretch ratio λ (9). A molecular sub-unit entering a crystallite from a strand in a stretched molecular network undergoes a smaller loss of configurational entropy than from the unstretched state because its configurational entropy has already been lowered somewhat by stretching. The reduction Δs in the entropy of fusion can be evaluated from the statistics of deformed and undeformed molecular networks. The result leads, by means of equation 13, to a predicted increase in the melting temperature on stretching (9):

$$(hqM_o/R) \ (T_m^{-1} - T_{m,\lambda}^{-1}) = (6/\pi n)^{\frac{1}{2}}\lambda - (\lambda^2 + 2\lambda^{-1})/2n \qquad (14)$$

where h is the latent heat of fusion per gram, R is the gas constant, and λ is the tensile stretch ratio applied to the network. This relation is found to give reasonably satisfactory predictions of the melting temperatures $T_{m,\lambda}$ at moderate extensions, in the range 100 to 400 per cent (λ = 2 to 5) and for different degrees of crosslinking, represented by different values of the molecular strand length n. Some typical results are shown in Figures 7 and 8 (10,11).

Both calculated and observed increases in melting temperature on stretching are found to be larger for some elastomers than for others, and for natural rubber the effect is largest of all. The reason for this lies in the unusually small value of the latent heat of fusion h for cis-polyisoprene, Table 1. As equation 14 indicates, the increase in T_m on stretching is inversely related to h. Thus, the smaller the value of h the greater will be the tendency to exhibit strain-induced crystallization. An abnormally low value of h for natural rubber appears to be associated in part with the relatively small change in density that accompanies crystallization and in part with the absence of strong interatomic binding in the unit cell. Whatever the exact cause, the low value of h is clearly responsible for the facility with which natural rubber crystallizes on stretching.

5. Abrasive Wear

Wear of rubber under sliding conditions resembles small-scale tearing (12). Indeed, it has been treated as cumulative tearing - a mechanical fatigue process - taking place under the repeated action of frictional forces. A quantitative relationship has been derived in this way for the rate of wear in terms of the rate of crack growth under repeated stressing (13,14). When the rubber is rather tough and wear resistant, however, there is evidence of chemical deterioration during sliding, in addition

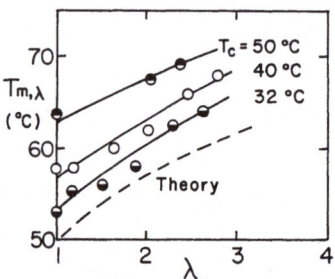

Fig. 7. Melting temperatures for crosslinked <u>trans</u> 1, 4-polychloroprene, held at various stretch ratios $\underline{\lambda}$ and crystallized at various temperatures $\underline{T_c}$ (10). Theoretical relation from equation 14 for increase in $\underline{T_m}$ with $\underline{\lambda}$.

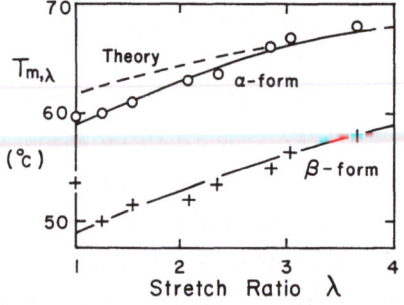

Fig. 8. Melting temperatures for crosslinked <u>trans</u> 1, 4-polyisoprene, held at various stretch ratios $\underline{\lambda}$ (11). Theoretical relation from equation 14 for increase in $\underline{T_m}$ with $\underline{\lambda}$.

to tearing (15,16). When this deterioration is extensive the rubber and abrading surface become covered with an oily decomposition product and the tearing process is altered, if not stopped altogether.

An example of a particle of wear debris torn from a rather weak material, an unfilled butadiene-styrene (SBR) vulcanizate, is shown in Figure 9. It has characteristically rough, torn surfaces. In contrast, the particle shown in Figure 10, obtained from a carbon-black-filled SBR vulcanizate, has a smooth, shiny appearance and the surface is sticky, as if covered with an oily or tarry film. The debris from carbon-black-filled natural rubber vulcanizates is even more highly degraded, so that the individual particles can hardly be distinguished in this case. On the other hand, the debris from carbon-black-filled cis 1, 4-polybutadiene materials is finely-divided and particulate showing no signs of decomposition and every indication of having been mechanically torn away from the rubber surface. Thus, the wear process for reinforced elastomeric materials of roughly equal hardness and friction coefficient, and of comparable tear strength and tensile strength, differs strikingly in character from one polymer to another. These differences must be ascribed to different chemical features of the elastomers.

It should be pointed out at this stage that the formation of an oily degraded surface layer is not necessarily a beneficial feature. If the layer is readily removed from the rubber, then further deterioration can proceed rapidly. Indeed, if in the early stages of decomposition, the rubber is rendered softer and weaker, it will be torn away more easily and the rate of wear will be correspondingly greater than in the absence of general molecular scission. On the other hand, if the liquidlike film is viscous, tarry, and adhesive, it appears to be retained on the rubber surface to act as a protective layer. The rate of wear is then much reduced.

In order to account for the formation of a degraded surface film in some instances, a number of possible chemical processes can be hypothesized:

(i) Thermal decomposition, as a result of frictional heating.

(ii) Oxidative scission of the molecular network, possibly accelerated by frictional heating.

(iii) Mechanical rupture of the molecular network, followed by internal and external reactions of the polymer radicals generated in this way.

The first process need not be considered further here, because all of these elastomers are more or less equally susceptible to thermal decomposition whereas they do not all degrade during sliding. The second process is also probably not the main mechanism of decomposition because some elastomers show frictional decomposition even in inert atmospheres (15).

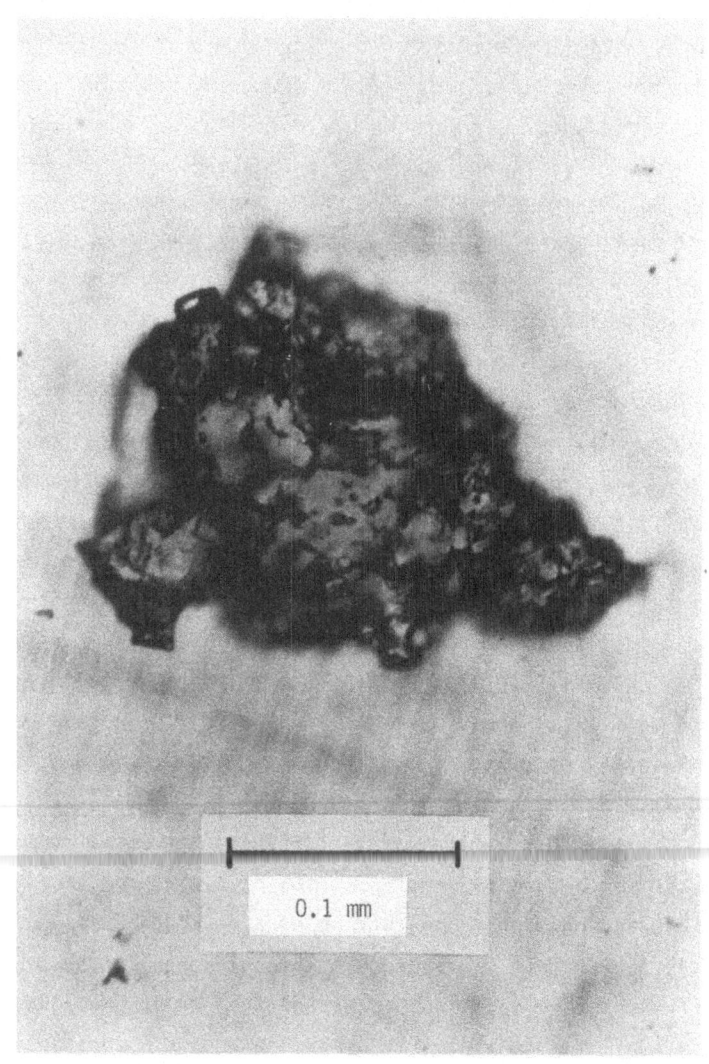

0.1 mm

Fig. 9.   Photograph of wear debris from an unfilled SBR vulcanizate.

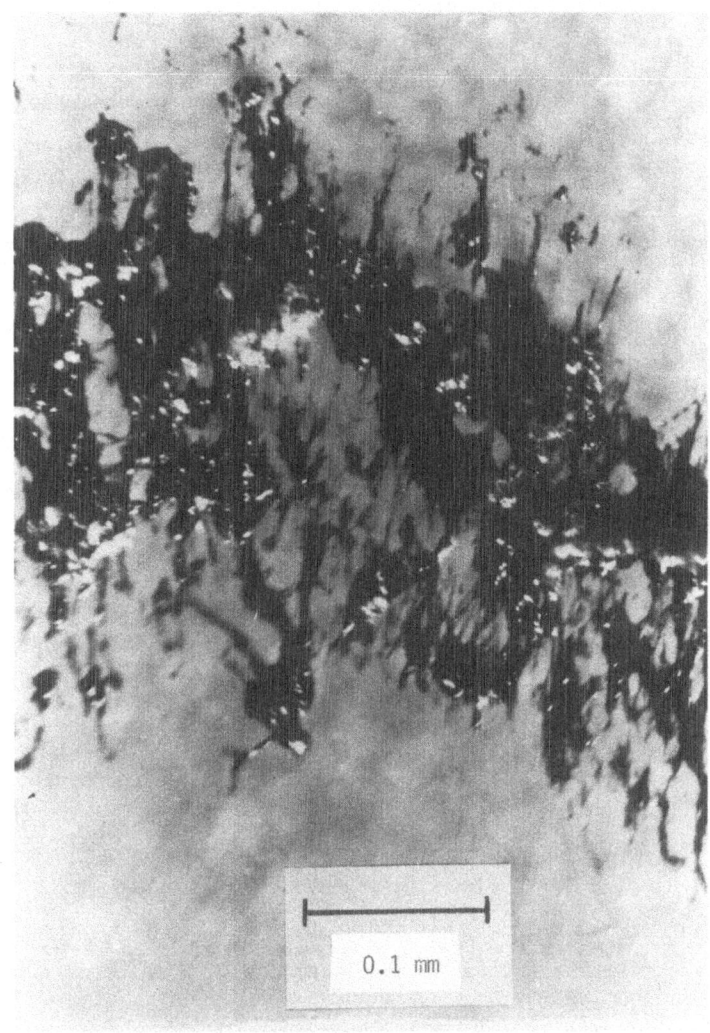

0.1 mm

Fig. 10.   Photograph of wear debris from a c rbon-black-filled SBR
vulcanizate.

The third process, however, does appear to be the basic mechanism of molecular decomposition during sliding. A rather good correlation is found to hold between the degree of decomposition observed during frictional sliding, both in air and in an inert atmosphere, and corresponding changes in molecular weight when the original elastomer (before crosslinking) is subjected to continuous mechanical shearing (15,16). For example, polybutadiene forms rather reactive macroradicals by molecular scission, and then undergoes crosslinking reactions, so that both in the shearing of the uncrosslinked polymer and the frictional sliding of the reinforced and crosslinked polymer, the product of mechanochemical processes tends to become a crosslinked solid. In contrast, natural rubber forms a resonance-stabilized macroradical by molecular fracture, which, in the presence of oxygen, forms a peroxy radical and then a hydroperoxide by H abstraction so that the original chain fracture is rendered permanent. Indeed, subsequent oxidation steps may cause scission of other chains as well. Thus, the product of mechanical rupture of the molecular network in this case tends to become liquidlike rather than solid.

These considerations account for the formation of a viscous liquid film on certain materials, and not on others, during frictional sliding. Moreover, the properties of the film, its viscosity and adhesiveness, will clearly depend upon the detailed reactions initiated by mechanical rupture of the elastomer molecules. They will therefore differ from one elastomer to another and they will also depend upon the particular ingredients used in the rubber mix formulation, especially when these substances are themselves able to participate in free-radical reactions.

Many of the differences encountered in the wear behavior of practical rubber compounds can thus be accounted for in a qualitative way when the specific chemical process involved in wear is recognized.

Acknowledgements

This review was prepared in the course of a research program supported by the Office of Naval Research (Contract No. N00014-85-K-0222). An earlier version was presented at the conference Natural Rubber: Towards The Year 2000, held in Saltillo, Mexico, July 7-11, 1980, under the auspices of the Centro de Investigacion en Quimica Applicada. The author is indebted to Dr. C.T.R. Pulford of the Goodyear Tire and Rubber Company Research Laboratories for his helpful comments.

References

1. L. R. G. Treloar, "Physics of Rubber Elasticity", 2nd. Ed., Oxford University Press, London, 1958.

2. G. J. Lake and A. G. Thomas, Proc. Roy. Soc. London A300, 108 (1967).

3. A. N. Gent and R. H. Tobias, J. Polymer Sci., Polymer Phys. Ed. 20, 2051 (1982).

4. J. D. Ferry, "Viscoelastic Properties of Polymers", 3rd. Ed., Wiley, New York, 1980.

5. L. Mullins, Trans. Inst. Rubber Industry 35, 213 (1959).

6. H. W. Greensmith and A. G. Thomas, J. Polymer Sci. 18, 189 (1955).

7. G. J. Lake and P. B. Lindley, Rubber J. Internatl. Plastics 146 (10), 24 (1964); 146 (11), 30 (1964).

8. L. Mandelkern, "Crystallization of Polymers", McGraw-Hill, New York, 1964.

9. P. J. Flory, J. Chem. Phys. 15, 397 (1947).

10. A. N. Gent, J. Polymer Sci. A3, 3787 (1965).

11. A. N. Gent, J. Polymer Sci., Part A2 4, 447 (1966).

12. A. Schallamach, Rubber Chem. Technol. 41, 209 (1968).

13. A. G. Thomas, J. Polymer Sci: Symposium No. 48, 145 (1974).

14. E. Southern and A. G. Thomas, Plastics and Rubber: Materials and Applications 3 (4), 133 (1978).

15. A. N. Gent and C. T. R. Pulford, J. Materials Sci. 14, 1301 (1979).

16. A. N. Gent and C. T. R. Pulford, J. Appl.Polymer Sci. 28, 943 (1983).

THE STATISTICAL MECHANICS OF ENTANGLED NETWORKS

S.F. Edwards

Cavendish Laboratory
University of Cambridge
Madingley Road, Cambridge CB3 OHE, U.K.

It is shown that the problem of formulating the statistical thermodynamics of networks can be resolved by the use of ideas from quantum statistical mechanics by appropriate generalization. Examples are given in terms of gaussian and liquid crystal polymer networks.

1. Introduction

The problem of understanding networks is a central one in chemical physics, and at each point poses problems. The way networks are formed is not clear and a whole new discipline of aggregation theory is coming into being. Once formed, it is clear that the standard formulae of Gibbsian statistical mechanics do not apply since a series of permanent, but statistically defined, constraints apply. Although these constraints are intuitively obvious, the problem of formally writing them down is unsolved and at a deep mathematical level is likely to remain so. Nevertheless important progress is being made both by theoretical advances, by the creation of well characterized networks, and by accurate experiments on them.

This paper will concentrate on theoretical aspects when the network is specified i.e. we do not enter the problem of predicting what kind of network appears in particular experimental circumstances, simply accepting simple specifications as given, and exploring the consequences.

2. Statistical Mechanics

The standard formulation of statistical mechanics perfected by Gibbs writes the free energy in the form

$$ e^{-F/k_B T} = \int e^{-H/k_B T} \tag{2.1} $$

where H is the energy of the system and the integral is over all degrees of freedom. However in a network problem the cross links of the network are permanent and it follows that the free energy as obtained from (2.1) must be a function of their specification. In practice there will be so many cross links that one will observe the average F, so (2.1) is modified to

$$\bar{F} = -k_B T \sum_c p_c \log \int e^{-H/k_B T} \delta_c \qquad (2.2)$$

where δ_c is a symbol meaning that the integration is over a particular network, and p_c is the probability of this network occuring. Formula (2.2) is well known in statistical physics other than in rubber theory, for example the properties of an alloy, quenched rapidly from a very high temperature to one where the components do not diffuse, has 'c' to mean the random positions of the atoms and p_c is a constant in the case of the initial temperature high and final low. Although this formula (2.2) is implied in all studies of molecular networks it is seldom stated explicitly. From it approximations can flow.

A particularly simple specification of the network is to suppose that up to some initial time the cross links are sliding freely along the chains and the chains are transparent to each other and they are then frozen. The probability of finding a particular configuration is then just the Gibbs function in the initial configurations i.e.

$$p_c = \int_{\text{original}} \delta_c \, e^{(F_0-H)/k_B T} \qquad (2.3)$$

$$\bar{F} = -k_B T \sum_c \int_{\text{orig}} \delta_c \, e^{(F_0-H)/k_B T} \log \int_{\text{final}} e^{-H/k_B T} \delta_c , \qquad (2.4)$$

where 'final' means that the integral is taken over the system in its final state which in general is strained into a new configuration, without loss of generality a box with sides a originally and strained to $\lambda; a$ finally, where a^3 is the volume of the system. But the formulation in (2.2) is quite general and other specifications are possible, for example the network could be made up of chains of equal length or of some given distribution of lengths (of which (2.3) is a particular sample), and different techniques are appropriate for different cases. One general observation is however valuable. The fact that

$$A^n = 1 + n \log A + O(n^2) \qquad (2.5)$$

suggests that it might be fruitful to consider

$$\log \int e^{-H/k_B T} \delta_c = \text{coeff. } n \text{ in}$$

$$\left(\int e^{-H/k_B T} \delta_c \right)^n = \int e^{-\sum_1^n H^{(\alpha)}/k_B T} \prod_1^n \delta_c^{(\alpha)} , \qquad (2.6)$$

a point further emphasized in the case of (2.3) and (2.4) where one can write [1]

$$F(n) = F_0 + n\bar{F} + O(n^2) \tag{2.7}$$

$$F(n) - F_0 = -k_B T \log \sum_c \int\int_{(o)(F)} \cdots \int_{(F)} e^{-\sum_o^n H^\alpha/k_B T} \prod_o^n \delta_c^{(\alpha)} \tag{2.8}$$

where the integral over the (o) degrees of freedom is the original configuration, and the (1)...(n) configurations are strained. This
amounts to saying that if we consider the free energy of the system of the
initial network and n strained networks as a system in 3 + 3n dimensions
in a box of sides (1,1,1) and $(\lambda_1, \lambda_2, \lambda_3)n$ times, then the coefficient of n
is required energy. This 'replica' method is built on shaky mathematical
foundations, but for polymer systems seems able to produce answers far
more quickly than other methods. It is to be emphasized however that almost
all problems in the end are attacked by approximations, and these can be
applied at the level of (2.2).

We now consider the physical content to be put into (2.2) and the
mathematics of the various models available.

3. Polymers and Constraints

In this section we study what specifications the network may have.
It will be made of polymer molecules between cross links. In free space
the chains will be specified either by all the links

$p(R_1 \cdots R_N)$ or just the end points

$p(R_1, R_N)$,

but the possibility of rigidity implies that 'R' has to be interpreted as
coordinate and tangent and curvature. Two extreme cases emerge, the
fully flexible chain

$$p = \left(\frac{3}{4\pi N l^2}\right)^{3/2} exp\left(-\frac{3(R_1 - R_N)^2}{2 N l^2}\right) \tag{3.1}$$

and the rigid rod

$$p = \left(\frac{1}{4\pi l^2}\right) \delta\left(|R_1 - R_N| - l\right) \tag{3.2}$$

If there are interactions (3.1) will not be correct but in this
paper we will assume melt or θ conditions which are fully accessible
physical states. Form (3.2) is a full specification of a rod, but (3.1)
is not a full specification of a flexible chain R_1, $R_2 \cdots R_N$.

This might be

$$\mathcal{N} exp\left[-\frac{3}{2 l^2} \sum_1^{N-1} (R_i - R_{i+1})^2\right] \tag{3.3}$$

which permits the transition to a path integral form

$$exp \left[-\frac{3}{2\lambda} \int_0^L \left(\frac{\partial \underline{R}}{\partial s}\right)^2 ds \right] \qquad (3.4)$$

which can be generalized easily to a worm like chain, or a chain with torsional rigidity by adding $\ddot{\underline{R}}^2$ or $\dot{\underline{R}}^2$ terms respectively [2]. Such formulations, though still treated as a novelty in polymer science, permit a rapid transition to differential equation forms. (Beware however: forms like (3.2) do not).

The chains have cross link points and entanglements. Cross links which literally have cross linked chains are easily specified. If chain R_1 (S_1) meets chain $R_2(S_2)$ at S_1^2, and S_2^1 equivalently the "δ_c"

$$\delta \left(\underline{R}_1 \left(S_1^2 \right) - \underline{R}_2 \left(S_2' \right) \right) \qquad (3.5)$$

If there are multifunctional units of chains whose ends are R_{10} and R_{1f} then

$$\delta \left(\underline{R}_{10} - \underline{R}_{20} \right) \delta \left(\underline{R}_{10} - \underline{R}_{30} \right) \qquad (3.6), \text{ has chains 1,2,3 meeting.}$$

Entanglements are more difficult since there seems no comprehensive way of describing them. A fruitful model is that of slip links i.e. suppose a cross link (3.5) can slip an amount η , then the constraint (3.5) is replaced by (3.6)

$$\frac{1}{\eta^2} \int_{-\eta/2}^{\eta/2} \int_{-\eta/2}^{\eta/2} \delta \left(\underline{R} \left(S_1^2 + \delta_1 \right) - \underline{R} \left(S_2' + \delta_2 \right) \right) d\delta_1 d\delta_2 \quad (3.7)$$

Rather remarkably for a network of a given chain density one can actually calculate η since when a chain moves in a 'lattice' of other chains the paths available to it are equivalent to a Cayley tree i.e. to a random walk which always retraces its steps. Details are given in refs 4 and 5 (see also ref 6). Since it appears the only tractable representation available to us, we adopt (3.7) and take entanglements to be equivalent to slip links i.e. are characterized by an η and are quadrifunctional.

Now we are in a position to create mathematical formalisms for the problem. The problem of the network has a strong affinity to the many body problems of quantum mechanics and we can adopt some of the language of that discipline. In particular it is well known that the deepest level of quantum theory demands second quantisation, but under certain lucky circumstances one may get away with first quantisation (note that one can always write problems properly expressed in second quantisation by an infinitely elaborate first quantisation. I don't mean this, but refer to problems like a single electron in the field of fixed random scatterers which is a true first quantisation problem)

4. Quantization methods

(4.1) First quantisation methods These only work well for quadrifunctional networks, for in that case one can regard the whole network as one vast chain with simple cross links of type (3.5) and slip links of type (3.7) (chain ends are easily added if required). This chain is called $\underline{R}(s)$ and has a length L equal to the total length of all the original polymers in the network. If there are N_c cross links, there are N_c constraints of type (3.5) The replica formula now comes into its own since

$$\Pi \delta_c = \prod_{\{ij\}} \prod_{\alpha=0}^{n} \delta\left(\underset{\sim}{R}^{(\alpha)}(s^i) - \underset{\sim}{R}^{(\alpha)}(s^j)\right) \quad (4.1)$$

where "C" is the incidence labels, s^i, s^j is the cross link which links monomer s^i with monomer s^j, and whereas in formula (2.2) the pairs s^i, s^j have to be averaged over _after_ integrating over the degrees of freedom of the chain, we can now invert the order of averaging to obtain

$$\sum \prod \delta \longrightarrow \int d\mu \, exp\left\{\mu \prod_{\alpha=0}^{n} \iint \delta(R^{(\alpha)}/s) - R^{(\alpha)}/s')ds\,ds' - N_c \log \mu\right\} \quad (4.2)$$

where μ is a fugacity variable which in practice is obtained by minimizing the free energy i.e. all the value of the integral has at the saddle point, so no integral needs to be shown. Thus finally[1]

$$e^{-F(n)/k_BT} = \int_{\cdots}\int \prod \delta R^{(\alpha)} exp\left[-\frac{3}{2\ell}\int\left(\frac{\partial R^{(\alpha)}}{\partial s}\right)^2 ds + \mu \iint \rho(R^{\alpha}-R^{(\alpha)})ds\,ds' - N_c \log \mu\right] \quad (4.3)$$

where δR is shorthand for integrating over every monomer position i.e. the path integral for $R(s)$. To allow for slip links one adds a further

$$\frac{\tilde{\mu}}{\eta^2}\int_{-\eta/2}^{\eta/2}\int_{-\eta/2}^{\eta/2}\delta\left(R^{(\alpha)}(s_1+\delta_1^{(\alpha)}) - R^{(\alpha)}(s_2+\delta_2^{(\alpha)})\right)\prod_\alpha d\delta_1^{(\alpha)}d\delta_2^{(\alpha)}ds_1 ds_2 \quad (4.4)$$

in the exponent. Note the (α) label on the slip variable $\delta^{(\alpha)}$ which is crucial. The problem is how concisely and exactly formulated. If $\mathcal{R}$ is the $3 + 3n$ dimensional vector $R^{(0)} \ldots R^{(n)}$

we have
$$e^{-F(n)/k_BT} = \int_{\cdots}\int \delta\mathcal{R} \, e^{-\mathcal{L}([\mathcal{R}])} \quad (4.5)$$

where $\mathcal{L}$ is explicit.

If the functionality of the cross links is not four, or if there is some given distribution of lengths, or if the chains are remote from gaussian, this method is weak. The other method is that of:

(4.2) Second Quantization: This technique appeared long ago with the work of Fixman. The first use in network problems was also some time ago,[7] but this method proved cumbersome and here we present a new approach. It was stimulated by the experimental observation of Bastide and Boué [8] that a possible, if rather strained, explanation of the neutron scattering from networks is to suppose that the chains are rather stiff and the straining process is analogous to a firetongs wherein the length of segments remains the same, but the length of the pantograph increases. (A much more plausible explanation is heavy clumping, a phenomenon already well known[9], but difficult to quantify). But our present purpose is to discuss possible situations rather than specialise to that experiment. If we have a system of rods, provided their functionality is less than (or marginally equal to) four, the network has an entropy; what is it? Another problem is that of a given distribution of segment lengths, which is awkward in the first quantisation formulation. We adopt the functional integration structure which generalizes the simple formulae[2]

273

$$\frac{\int x^2 e^{-x^2/2a}\, dx}{\int e^{-x^2/2a}\, dx} = a, \quad \frac{\int x_i x_j\, e^{-\sum x_m^2/2a_m}\, dx}{\int e^{-\sum x^2/2a}} = a_i\,\delta_{ij}$$

$$\frac{\int x_i x_j\, e^{-\frac{1}{2}\sum\sum x_m\, a_{ml}^{-1} x_l}\,\Pi dx}{\int e^{-\frac{1}{2}\sum\sum x a^{-1} x}\,\Pi dx} = a_{ij}, \quad \frac{\int x_i x_j\, x_p x_q \cdots\, e^{-\frac{1}{2}\sum x a^{-1} x}\,\Pi dx}{\int e^{-\frac{1}{2}\sum x a^{-1} x}} = \sum_{perms} a_{ij}\, a_{pq} \cdots$$

$$(4.6)$$

to the complex form

$$\mathcal{N}\int z_i z_j^* z_p z_q^* \cdots e^{-\sum z a^{-1} z^*}\,\Pi\, dx\, dy = \sum_{\substack{perms \\ i,j}} a_{ij}\, a_{pq} \cdots \qquad (4.7)$$

and hence to the continuous extension of i, j

$$\mathcal{N}\int \phi(r_i)\, \phi^*(r_j)\, \phi(r_p)\, \phi^*(r_q)\, \exp\left(-\frac{1}{2}\int \phi(r)\, \phi^*(r)\, d^3r\right)\, \delta\phi$$

$$= \sum_{\substack{perm \\ i\,p\cdots \\ j\,q\cdots}} \delta(r_i - r_j)\,\delta(r_p - r_q) \cdots \cdots \int \exp\left(-\frac{1}{2}\int \phi\phi^*\right)\,\delta\phi$$

$$(4.8)$$

Thus if we consider N_p polymers which have g configurations when their end points are at r and r', $g = g(r-r')$ and these ends are cross linked by N_x cross links with functionality m, the number of configurations (and hence the probability) is given by

$$\mathcal{N}\int (\delta\phi)\left(\int\int \phi(r)\, g(r-r')\, \phi(r')\, d^3r\, d^3r'\right)^{N_p}\left(\int \phi^{*m}_{(r)}\, d^3r\right)^{N_x}$$

$$\exp\left(-\frac{1}{2}\int \phi\phi^*\, d^3r\right) \qquad (4.9)$$

where $\quad \mathcal{N}^{-1} = \int \delta\phi\, \exp\left(-\frac{1}{2}\int \phi\phi^*\, d^3r\right) \qquad (4.10)$

and $\delta\phi$ corresponds to the integral over all functions ϕ; simply understood by thinking of the space of $\underline{r}$ replaced by a lattice r_n, $\delta\phi$ is $\Pi\, d\phi_n$ where ϕ_n is $\phi(r_n)$ and then let the lattice spacing go to zero. Integrals of this type are much discussed in the literature. Our use of them will be simple. Notice that unless the number of ϕ and ϕ^* are the same, the integral vanishes. Thus we must have $2N_p = m N_x$. One can picture

$$\int \phi g \phi : \quad \xleftarrow{\quad g \quad} \qquad \int \phi^{*m} \qquad (m = 3, etc)$$

$$(4.11)$$

The integral joins them up

If we have rods or gaussian coils we just use the appropriate g. If there is a mixture of $N_p^{(1)}$ of g_1 and $N_p^{(2)}$ of g_2, then one does the trivial generalization. Thus we are faced with

$$\int \delta\phi\,\delta\phi^*\, \exp\left[A([\phi]) + B([\phi^*]) - C([\phi\phi^*])\right] \qquad (4.12)$$

and can in principle calculate correlation functions like

$$\int \phi(R_1) \phi^*(R_2) \, e^{\mathcal{L}(\phi,\phi^*)} \, \delta\phi \, \delta\phi^* \qquad (4.13)$$

which will give the freedom of the network. Thus first quantisation takes the monomer points of the polymer $R_1 \cdots R_N$ and turns them into $\underline{R}(s)$ which gives path integrals (which, though we have not explored it here gives differential equations). Second quantisation has lost the internal variables altogether, but is very flexible in its generality. Notice that unlike quantum field theory our formulation is <u>not</u> hermitian. This is permitted for our network problem and greatly enhances the power of the method. As an illustration consider a trifunctional network of rods. The entropy is given by

$$e^{S/k_o} = \mathcal{N} \int \delta\phi \, \delta\phi^* \, exp\left[N \log \int \phi g \phi + \frac{2N}{3} \log \int \phi^{*3} - \frac{1}{2} \int \phi \phi^* \right] \quad (4.14)$$

where

$$g = (4\pi l^2)^{-1} \, \delta\left(|r - r'| - l\right) \qquad (4.15)$$

which in fourier transform is

$$g(k) = \frac{\sin kl}{kl} \qquad (4.16) \text{ so that}$$

$$\int \phi g \phi = \frac{1}{(2\pi)^3} \int \phi_k \phi_{-k} (\sin kl)/kl \, d^3k \qquad (4.17) \qquad (4.17)$$

For a mixture of N_1 rods length l_1 and N_2 of length l_2 one replaces

$$N \log \int \phi g \phi \quad \text{by} \quad N_1 \log \int \phi \, g_1 \phi \; + N_2 \log \int \phi g_2 \phi$$

where

g_1 contains l_1 and g_2, l_2 and replaces $\frac{2N}{3}$ by $\frac{2}{3}(N_1 + N_2)$.

The reader will see how this readily generalizes to any mixture of any kinds of chains and cross links.

We now consider particular examples of polymer systems and of the methods.

5. Applications and models

One makes progress in modelling by studying extreme cases. Let us therefore consider the intuitive picture of a network and see what extreme cases are available. The general problem is that of given chain and given cross link contributions, but entanglements are imposed i.e. we cannot remove them or control them in anyway. The first model is to ignore the entanglements and consider phantom chains. This still leaves a complex problem, so assume the chains are gaussian. We then reach the classic problem of James and Guth[10]. It can be resolved by both our 4.1 and 4.2 methods very easily. The 4.1 method is given in ref(/) and 4.2 method will be reached later. There is often a discussion of the gaussian network in terms of the entropy of the chain i.e. for fixed cross link points, and the entropy involved when the cross link points move. The simplest version of the network is the Wall theory, developed by Flory[11], in which the cross link points are taken to move affinely and the entire entropy lies in the chains. It is well known to under estimate the entropy by a factor of two. It is not a consistent model for there is no way that one can fix cross links whilst leaving entropy in the chains. However a reverse system is possible: by taking polymers to be rods, all the entropy lies in the freedom of the cross links and none in the polymers, and this problem is available by method 4.2. Finally to include entanglements one can employ the slip link method, but this fits in much better with method 4.1, and although there are ideas possible for 4.2 (i.e. it does not seem impossible to include[12] entanglements it has yet to be done). It has been suggested by Flory[12] that the model of most of the entropy lying in the freedom of the cross link is a way to handle entanglements, but this is not consistent, and if one can succeed in doing the slip link calculation there seems no reason to restrict oneself.

5i. Slip link models. The analysis of this problem involves quite complex algebra, but since it is published elsewhere[13] we content ourselves with the physical arguments. The phenomenon being addressed is that whereas a gaussian network always gives the classical form of the free energy proportional to $\sum \lambda_i^2$, a real network, relative to

this, softens on straining at first, but then strengthens. A simple model of this is to note that if a chain has a slip link representing its entanglement with another chain, on straining the amount of slip available increases, and therefore the stress decreases[3]. To be precise, the formula for the free energy becomes modified to

$$\frac{1}{2} \sum_i \left\{ \frac{\lambda_i^2 (1+\eta)}{1+\eta \lambda_i^2} + \log \left(1 + \eta \lambda_i^2\right) \right\}$$

The segments between slip links are quite small, a fact which can be ascertained from the plateau modulus which may be taken to give the total number of links, cross and slip. It follows that at large deformation one reaches the region of fully extended segments and a resulting stiffening. The complete approximation gives[13]

$$\frac{F}{k_B T} = \frac{1}{2} N_c \left\{ \frac{\sum (1-\alpha^2)\lambda_i^2}{1 - \alpha^2 \sum \lambda_i^2} + \log \left(1 - \alpha^2 \sum \lambda_i^2\right) \right\}$$

$$1 \frac{1}{2} N_s \left[\sum_i \left\{ \frac{\lambda_i^2 (1+\eta)(1-\alpha^2)}{(1+\eta \lambda_i^2)(1-\alpha^2 \sum \lambda_i^2)} + \log \left(1 + \eta \lambda_i^2\right) \right\} \right.$$

$$\left. + \log \left(1 - \alpha^2 \sum \lambda_i^2\right) \right]$$

This formula agrees well with experiment as is shown in ref. 13.

5ii. Rod models It is well known that non quadratic functional integrals are difficult if not impossible to integrate accurately, but it is also known that quite crude methods give effective approximations. For example the statistical mechanics of an AB alloy can be transformed into the evaluation of

$$\int \delta \phi \exp \left[\sum_{i \, on \, lattice} \log \cosh \phi(r_i) - \frac{k_B T}{2} \int \phi \, V^{-1} \phi \right] \tag{5.1}$$

where r_i are lattice points and V^{-1} is the operator which is the inverse of the interaction potential, for example if the potential is $v \exp(-\alpha r)/r$ then $V^{-1} = \frac{1}{(2\pi)^3} v^{-1} (k^2 + \alpha^2)$.

The mean field method just looks at the saddle point

$$\left(\nabla^2 - \alpha^2\right)\bar{\phi} = v \tanh\frac{\bar{\phi}}{k_a T},\tag{5.2}$$

where the lattice has been blurred out. If $\bar{\phi}$ is taken a constant and small we have

$$-\alpha^2\bar{\phi} = \frac{v}{kT}\left(\bar{\phi} - \frac{1}{3}\bar{\phi}^3 \cdots\right)\tag{5.3}$$

$$\bar{\phi} = 0 \quad \text{or} \quad \bar{\phi} = \alpha\left(1 - \frac{T_c}{T}\right)^{1/2},\tag{5.4}$$

i.e. we have the Bragg-Williams alloy theory. Apply this to our rod lattice we get

$$N\int g\,\phi = \phi^*\left(\int \phi g\phi\right)\tag{5.5}$$

$$N\,\phi^{*\,m-1} = \phi\left(\int \phi^{*m}\right).\tag{5.6}$$

$$\left\{\begin{matrix}\phi = S \\ \phi^* = \psi\end{matrix}\right\}\tag{5.7}$$

If the solutions $\left\{\begin{matrix}\phi = S \\ \phi^* = \psi\end{matrix}\right\}$ are taken to be constant, since $\int g = 1$

we have

$$S\psi = N\tag{5.8}$$

(Note that S is not in general the complex conjugate of ψ for our structure is not hermitian). It will be seen that S, ψ are not separately available. If one now considers two lattices, cross linked in the identical topology, one can analyse this situation by extending to $\phi(r_1, r_2)$, g to $g(r_1 r_1')g(r_2)$, and ϕ^* to $\phi^*(r_1, r_2)$. One then gets

$$\iint g(r_1 r_1')\,g(r_2 r_2')\,\phi(r_1' r_2')\,d^3r_1'\,d^3r_2'$$
$$= \gamma\,\phi^{(m-1)^{-1}}(r_1, r_2).\tag{5.9}$$

Look for a solution depending only on $\phi(r_1 - r_2)$ i.e. see if the two lattices can be found to approximate to each other. This gives a solution S, ψ as before

$$\int g_k^2 S_k\, e^{ik\cdot r}\, d^3k = \gamma\, S^{(m-1)^{-1}}$$

At this point we notice that (5.10)

$$g_k^2 = \frac{\sin^2 kl}{kl} \simeq \frac{1}{2k^2l^2 + 1}.\tag{5.11}$$

So that, using ψ , and taking the most free functionality $m = 3$

$$\left(-2l^2\nabla^2 + 1\right)\psi = \beta\psi^2.\tag{5.12}$$

The solution $\psi =$ constant corresponds to two independent lattices but there is a second solution (which in one dimension would be an elliptic

277

function) in which the two lattices are close to each other with a distribution $\psi(r)$ which scales in r with ℓ, as one might suppose. An approximate solution of (5.12) is

$$\psi \sim e^{-\alpha r/\ell}$$

where $\alpha \sim 5$,

(Note that for $m = 4$ the network still has freedom and therefore entropy, but for $m > 4$ it is locked, and the equivalent of equation (5.12) has no sensible solution).

For the free energy one must take $\phi(r_0, r_1 \cdots r_n)$. Although working in $3 + 3n$ dimensional space is complicated, it turns out that the gaussian chain model is soluble without difficulty, and that gives one a lead on how the rod or other networks can be solved. This is too long a calculation to be given here but will be published elsewhere. (14)

(5iii) Important Networks Having studied gaussian and rod networks, the formalism above offers the opportunity of studying the kinds of network which result from biological macromolecules. For example polysaccharide networks have considerable stiffness, and variable functionality. Studies of them in the past have been hampered by the absence of an appropriate formalism, and the author hopes that the present work will be sufficiently robust to make progress with these important problems.

6. Acknowledgements

The author has done the work on slip links in collaboration with Th. Vilgis and that on rods with Francois Boue, and will be publishing technical details of this work in a co-authored paper. Their stay at Cambridge has been most fruitful.

7. References

1.   Deam R.T. and Edwards S.F., Phil. Trans. 280, 317, (1976).
2.   Freed K.F., Adv. Chem. Phys. 22, 1, (1972).
3.   Ball R.C., Doi M., Edwards S.F. and Warner M., Polymer 22, 1010, (1981).
4.   Needs R.J. and Edwards S.F., Macromolecules 16, 1492, (1983).
5.   Helfand E.H. and Pearson D.S., J. Chem. Phys. 79, 2054, (1983).
6.   Graessley W.W. and Edwards S.F., Polymer 22, 1329, (1981).
7.   Edwards S.F. and Freed K.F., J. Phys. A 2, 145, (1969).
8.   Bastide J., Herz J. and Boué F., J. Phys. 46, 1967, (1985).
9.   Chompff A.J., Polymer Networks., Plenum Press, N.Y., N.Y. (1971).
10.   James H.M. and Guth E., J. Chem. Phys. 11, 455, (1943).
11.   Flory P.J., Principles of Polymer Chemistry., Cornell University Press, Ithaca, N.Y. (1953).
12.   Flory P.J. and Erman B., Macromolecules 15, 800, (1982).
13.   Edwards S.F. and Vilgis Th., Polymer 27, 483, (1986).

CALCULATION OF MOLECULAR DEFORMATION AND ORIENTATION IN

ELASTOMERS USING THE FLORY NETWORK MODEL

Burak Erman

School of Engineering, Bogazici University
Bebek 80815, Istanbul, Turkey

ABSTRACT

The molecular model of an elastomeric network with local intermolecular correlations, given by Flory, is used to calculate the components of the molecular deformation tensor and molecular orientation. Effects of molecular parameters such as severity of entanglements, network inhomogeneities and conditions during cross-linking are discussed. Components of molecular deformation and orientation are calculated for a network under uniaxial stress.

INTRODUCTION

Progress in the understanding of the constitution and behavior of real networks in the last decade owes much to the treatment of the statistical thermodynamics of random networks by Flory[1] in 1976. Properties of the now widely used phantom network model were first rationally analyzed in that paper[2]. Two important features of the phantom network were deduced, stating that (i) the fluctuations of a chain from its mean in the network are substantial and are of the order of the mean squared chain vector, and that (ii) the instantaneous distribution of fluctuations from the mean are independent of the macroscopic state of strain. Only the mean vectors transform affinely with macroscopic strain.

The second of the above properties of the phantom network was shown[1] to be a consequence, but not a hypothesis of the James and Guth theory[3]. The chains of the phantom network depicted in this manner are locally uncorrelated. A group of n spatially neighboring, closely interspersed and entangled chains capable of moving through each other in a phantom-like manner, simultaneously conserving the connectivity of the network thus forms a hypothetical system.

The phantom network, with its constitution and structure well defined, became a reference system on which subsequent analyses of real networks are based. In a sense, this became a change of the reference system, because until then various properties of networks were being interpreted in terms of the affine network model[4] according to which the end-to-end chain vectors transform affinely with macroscopic strain. The fact that real networks should exhibit properties between phantom and affine network models was first stated explicitly in the 1976 paper of Flory.

Introduction of the effect of local correlations among chains to obtain the real network behavior was first made by Ronca and Allegra[5], successfully explaining the experimental results of uniaxial tension and compression of amorphous networks. A more comprehensive analyis of the effect of local correlations among neighboring chains was then presented by Flory[6]. The formulation of Flory as well as that of Ronca and Allegra rest on extensive experimental studies of swelling and deformation of different network systems[7]. As later indicated by Flory[8], two experimental investigations on natural rubber[9,10] played crucial role in consolidating the present understanding of real networks. The classic experiments of Gee[9] showed that the elastic modulus, or the reduced stress, decreased with increasing tensile strain (the C_2 effect) and also with swelling. Removal of the solvent restored the modulus. The investigations of Allen et al., indicated, further, that the modulus of a network obtained by extrapolating to infinite stain was independent of the degree of swelling. These experimental observations led to the postulates that the modulus of the phantom network coincides with the modulus of the highly swollen or deformed network and that local correlations contribute to the modulus at intermediate strains. The correlations are the results of diffuse elastic entanglements of chains with their neighbors. For convenience of mathematical formulation, the junctions at the termini of chains were shown as the sites of entanglements. The term 'constraints on junctions' was used with the understanding that these constraints were derived directly from local intermolecular entanglements among chains. The dependence of these constraints on strain, obtained by a statistical analysis of the local geometry about the junctions, forms the basis of the Flory theory of elasticity of real networks. The predictions obtained by using the theory are shown to agree with results of experiments of diverse nature[11]. Mechanical behavior under simple and multiaxial states of stress,

swelling, birefringence and segmental orientation are some of the phenomena for-which the theory is observed to agree satisfactorily with experiment.

The contribution of entanglements to the modulus, and to the nonaffine
nature of strain at the microscopic level seems to be the focus of current
research on elastomers. Several papers presented at this symposium are
devoted to the study of the various effects of entanglements in networks.
Among these are the statistical analysis of the effects of entanglements[16,17],
the calculation and measurement of segmental orientation[18], the analysis of
the dynamics of hindered diffusion among entanglements[19], assessment of modulus in quantitatively cross-linked networks[20], the relation of nonlinear
dynamic modulus to entanglements[21], and the effect of entanglements on microscopic deformation measured by small angle neutron scattering[22].

Knowledge of transformations of molecular dimensions under a given
macroscopic state of strain is of central importance to the understanding
of the action of entanglements in the deformed network. In the following,
we analyze the state of microscopic deformation and orientation in a network
using the model given by Flory[6] which was later extended to treat the related phenomena of birefringence[12] and segmental orientation[13]. In the first
section below, we summarize the equations for microscopic deformation and
orientation that have been given previously[6,12,13]. For further details the
reader should refer to these papers. In the second section, we present
sample calculations of local deformation and orientation and relate the
predicted phenomena to the molecular constitution of the network.

THEORY

Let $\underset{\sim}{r}_i$ represent the instantaneous end-to-end vector for chain i in a
deformed network. Let $\bar{\underset{\sim}{r}}_i$ and δr_i denote, respectively, its time averaged
mean and instantaneous fluctuation from this mean. Hence,

$$\underset{\sim}{r}_i = \bar{\underset{\sim}{r}}_i + \delta r_i \tag{1}$$

The mean vector $\bar{\underset{\sim}{r}}_i$ may be expressed as the sum of two means,

$$\bar{\underset{\sim}{r}}_i = \bar{\underset{\sim}{r}}_{i,ph} + \Delta\bar{\underset{\sim}{r}}_i \tag{2}$$

where $\bar{\underset{\sim}{r}}_{i,ph}$ is the mean that would obtain in the absence of intermolecular
correlations, i.e;, in the phantom network, and $\Delta\bar{\underset{\sim}{r}}_i$ indicates the additional contribution to the phantom mean, due to effects of local intermolecular

correlations. Denoting the macroscopic strain tensor by $\underset{\sim}{\lambda}$, the phantom mean may be expressed as $\bar{r}_{i,ph} = \underset{\sim}{\lambda} \bar{r}_{i,ph,o}$, where $\bar{r}_{i,ph,o}$ is the mean in the undeformed phantom network. Eq (1) then reads as,

$$\underset{\sim}{r}_i = \underset{\sim}{\lambda} \bar{r}_{i,ph,o} + \Delta \bar{r}_i + \delta \underset{\sim}{r}_i \tag{1}'$$

Both $\Delta \bar{r}_i$ and $\delta \underset{\sim}{r}_i$ depend on the state of macroscopic deformation of the network. The x-component of eq. (1)', where x denotes a principal coordinate direction, may be written as

$$x_i = \lambda_x \bar{x}_{i,ph,o} + \Delta \bar{x}_i + \delta x_i \tag{1}''$$

where, λ_x denotes the principal component of $\underset{\sim}{\lambda}$ along x.

The quantities on the right hand side of eq (1)'' are statistically independent. Hence, the average of x_i^2 over the ensemble may be obtained as

$$\langle x^2 \rangle = \lambda_x^2 \langle \bar{x}_{ph}^2 \rangle_o + \langle (\Delta x)^2 \rangle \tag{3}$$

where $\langle (\Delta x)^2 \rangle = \langle (\Delta \bar{x}_i)^2 \rangle + \langle (\delta x_i)^2 \rangle$ is the average of fluctuations from phantom means and the brackets denote the ensemble average. For a φ functional network, $\langle \bar{x}_{ph}^2 \rangle_o = (1 - 2/\varphi) \langle x^2 \rangle_o$, where $\langle x^2 \rangle_o$ denotes the average for a free chain[1]. Using this equality, eq. (3) may be written as

$$\langle x^2 \rangle = \left[\lambda_x^2 (1 - 2/\varphi) + (2/\varphi) \langle (\Delta x)^2 \rangle / \langle (\Delta x)^2 \rangle_{ph} \right] \langle x_o^2 \rangle \tag{4}$$

where, the equality[1] $\langle (\Delta x)^2 \rangle_{ph} = \frac{2}{\varphi} \langle x^2 \rangle_o$ has been used.

The dependence of the ratio $\langle (\Delta x)^2 \rangle / \langle (\Delta x)^2 \rangle_{ph}$ in eq. (4) on deformation is required in order to establish the value of $\langle x^2 \rangle$ in the deformed network. It is assumed that[12],

$$\langle (\Delta x)^2 \rangle / \langle (\Delta x)^2 \rangle_{ph} = \langle (\Delta X)^2 \rangle / \langle (\Delta X)^2 \rangle_{ph} \tag{5}$$

where $\langle (\Delta X)^2 \rangle$ is the average square of instantaneous fluctuations of junctions in the real network from their phantom centers. $\langle (\Delta X)^2 \rangle_{ph}$ is the average squared junction fluctuations of the phantom network. According to the theory[6,14], the ratio on the right hand side of eq. (5) is given as $(1 + B_t)$, where,

$$B_t = (\lambda_t - 1)(\lambda_t + 1 - \zeta \lambda_t^2)/(1 + g_t)^2 \tag{6}$$

with t=x, y, z and

$$g_t = \lambda_t^2 [\chi^{-1} + \int (\lambda_t - 1)] \tag{7}$$

Here, χ denotes the severity of intermolecular correlations, given by the ratio

$$\chi = \langle (\Delta x)^2 \rangle_{ph} / \langle (\Delta s_x)^2 \rangle_o \tag{8}$$

where $\langle (\Delta s_x)^2 \rangle_o$ denotes the mean squared fluctuation of junctions from their average locations in the undeformed network. Defined in this manner, χ varies between 0 and ∞. The fluctuations of junctions from their constraint centers are assumed to transform affinely under strain according to the original theory[6]. However, due to various factors, such as inhomogeneities in cross-linking or in the distribution of entanglements, the transformations of these fluctuations may be non-affine. The coefficient $\int$ which appears in eqs. (6) and (7) is introduced in recognition of the nonuniformity of constraints. The effect of $\int$ seems not to be important in the interpretation of stress-strain data of dry networks[8,15]. It may, however, assume significant importance in the explanation of results of scattering and spectroscopy experiments. The ratio $\langle x^2 \rangle / \langle x^2 \rangle_o$ represents a measure of the average deformation of molecules in the network. Alternatively, it may be regarded as the average deformation experienced by a chain in the network. This ratio is the x-component of the molecular deformation tensor[12], $\underset{\sim}{\Lambda}^2$. Using eqs. (4) and (6), the components may be written as

$$\Lambda_t^2 = (1 - 2/\varphi) \lambda_t^2 + (2/\varphi)(1 + B_t) \tag{9}$$

Inasmuch as the entanglements are assumed to be elastic, the additional deformation in domains affected by them should also be taken into consideration[12]. This local deformation of the local entanglement field is represented according to the theory as

$$\Theta_t^2 = \langle (\Delta s_{*t})^2 \rangle_{\lambda_t} / \langle (\Delta s_t)^2 \rangle_{\lambda_t} = 1 + g_t B_t \tag{10}$$

Here, $\langle (\Delta s_{*t})^2 \rangle_{\lambda_t}$ is the t-component of the mean squared junction fluctuations from centers of constraint at deformation λ_t. $*$ signifies the presence of the effect of network connectivity. $\langle (\Delta s_t)^2 \rangle_{\lambda_t}$ is the corresponding average in the absence of connectivity.

The orientation function, S'_x , of vectors affixed to a known location on each chain of the network stretched along the x-axis may be written in terms of the components of the molecular deformation tensor as

$$S'_x = D_o\left[\Lambda_x^2 - (\Lambda_y^2 + \Lambda_z^2)/2\right] \tag{11}$$

where D_o is given as[13]

$$D_o = (3\langle r^2 \cos^2\Phi \rangle_o / \langle r^2\rangle_o - 1) /10 \tag{12}$$

and the prime signifies that the orientation is due to change in the dimensions of the chain. The averages in eq. (12) are performed over all configurations of a single chain with two ends fixed. Φ denotes the angle between the chosen vector of the chain and the vector joining the two ends. D_o is related to the constitution of the chain.

A second component, S''_x, in the total orientation has to be considered, due to the effect of further deformations of constraint domains[12,13] on the orientations of the segments. The segment of interest, trapped in the elastic domain of constraints, will undergo an orientation, that may be represented as[13]

$$S''_x = D_1\left[\Theta_x^2 - (\Theta_y^2 + \Theta_z^2)/2\right] \tag{13}$$

Here, D_1 stands as an empirical parameter signifying the proportionality of S''_x to the microscopic strains in the manner shown. Its rational relation to network parameters awaits further theoretical and experimental work. Letting $D_1 = eD_o$, where e is a constant representing the strength of coupling of the segment to its surrounding, the total orientation function may be expressed as the sum of S'_x and S''_x:

$$S_x = D_o\left\{\Lambda_x^2 -(\Lambda_y^2 +\Lambda_z^2)/2 + e\left[\Theta_x^2 -(\Theta_y^2 +\Theta_z^2)/2\right]\right\} \tag{14}$$

SAMPLE CALCULATIONS

Let a network be stretched along the x-direction. Let λ_x denote the component of the displacement gradient tensor in the direction of stretch. λ_x may be decomposed into a volumetric and a distortional part as

$$\lambda_x = (v_{2,o}/ v_2)^{1/3}\alpha \tag{15}$$

Here, λ_x is the ratio of the final length of the network to its length during cross-linking. $v_{2,o}$ and v_2 are the volume fractions of the polymer in the network during cross-linking and during the experiment, respectively. Accordingly, α is the ratio of the final length to the swollen but undistorted length. The measure of uniaxial deformation reported in experimental studies is α. Components of $\underset{\sim}{\lambda}$ along the direction perpendicular to the direction of stretch are obtained from the condition of incompressibility as

$$\lambda_y = \lambda_z = (v_{2,o}/v_2 \lambda_x)^{1/2} = (v_{2,o}/v_2)^{1/3} \alpha^{-1/2} \tag{16}$$

Substitution of eqs. (15) and (16) into eqs. (9) and (10) leads to the evaluation of the molecular deformation tensor in terms of α.

Results of calculations for $\Lambda_{//}$ and $\Lambda_\perp$ are presented in Figure 1. The ordinate values, $\Lambda_{//}$ and $\Lambda_\perp$, denote the square root of the components of the molecular deformation tensor parallel and perpendicular to the direction of stretch, respectively. Results are presented in terms of α for networks cross-linked and stretched in the dry state, i.e., $v_{2,o} = v_2 = 1$. The dotted lines show results for the affine ($\varkappa = \infty$) and phantom ($\varkappa = 0$) networks. Solid curves represent results for a network with $\varkappa = 8$. The curve for $\varsigma = 0$ on the upper portion of the figure, representing molecular deformation parallel to the direction of extension, lies closer to that for the affine network. The corresponding curve for $\varsigma = 0.5$, however, is significantly closer to the phantom network result. In the lower portion of the figure, values of molecular deformation transverse to the direction of extension coincide for $\varsigma = 0$ and 0.5, as shown by the solid curve. In this region, $\Lambda_\perp$ values for the real network are closer to those for the affine network.

In Figure 2, the effect of conditions of cross-linking on molecular deformation is shown. For a sample cross-linked in solution, $\Lambda_{//} \neq 1$ and $\Lambda_\perp \neq 1$ when $\alpha = 1$. The ordinate values of Figure 2 are normalized as $\Lambda_{//}/\Lambda_{//}^o$ and $\Lambda_\perp/\Lambda_\perp^o$, where the superscript refers to molecular deformation when $\alpha = 1$, i.e., when there is no externally applied stress. The dotted lines denote results for the affine and phantom networks. The solid lines are calculated for three phantom networks cross-linked at the indicated values of $v_{2,o}$. Results indicate that molecular deformation may be substantially less (both in parallel and perpendicular directions to the direction of stretch) than that for a phantom network cross-linked in the dry state.

In Figure 3, the ratio of the orientation function of a network (with $\varkappa = 8$ and $\varsigma = 0.1$) to that of a phantom network is plotted as a function of α^{-1}, for various values of the empirical constant e. Calculations show

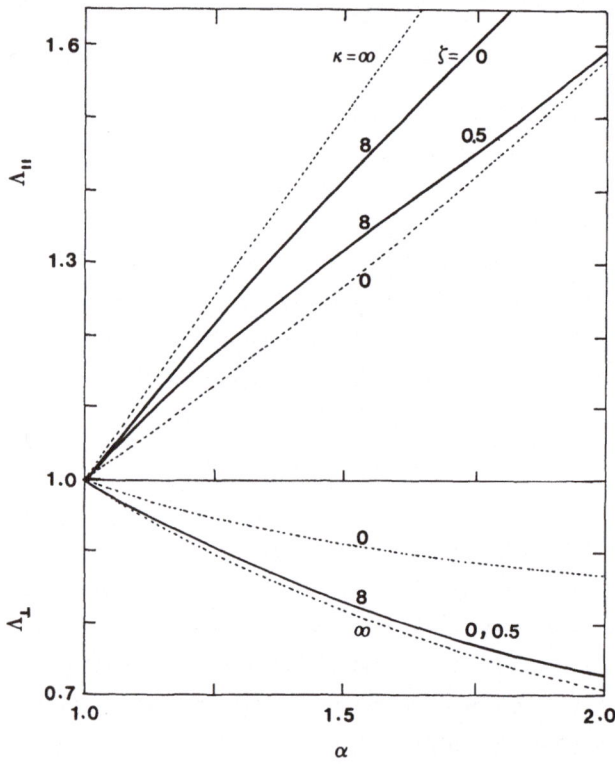

Figure 1. Effect of χ and ζ on molecular deformation. α represents the macroscopic extension ratio. $\Lambda_{\parallel}$ and $\Lambda_{\perp}$ along the ordinate represent square root of the components of molecular deformation tensor $\underset{\sim}{\Lambda}^2$ parallel and transverse to the direction of extension, respectively. Dotted curves indicate results for the affine ($\kappa = \infty$) and the phantom ($\chi = 0$) networks. Solid curves represent results for a real network with $\chi = 8$. Molecular deformation along the direction of stretch is affected considerably by ζ as seen from the two solid curves for $\zeta = 0$ and 0.5 in the upper part of the figure. Molecular deformation in the perpendicular direction is insensitive to ζ as shown by the single curve in the lower part of the figure, approximating the results for both $\zeta = 0$ and 0.5 .

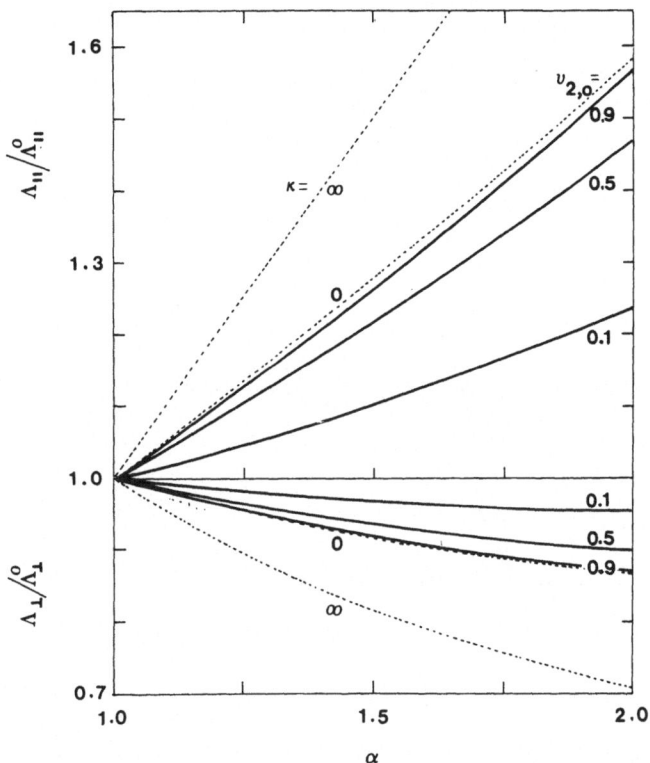

Figure 2. Effect of cross-linking in solution on molecular deformation. $\Lambda_{||}^{\circ}$ and $\Lambda_{\perp}^{\circ}$ correspond to values when $\alpha = 1$. Dotted lines refer to affine ($\kappa = \infty$) and phantom ($\kappa = 0$) network results. Solid curves represent results for three networks with $\zeta = 0$ and different degrees of dilation during cross-linking. Corresponding values of $v_{2,o}$ are shown on each curve.

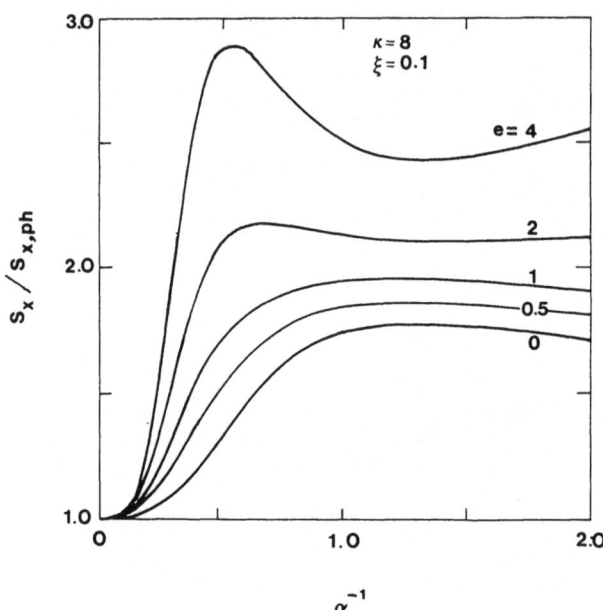

Figure 3. Segmental orientation in a uniaxially deformed network. $S_x/S_{x,ph}$ denotes the ratio of the orientation function for a real network to that of a phantom network. Values of the ratio are expressed in terms of the reciprocal extension ratio α^{-1}. Calculations are made for a network with χ =8 and $\int$ =0.1 for various values of e, where e represents the strength of coupling of a segment to its environment.

that orientation is strongly affected by the introduction of e. In the compression region, the ratio $S_x/S_{x,ph}$ seems to be insensitive to deformation, whereas strong dependence is observed in the tension region with sharp drop for elongations above $\alpha = 2$.

CONCLUSION

The Flory model of a real network with intermolecular correlations is used to represent the components of the molecular deformation tensor and the orientation function in an elastomer under uniaxial stress. The formulation makes it possible to identify the molecular parameters contributing to local deformation and orientation. In the present paper, effects of the degree of entanglements, inhomogeneities and conditions during cross-linking are illustrated in Figures 1 and 2. The fact that molecular deformation parallel to direction of stretch tends to approach that of the phantom network, while deformation along perpendicular direction is closer to affine behavior, is of significance as indicated by recent neutron scattering experiments.

Results of calculations of orientation shown in Figure 3 indicate significant deviations from phantom behavior. Contributions to orientation from local coupling of a segment to its environment, represented by the parameter e, place strong emphasis on the role of probes used in the measurement of orientation.

ACKNOWLEDGEMENT

Financial support by the American Chemical Society for the author's participation in the Chicago, Illinois Meeting, Sept. 1985, is gratefully acknowledged.

REFERENCES

1. P. J. Flory, Proc. R. Soc. London Ser. A, 351, 351 (1976).
2. To be more exact, most of the contents of the paper appeared in 1964 in a manuscript by Flory, titled "The Theory of Elasticity of Gaussian Networks". The manuscript was widely circulated but not published.
3. H. M. James, and E. Guth, J. Chem. Phys., 15, 669 (1947).
4. P.J. Flory, "Principles of Polymer Chemistry", Cornell Univ. Press, 1953.
5. G. Ronca, and G. Allegra, J. Chem. Phys., 63, 4990 (1975).
6. P. J. Flory, J. Chem. Phys., 66, 5720 (1977).
7. J. E. Mark, Rubber Chem. Technol. 48, 495 (1975).

8. P. J. Flory, British Polym. J. $\underline{17}$, 96 (1985).

9. G. Gee, Trans. Faraday Soc., $\underline{42}$, 585 (1946).

10. G. Allen, M. J. Kirkham, J. Padget, and C. Price, Trans. Faraday Soc., $\underline{67}$, 1278 (1971).

11. B. Erman, British Polym. J. $\underline{17}$, 140 (1985).

12. B. Erman, and P. J. Flory, Macromolecules, $\underline{16}$, 1601 (1983).

13. B. Erman, and L. Monnerie, Macromolecules, $\underline{18}$, 1985 (1985).

14. P. J. Flory, and B. Erman, Macromolecules, $\underline{15}$, 800 (1982).

15. P. J. Flory, Polymer J. $\underline{17}$, 1 (1985).

16. S. F. Edwards, and T. Vilgis, this symposium.

17. R. Ullman, this symposium.

18. J. P. Queslel, B. Erman, and L. Monnerie, this symposium.

19. R. J. Gaylord, this symposium.

20. R. W. Brotzman, and P. J. Flory, this symposium.

21. B. J. R. Scholtens, and P. J. R. Leblans, this symposium.

22. H. Yu, T. Kitano, C. Y. Kim, E. J. Amis, T. Chang, M. R. Landry, J. A. Wesson, C. C. Han, T. P. Lodge, and C. J. Glinka, this symposium.

RUBBER ELASTICITY, ENTANGLEMENT CONSTRAINTS AND
THE MEMORY-LATTICE MODEL

Robert Ullman
Department of Nuclear Engineering
University of Michigan
Ann Arbor, MI 48109

INTRODUCTION

The conformational freedom of a polymer chain in an elastomeric
network is severely restricted by the presence of neighboring chains.
Two monomer units cannot occupy the same space, and two chains cannot
pass through each other. These entanglement constraints are not easy to
introduce into the theory of rubber elasticity though many attempts have
been made.

In our opinion, a fruitful procedure can be found in modifications
of the classical theory. The description of an elastomer as an assembly
of polymer chains has reproduced qualitatively most of the physical
properties normally associated with a rubbery material.

For example, the phantom network model of James and Guth (1,2) gave
a recipe for predicting the deformation of a polymer network by an
applied stress, and allowed predictions of the change in chain dimensions
as a function of network expansion or distortion. In an effort to make
the phantom model more realistic, and to fit the model to a variety of
experimental results, P.J. Flory and collaborators (3,4,5) proposed that
the fluctuation of crosslink junction points calculated by the James-
Guth method should be very much restricted by chain entanglements.
Ronca and Allegra (6), and Flory, and Flory and Erman (5) pointed out the
probable anisotropy of these fluctuations in an oriented elastomer, and
these considerations have been in good accord with a considerable number
of experimental investigations (7).

Nevertheless, the Flory-Erman theory has left a number of questions
unanswered, and it has become evident that it is not possible to take
account of all the physical effects of chain entanglements by considering

fluctuations of crosslink junctions alone. We discuss some of the problems below.

EXPERIMENTAL ANOMALIES

High Moduli

The elastic modulus of a rubber according to the phantom network theory is much lower than the modulus of the same network with all junction fluctuations suppressed. If the fluctuations are partially suppressed, the calculated modulus lies between these limits. In fact, in many cases, the measured modulus is many times greater than predicted by fixed junction models (8,9).

Memory of the Crosslinking Condition

Alfrey and Lloyd (10,11) have studied the swelling of networks prepared in solution in a variety of solvents. The swelling of these networks prepared at several dilutions differed from each other as well as from networks prepared in bulk.

Kramer, Ferry and collaborators (12,13) prepared networks from linear polymers which had been stretched, frozen, crosslinked and reheated. The anisotropy of the network crosslinked in the stretched state remained in part in the crosslinked polymer, and the mechanical properties of the network depended to a considerable degree on the degree of orientation of the polymeric sample when the network was formed.

Both the experiments of Alfrey and Lloyd and the studies of Ferry and Kramer are inexplicable by a phantom network model. Incorporation of constraints on junction fluctuations cannot account for the memory of the conditions of crosslinking. Ferry and Kramer use a two network model proposed earlier by Flory (14) to explain their results.

Small Angle Neutron Scattering (SANS) of Polymer Networks

Direct measurements of chain deformation in rubbers or other polymer networks can be carried out in a SANS experiment on a material containing some labeled (by deuterium) chains. The radius of gyration of the chain is obtained, and if the chain is assumed to follow random flight Gaussian statistics, the change in radius of gyration with network deformation is easy to compute from the usual assumption that the vector connecting the chain ends is distorted in the same manner as the macroscopic distortion. The result depends on whether junction points fluctuate or not, but, given a prescription for the magnitude of the fluctuation, the calculation can be performed. There have been many experiments, not all of which are reliable, but it has been become clear that no simple relation between network deformation and chain expansion can be found (15,16,17,18,19).

The inconsistency between SANS experiments and the phantom or fixed junction models of rubber elasticity originates, in our view, in chain rearrangement which takes place when the network is stretched or swollen. Consider a polymeric sample which is lightly crosslinked to form a rubber. The crosslink junctions are uniformly distributed, connected to each other by a highly entangled swarm of polymer chains. It is convenient to classify pairs of junctions according to their relative positions in the network. Pairs which are directly connected by a single polymer chain with no other intervening junction points are called topological neighbors (TN), all other junction pairs are topologically remote (TR) A simple calculation on such a rubber shows that most geometric near neighbors in a network are TR. The various chains are sufficiently intertwined that only rarely will a TN be a geometric neighbor. This depends on the molecular weight between crosslinks (M_c), and as M_c becomes lower, the percent of geometric neighbors which are TN increases.

Let us suppose that a network is stretched or swollen. The free energy change of stretching a polymer chain is positive, and as a consequence the free energy change of the network will be minimized if the individual chains are stretched as little as possible. This result may be accomplished if TN junctions separate less and TR junctions compensate by separating more when the network is deformed such that the uniformity of distribution of junction points is maintained. The chain rearrangement which takes place is limited by the entanglement structure; clearly the rearrangement possibilities are decreased if the extent of chain entanglement is greater. Note that both the modulus of the rubber, and the changes in chain dimensions are very much dependent on the amount of junction rearrangement which takes place.

Theoretical Model and Experimental Anomalies

High moduli, memory effects, and SANS results which are inconsistent with classical theories of rubber elasticity provoke the need for a new theory. The ideas of junction rearrangement, if correct, require that none of the models of affine deformation should be expected to apply. A statistical mechanical partition function, properly formulated for a polymeric elastomer, should yield predictions of chain deformation, and additional assumptions relating macroscopic and molecular geometry are superfluous.

In the following sections a recipe for calculating a statistical mechanical partition junction of an elastomeric polymeric network is proposed. It was formulated with the above-mentioned anomalies in mind, nevertheless many of the ideas on which earlier models were based are incorporated in its structure.

THE MEMORY-LATTICE MODEL OF ENTANGLEMENTS IN A RUBBER
Structure of the Model

The network under discussion will be taken as "perfect" in the following sense. All chains are the same length, and there are N chains in the network. These chains are end-linked, ϕ chains at each crosslink junction. Before crosslinking, each chain obeys Gaussian statistics, and the chains are sufficiently long that the Gaussian model is not exceeded when the network is deformed.

Upon crosslinking, the chains are fixed in the network in the conformation at the moment of crosslinking. After this process, the freedom of motion of each chain is restricted, and the average behavior of chain i is no longer identical with chain j. To be specific, before crosslinking, the end-to-end vector $\underset{\sim}{R}$ of chain i changes with time such that its average value $\bar{\underset{\sim}{R}}$ equals zero. After crosslinking, the fluctuation of $\underset{\sim}{R}$ of chain i continues, but the network constraints require that, in general, $\bar{\underset{\sim}{R}}$ is not equal to zero for any chain, but the ensemble average $\langle\bar{\underset{\sim}{R}}\rangle$ over all chains is zero. This point of view is identical with the phantom network hypothesis.

After crosslinking, the initial probability that chains are separated by $\underset{\sim}{R}$, $P_0(\underset{\sim}{R})$, is modified by the constraints, and the modification is different for every chain in the network. The probability that a particular chain is has an end-to-end vector $\underset{\sim}{R}$ is given by:

$$P(\underset{\sim}{R};\underset{\sim}{R_c}) = P_0(\underset{\sim}{R})\exp\left[-V(\underset{\sim}{R};\underset{\sim}{R_c})/kT\right]/Q(\underset{\sim}{R_c}) \quad (1a)$$

$$Q(\underset{\sim}{R_c}) = \int P(\underset{\sim}{R},\underset{\sim}{R_c})\,d\underset{\sim}{R} \quad (1b)$$

$V(\underset{\sim}{R};\underset{\sim}{R_c})$ is a potential function which is at a minimum when $\underset{\sim}{R} = \underset{\sim}{R_c}$. R_c varies from chain to chain, $Q(\underset{\sim}{R_c})$ is the partition function associated with a particular R_c. The number of chains at a given $\underset{\sim}{R_c}$ is $N(\underset{\sim}{R_c})$, and of course, $\int N(\underset{\sim}{R_c})\,d\underset{\sim}{R_c} = N$.

The thermodynamic functions and geometric properties of the chain are calculated from $Q(\underset{\sim}{R_c})$ and $P(\underset{\sim}{R};\underset{\sim}{R_c})$ as a function of $\underset{\sim}{R}$. The free energy changes and mean chain dimensions in the network are obtained by summing over all values of $\underset{\sim}{R_c}$.

This constitutes the formal basis of the theory. What is crucial is the potential function $V(\underset{\sim}{R};\underset{\sim}{R_c})$ and how it changes as the network is deformed. $P_0(\underset{\sim}{R})$, $N(\underset{\sim}{R_c})$ and $V(\underset{\sim}{R};\underset{\sim}{R_c})$ contain important parameters. Chain geometry and crosslink fluctuations yield values of some parameters. It is at this point where experimental results provide considerable guidance.

Specification of Details of the Model

The probability that the end-to-end vector of chain in the polymer is given by $\underset{\sim}{R}$ before crosslinking is:

$$P_0(\underset{\sim}{R}) = (a/\pi)^{3/2} \exp\left[-a\,\underset{\sim}{R}^2\right] \tag{2a}$$

$$\langle\underset{\sim}{R}^2\rangle = 3/2a \tag{2b}$$

The potential energy of the network crosslinked in bulk in the absence of external stresses is taken to be:

$$V(\underset{\sim}{R};\underset{\sim}{R_c})/kT = b(\underset{\sim}{R} - \underset{\sim}{R_c})^2 \tag{3}$$

When this is substituted in Eq. (1) above, and $\langle\underset{\sim}{R}^2\rangle$ computed, it is assumed that after averaging over all values of $\underset{\sim}{R_c}$ that $\langle\underset{\sim}{R}^2\rangle$ is unchanged by crosslinking.

In the phantom network model the mean square fluctuation in $\langle\underset{\sim}{R}^2\rangle$ is given by:

$$\delta^2 = (2/\varphi)\langle\underset{\sim}{R}^2\rangle, \tag{4}$$

φ being the network functionality. Fluctuations are much less in a real network, and very much a function of network structure. We define a parameter A_f by

$$\delta_{\underset{\sim}{x}}^2 = A_f\langle X^2\rangle \tag{5}$$

where X is the x coordinate of $\underset{\sim}{R}$. In general, it is convenient to deal in cartesian coordinates where $\underset{\sim}{R} = (X,Y,Z)$ and $\underset{\sim}{R_c} = (X_c, Y_c, Z_c)$.

The number of chains with N_c at a given value $\underset{\sim}{R_c}$ is taken to be Gaussian

$$N(\underset{\sim}{R_c}) = N(c/\pi)^{1/2} \exp\left[-c\underset{\sim}{R_c^2}\right] \tag{6}$$

From chain geometry, and the specification of fluctuations, (eq.5), it can be shown that [20]

$$b/a = (1-A_f)/A_f \tag{7a}$$

$$c/a = 1-A_f \tag{7b}$$

From the Kramer and Ferry experiments on networks crosslinked in the stretched conformation, it is clear that the network retains a memory of the conformation of the chains at the time of crosslinking. This memory resides in the entanglement structure, and is expected to vary from one system to another. It is less obvious that this memory exists also in a network crosslinked in the isotropic relaxed state since a memory drives the system to the same isotropic conformation which arises from random Brownian motion even in the absence of such memory. It is our contention that the network retains a memory of the chain conformation at crosslinking for rubber prepared in both the oriented and isotropic state.

If the initial dimensions of the sample are L_x^0, L_y^0 and L_z^0, and these became L_x, L_y and L_z upon stretching, the deformations are defined by $\lambda_x = L_x/L_x^0$, $\lambda_y = L_y/L_y^0$ and $\lambda_z = L_z/L_z^0$. The x component of the potential function of the oriented rubber takes the form

$$V(X; X_c) = b_0 (X-X_c)^2 + b_1 (X-\lambda_x X_c)^2 \qquad (8)$$

memory   lattice

As shown in Eq. 8, the potential splits into two parts upon stretching, one which restores the chain to its initial conformation at crosslinking, the memory term, and, a second, the lattice potential, which tends to move the chain along with the macroscopic sample. The lattice term in the potential deforms affinely, the chain, itself, tends to deform affinely if b_1 is much greater than b_0.

The relative values of b_0 and b_1 are not known. The quantity b_0 is not a function of deformation. We follow the suggestion of Ronca and Allegra, and also Flory in that b_1, to a considerable extent is proportional to λ_x^{-2}. The dependence of b_1 on λ_x is based on the concept that the crosslink junctions act as foci of entanglement restrictions. Thus, one has

$$b_1 = b_1^0 \psi (\lambda) \qquad (9a)$$

$$b_1^0 = ab \qquad (9b)$$

$$b_0 = (1 - a)b \qquad (9c)$$

$$\psi (\lambda_x) = K + (1 - K)/\lambda_x^2 \qquad (9d)$$

Having specified the details, one can calculate free energies, retractive forces and molecular dimensions. General results will be set down below. Mathematical details will be presented elsewhere (20).

In extending the calculations to include all the cartesian components, it is necessary to keep in mind that the volume of a bulk rubber changes very little when the rubber stretched. It is sufficient to set $\lambda_x \lambda_y \lambda_z$ equal to unity.

Geometric Changes upon Uniaxial Deformation

The mean squared end-to-end distance in the X direction after network deformation, $\langle X^2 \rangle$ is related to the mean squared end-to-end distances before deformation by the formula:

$$\langle X^2 \rangle = \langle X^{0^2} \rangle \left[\frac{A_f}{E_1(\lambda_x)} + \frac{(1 - A_f)(1 - a + a \lambda_x \psi(\lambda_x))}{(E_1(\lambda_x))^2} \right] \quad (10a)$$

$$E_1(\lambda_x) = A_f + (1 - A_f)(1 - a + a\psi(\lambda_x)) \quad (10b)$$

Similar results are obtained for the Y and Z components.

In uniaxial deformation, what is measured in a scattering measurement is the projection of the radius of gyration in various directions. We define $R_{\parallel}^2$ and $R_{\perp}^2$, as the projections parallel and perpendicular to the direction of stretch. An increase in the mean squared end-to-end distance of a Gaussian chain from $\langle R^{\circ 2} \rangle$ to $\langle R^2 \rangle$ brings about a change in the radius of gyration according to

$$R_g^2 = 0.5 \, R_g^{\circ^2} \left[1 + \langle R^2 \rangle / \langle R^{\circ 2} \rangle \right] \quad (11)$$

Note that this yields the correct result for a Gaussian ring.

If the direction of stretch in uniaxial deformation is the x direction, then

$$R_{\parallel}^2 = 0.5 \, R_g^{0^2} \left[1 + \langle X^2 \rangle / \langle X^{0^2} \rangle \right] \quad (12a)$$

$$R_{\perp}^2 = 0.5 \, R_g^{0^2} \left[1 + \langle Y^2 \rangle / \langle Y^{0^2} \rangle \right] \quad (12b)$$

$$R_g^2 = (R_{\parallel}^2 + 2R_{\perp}^2)/3 \quad (12c)$$

It is interesting to consider two limiting cases of Eqs (10), that of the phantom network where $A_f = 2/\varphi$, $a = 1$, and $\psi(\lambda_x) = 1$, and the network with non-fluctuating junctions where $A_f = 0$, $a = 1$ and $\psi(\lambda_x)$ 1. We find

$$\langle X^2 \rangle = \langle X^{0^2} \rangle (1 - 2/\varphi) \lambda_x^2 + 2/\varphi \quad \text{(phantom network)} \quad (13a)$$

$$\langle X^2 \rangle = \langle X_0^2 \rangle \, \lambda_x^2 \qquad\qquad \{\text{fixed junctions}\} \qquad\qquad (13b)$$

These agree with the results of the classical calculations, and when substituted in Eqs. (12a) and (12b) yield $R_{\|}^2$ and $R_{\perp}^2$.

Free Energy and Retractive Forces

The free energy of deformation can be decomposed into the sum of three terms. For simplicity, we assume that crosslinking is performed on an isotropic material. We find:

$$\Delta A = \Delta A_x + \Delta A_y + \Delta A_z \qquad\qquad (14a)$$

$$\frac{\Delta A_x}{NkT} = \frac{.5 \, N_1(\lambda_x) + N_2(\lambda_x)}{E_1(\lambda_x)} - .5 \qquad\qquad (14b)$$

$$N_1(\lambda_x) = 1 - a + a\lambda_x^2 \, \psi(\lambda_x) \qquad\qquad (14c)$$

$$N_2(\lambda_x) = \frac{1 - A_f}{A_f} (1-a)\alpha(\lambda_x - 1)^2 \, \psi(\lambda_x) \qquad\qquad (14d)$$

ΔA_y and ΔA_z are computed in the same way. The retractive force is calculated by differentiating ΔA with respect to λ subject to the constraint of constant volume. The calculation is straightforward, results will be presented elsewhere (20).

An important consequence of the memory-lattice model is that high moduli can be accommodated provided that the memory term in the potential function does not vanish. In terms of the model, a memory effect is present if α is not equal to unity. This is evident in Eq(14d), low fluctuations corresponding to small A_f lead to large values of ΔA, the free energy of deformation, and corresponding large values of the retractive force. The coupling of the modulus to junction fluctuations does not appear in either the phantom network or fixed junction models. It arises here only because the minimum in the total potential is not exactly centered on the lattice.

It is interesting to ask for the limiting values of ΔA in the phantom case ($A_f = 2/\varphi$) and the fixed junction case ($A_f = 0$). In both of these, α equals unity and $\psi(\lambda) = 1$.

Calculating from Eq (14), we find

$$\Delta A/NkT = .5 \left[\lambda_x^2 + \lambda_y^2 + \lambda_z^2 \right] - 3/2 \qquad\qquad (15)$$

In this limit, the free energy of deformation is independent of junction fluctuations. The result agrees with that obtained for the fixed junction model, but is not in accord with that obtained for a phantom network.

Networks Prepared from an Oriented Sample

The memory-lattice model has the same structure here as for the isotropic material. The potential is changed. Consider a chain in an uncrosslinked polymer which is at a minimum potential energy before stretching at $R = R_0$. The sample is stretched so that R_0 changes to R_c where $X = \lambda_x^c X_0$. The sample is frozen, crosslinked at this elongation, and then reheated to the rubbery state. The potential associated with an arbitrary deformation is given by

$$V(X;X_0) = b_0(X - \lambda_x^c X_0)^2 + b_1(X - \lambda_x X_0)^2 \tag{16}$$

The modifications of Eq(8) leading to Eq(16), taken together with Eq (6) (with R_0 replacing R_c) generates a partition function appropriate for networks crosslinked in a deformed state. Using this formalism, one can show that the network remains partially oriented in the absence of an external force, and that stress-strain behavior is characteristic of that found for materials prepared in this way. Further details are given elsewhere (20).

CONCLUSION

A statistical mechanical model of an elastomeric network is proposed which takes into account that a chain in the network tends to follow the deformation of the macroscopic sample, but is restrained from doing so by the entanglement structures formed during crosslinking.

The theory is consistent with both low and high moduli, including cases where the moduli are higher than predicted by either phantom network or fixed junction models.

The possibility of chain rearrangement so that topological neighboring junctions move in concert to resist chain deformation is implicit in the structure of the theory. Both minor and major chain rearrangements can be expected depending on the extent of chain entanglement.

The fact that the entanglement structure retains a memory of the crosslinking procedure is a central tenet of the theory. This yields a simple explanation of properties of networks formed in the deformed state, and also plays an important role in the theory for networks formed from an isotropic polymer.

REFERENCES

1. James, H.M., _J. Chem. Phys._ 1947, _15_ 651.

2. James, H.M., Guth E. _J. Chem. Phys._ 1947, _15_ 689.

3. Flory, P.J., _Proc. Roy. Soc._ 1976, _A351_ 351.

4. Flory, P.J., _J. Chem. Phys._ 1977, _66_ 5720.

5. Flory, P.J., Erman, B., _Macromolecules_ 1982, _15_ 800.

6. Ronca, G.; Allegra, G., _J. Chem. Phys._ 1975, _63_ 4990.

7. A good review of much of this work is contained in
   Queslel, J.P.; Mark, J.E., _Adv. Polym. Sci._ 1984, _65_ 135.

8. Dossin, L.M.; Graessley, W.W., _Macromolecules_ _12_ 123.

9. Pearson, D.S.; Graessley, W.W., _Macromolecules_ 1980, _13_
   1001.

10. Alfrey, T., Jr.; Lloyd, W.G., _J. Polym. Sci._ 1962, _62_
    159.

11. Lloyd, W.G.; Alfrey, T., Jr., _J. Polym. Sci._ 1962, _62_
    301.

12. Carpenter, R.L.; Kramer, O; Ferry, J.D., _Macromolecules_
    1977, _10_ 117.

13. Hvidt, S.; Kramer, O; Batsberg, W.; Ferry, J.D.,
    Macromolecules 1980, _13_ 933.

14. Flory, P.J., _Trans. Faraday Soc._ 1960, _56_ 722.

15. Benoit, H.; Decker, D.; Duplessix, R.; Picot, C.; Rempp,
    P.; Cotton, J.P.; Farnoux, B.; Jannink, G.; Ober, R.,
    J. Polymer Sci. Polym. Phys. 1976, _14_ 2119.

17. Bastide, J.; Duplessix, R.; Picot, C.; Candau, S.,
    Macromolecules 1984, _17_ 83.

18. Beltzung, M.; Picot, C.; Herz, J., _Macromolecules_ 1984,
    _17_ 663.

19. Yu, H; Kitano, T.; Kim, C.Y.; Amis, E.J.; Chang, T.;
    Landry, M.R.; Wesson, J.A., _Polymer Preprints_ 1985, _26_
    (2) 60.

20. Ullman, R., _Macromolecules_ 1986, _19_ 1748.

ACKNOWLEDGEMENT

This work was supported by the National Science Foundation under grant No. DMR-8217460.

STRAIN-INDUCED CRYSTALLIZATION IN RUBBERS

Giuseppe Allegra

Dipartimento di Chimica del Politecnico
Piazza L. da Vinci 32, 20133 Milano (Italy)

INTRODUCTION

As it is well known, crosslinked samples of stereoregular polymers are able to crystallize at temperatures higher than T_m^o, i.e., the ideal melting point, if suitably deformed. Conformational entropy decrease of the molten chains, induced by deformation, is responsible for the effect[1], which is especially important in improving the ultimate properties of the material. It is the purpose of the present study i) to clarify the critical aspects of melting in strained rubberlike samples; ii) to discuss the favourable effect induced by relatively small values of the specific melting entropy upon the rubber's technical performance; iii) to point out the relationship between conformational disorder in polymer crystallites and the small entropic value referred to above, discussing few examples of practical relevance[2].

THE NATURE OF THE MELTING TRANSITION IN A UNIAXIALLY STRETCHED RUBBER SAMPLE

The following analysis will be focused on the thermodynamic behaviour at incipient crystallization at $T > T_m^o$ of a stretched rubberlike network consisting of long, Gaussian chains adjoining different crosslinks; dangling ends will be neglected. The temperature will be chosen so that the chains undergoing crystallization will be far from full elongation, and each of them will be assumed to (partly) crystallize within just one crystallite. Crystallites may be formed with their stems inclined at any angle ϕ with respect to the stretching direction (see Fig.1) and no chain folding will be allowed[1,3]. To maximize the amount of their crystal-packing interactions, all the stems belonging to a single crystallite will be assumed to comprise the same number ζ of chain statistical segments[3]. Besides, the number of stems within each crystallite will be supposed as sufficiently large that formation or melting of the crystallite may be regarded as a phase transition in the usual thermodynamic sense. Accordingly, the length of the crystalline stems $\zeta(\phi)$ will be the same for all the crystallites with the same angle ϕ, at a given temperature T and stretching ratio λ. No interaction between crystallites will be considered; the average displacement of the crosslinks under stretching will be regarded as affine. Although equilibrium is assumed, hence reversible formation of the crystallites, their average lifetime will be long with respect to the longest chain relaxation time. Accordingly, it will be possible to associate a free energy contribution to each chain, considering its crystallized stem as frozen while the amorphous part may effectively attain all its degrees of freedom.

Let us consider a single, partly crystallized chain in the stretched sample (see Fig.2). Neglecting considerations of contact probability between co-crystallizing chains and apart from a λ-dependent contribution $\mathcal{A}(\lambda)$ that arises from the co-operative behaviour of the network and is influenced by the packing constraints on the crosslink fluctuations, the chain contribution to the overall free energy may be expressed as[3]

$$\mathcal{A} = \mathcal{A}(\lambda) + N \left\{ \frac{3}{2} k_B T \left(\lambda_c^2 \langle \gamma^2 \rangle + w^2 - 2\lambda_c \langle \gamma \rangle w\cos\theta \right) / (1-w) - w(h_f - T s_f) \right\} \quad (1)$$

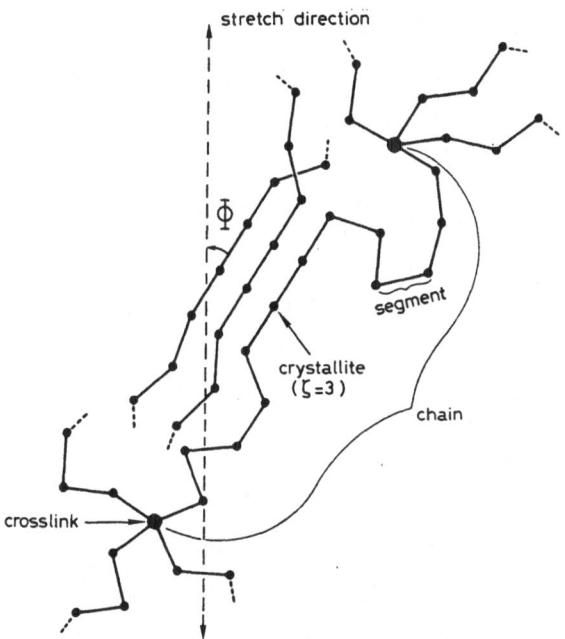

Fig. 1. Schematic drawing of a portion of a crystallizing rubber
network. A chain connecting two crosslinks comprises
$N = 14$ statistical segments, $\zeta = 3$ of which are crystallized.

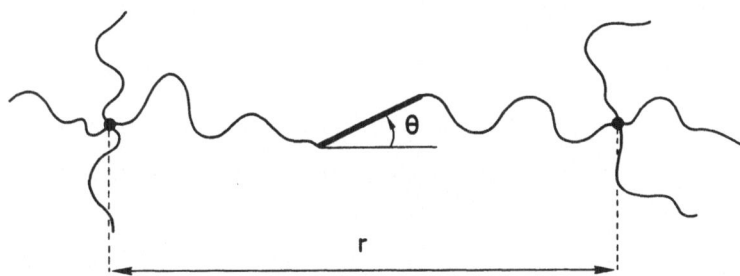

Fig. 2. Sketch of a chain connecting two crosslinks. The end-to-
end length is r, the solid line stands for a crystalline
stem at an angle θ with the direction represented by r.
Statistical-segment discontinuities are disregarded.

where λ_c is the average strain ratio applying to the particular chain, $N(\gg 1)$ and ℓ are the number of chain segments and the effective segment length, $\gamma = r/N$ where r is the chain end-to-end distance while h_f, s_f and w respectively are the melting enthalpy and entropy per chain segment and the degree of chain crystallization, i.e.,

$$h_f/s_f = T_m^o \; ; \quad w = \gamma/N \qquad . \qquad (2)$$

Using the affine assumption, the relationship between λ_c and the overall stretching ratio λ is [3]

$$\lambda_c = \lambda/(\cos^2\Phi_c + \lambda^3\sin^2\Phi_c)^{1/2} \; , \qquad (3)$$

where Φ_c is the average angle between the end-to-end chain vector and the stretch direction. From eqs. (1)-(3) we see that the free energy of a partly crystallized chain attains a relative minimum for $\cos\Theta = 1$ (i.e., $\Theta = 0$) while reaching the absolute minimum if the requirement $\lambda_c = \lambda$ (i.e., $\Phi_c = 0$) is also met, all other parameters being given. In other words, the chain crystallizes best if both its average end-to-end vector and its crystallized stem are parallel to the direction of stretch; in case the ete vector is at an angle with this direction, then the crystallized stem tends to stay parallel to the vector itself. It is an immediate consequence that the last crystallites to disappear in a sample held at constant λ and increasing temperature, will have their stems parallel to the direction of stretch; in fact, any crystallite with the same stem length and a slightly different orientation will contain a large number of chains, each having a slightly higher free energy (on the average), thus producing a lower thermodynamic stability.

For sake of simplicity we shall temporarily assume that all the chains crystallizing within a single crystallite have their end-to-end vectors parallel to the crystalline stems;

this amounts to take $\cos\theta = 1$ in eq.(1), thus minimizing the free energy for all crystallites so long as contact probabilities for co-crystallizing chains are disregarded (see remark just before eq.(1)). After putting $\partial \mathcal{A}/\partial w = 0$ in eq.(1), the degree of crystallization $w(\Phi)$ for a stem direction at an angle Φ with respect to sample elongation is

$$w(\Phi) = \langle\gamma\rangle \lambda/(\cos^2\Phi + \lambda^3\sin^2\Phi)^{1/2} - \frac{1}{2}\mathcal{E}(T) , \qquad (4)$$

where λ_c is now replaced by λ and

$$\langle\gamma\rangle = \langle r\rangle/N\ell \ (\sim \frac{1}{\sqrt{N}}) ; \qquad \mathcal{E}(T) = \frac{2}{3}\frac{s_f}{k_B}(1-T_m^\circ/T) \qquad (5)$$

are both quantities of order $N^{-1/2}$; higher-order terms are disregarded. Let us now define the temperature T_λ as the upper limit above which no crystallization takes place even at $\Phi = 0$, i.e.,

$$\mathcal{E}(T_\lambda) = 2\langle\gamma\rangle\lambda ; \quad T_\lambda = T_m^\circ/(1 - \frac{3k_B\langle\gamma\rangle}{s_f}\lambda) , \qquad (6)$$

and assume that the actual temperature T tends to T_λ from below, so that

$$0 \leqslant \left(T_\lambda - T\right) \ll \left(T_\lambda - T_m^\circ\right) . \qquad (7)$$

It is our goal to ascertain the resulting type of transition. From eq.(4), the limiting angle Φ_{lim} beyond which no crystallites may form under the actual values of T and λ , is obtained from the equation $w(\Phi_{lim}) = 0$. In view of eqs.(4) to (7) we get

$$\sin^2\Phi_{lim} = \frac{2}{\lambda^3-1} \frac{T_\lambda - T}{T_\lambda - T_m^\circ} , \qquad (8)$$

while the maximal degree of crystallization, corresponding to the chains with $\Phi = 0$, is

$$w(0) = \langle\gamma\rangle\lambda \frac{T_\lambda - T}{T_\lambda - T_m^\circ} . \qquad (9)$$

It is easy to prove that the overall amount, or degree, of crystallinity is proportional to the product of $w(0)$ times the area comprised between $\Phi = 0$ and $\Phi = \Phi_{lim}$ on the sphere of unit radius (see Fig.3). In turn, for small values of Φ_{lim} this area is proportional to $\Phi_{lim}^2 \simeq \sin^2 \Phi_{lim}$ and, since the crystallization heat content H_c is proportional to the crystallinity degree, we have from eqs.(8) and (9)

$$H_c \propto w(0) \cdot \sin^2 \Phi_{lim} \propto (T_\lambda - T)^2 \qquad (10)$$

Correspondingly, at $T = T_\lambda$ we have a discontinuity in the second-derivative of H_c, which amounts to a <u>third-order transition</u>. However, this conclusion is only true if $\lambda > 1$; for the isotropic sample with $\lambda = 1$, crystallites always have the same probability of existence at any angle Φ, so that $\Phi_{lim} (= \pi/2)$ is temperature-independent. Accordingly

$$H_c (\lambda = 1) \propto (T - T) \qquad (11)$$

and we have a <u>second-order transition</u>, instead of the ideal first-order transition expected for a non-crosslinked polymer sample. (It should be noted in this context that even at $\lambda = 1$, T_λ is different from T_m°, see eq.(6); the relative increase $(T - T_m^\circ)/T_m^\circ$ is of order $N^{-1/2}$.)

A comment on the previously stated assumption that all the chains participating to the same crystallite have parallel end-to-end vectors may be in order. All other features of the present model staying the same, if we relax the above assumption the following consequences are expected i) the temperature T_λ should still represent a transition, in that at $T > T_\lambda$ no crystallites may form while at $T < T_\lambda$ they may, provided most of their chain vectors are properly aligned along the stem direction; ii) on intuitive grounds, the transition should be even smoother than obtained above, therefore suggesting that its order should be no less than 3.

EFFECT OF THE MELTING ENTROPY

From eqs.(4)-(6) we see that the melting entropy per chain atom s_f plays an interesting role, in that its decrease leads to an increase of both $w(\Phi)$ and of T_λ if all other parameters do not change. It should be noted in this connection that keeping T_m^o fixed while lowering s_f, implies that h_f must also be lowered in the same proportion. These remarks suggest that the technical performance of a rubber is better the lower its melting entropy, provided the temperature of melting T_m^o does not depart too much from the usual range, roughly centered around 0°C. However, a good overall performance also requires that the sample should only crystallize to a limited extent, if any, at low temperatures and in the absence of strain (i.e., $T<T_m^o$, $\lambda = 1$). As a result of a thermodynamic investigation of crystallized networks, Wu obtains the following result for the overall crystallinity degree under these conditions[4]:

$$w(T) = 1 - \left[\frac{3/2 - \Upsilon^{-1}}{-3/2N\mathcal{E}(T)} \right]^{1/2} \qquad (12)$$

(Note that $\mathcal{E}(T) < 0$ if $T < T_m^o$, see eq.(5). More recently, Smith reaches a similar result through a different approach[5].) We see that in the present case $w(T)$ <u>decreases</u> with a decrease of s_f, since $\mathcal{E}(T) \propto s_f$ if T_m^o/T does not change. Accordingly, we may conclude that lowering s_f with a proportional change of h_f so that $h_f/s_f = T_m^o$ stays roughly constant, is beneficial to rubber performance in that i) it increases stress-induced crystallization at current working temperatures $(T > T_m^o)$, ii) it decreases temperature-induced crystallization at $T<T_m^o$, thus reducing the hazard of sample brittleness at low temperatures[2].

LOW MELTING ENTROPY AND CONFORMATIONAL DISORDER

The most significant factors that may lead to a low value of $s_f = s_{(melt)} - s_{(crystal)}$ are either a poor conformational freedom of the molten chains or some sort of chain disorder within the crystallites. The former case should be represented by stiff polymers, which are likely to melt at relatively high temperatures, perhaps forming liquid-crystalline phases. By far more relevant to rubber elasticity should be the latter case, where the most effective type of crystalline disorder is related to some amount of conformational freedom. An important example is natural rubber, i.e., cis-poly-1,4-isoprene, which gives s_f=3.45 e.u. per monomeric unit[6], to be compared with s_f=8.07 e.u. for cis-poly-1,4-butadiene[7] (one monomeric unit is roughly identifiable with one statistical segment[3]). On the basis of X-ray evidence, Corradini shows that the former polymer is substantially disordered in the crystalline state (see Fig.4)[8], contrary to the latter. In close analogy, the good rubberlike behaviour of trans-1,4 enchained copolymers of 1,3-butadiene and 1,3 pentadiene[9] may also be partly related with conformational disorder in the crystalline state. In fact, X-ray investigation of stretched copolymer samples unambiguously indicates that their crystal structure at room temperature is similar to that observed for the high-temperature (T>70°C), conformationally disordered polymorph of trans-poly-1,4-butadiene;[10] it is not surprising that random insertion of pentadiene units in the polybutadiene chains may promote the same type of disorder even at lower temperatures.

CONCLUSION

On the basis of a relatively simple statistical model we have shown that a uniaxially stretched rubberlike sample should undergo a transition of the third or higher order upon melting at constant deformation; obviously enough, terminal melting takes place above the normal melting temperature of the polymer, i.e., at $T_\lambda > T_m^\circ$. To this conclusion the following assumptions appear to be especially relevant: i) only extended chain crystallization is present in the vicinity of T_λ, and the crystallites consist of a large number of stems of uniform length $\xi\ell$

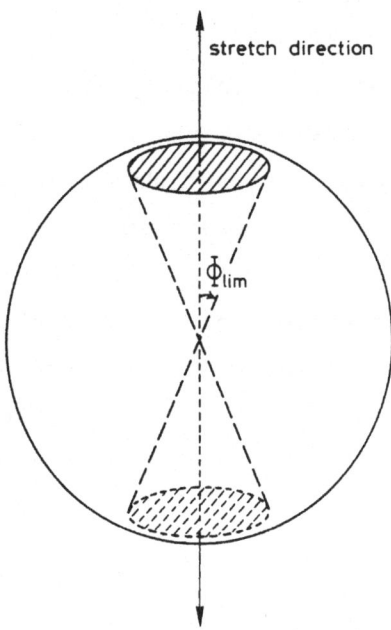

stretch direction

Φ_{lim}

Fig. 3. Sketch showing the angular region permitted to the crystalline stems in a rubber sample at given T and λ (see
text).

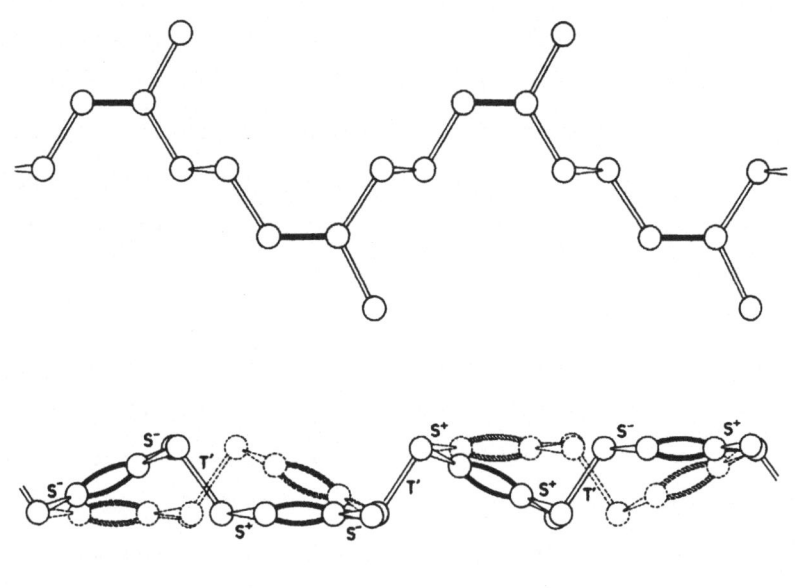

2 Å

Fig. 4. Two side views at right angles of cis-poly-1,4-isoprene
molecule in the crystalline state, showing conformational disorder as suggested by Corradini[8],[10]; hydrogen
atoms omitted (Courtesy of Macromolecules, Vol.16, p.
1169 (1983).)

($\zeta \to 0$ for $T \to T_\lambda$); ii) under the same conditions, crystallites having different stem orientation do not interact with one another; iii) in spite of the localized inhomogeneities represented by the permanent crosslinks, the rubberlike network is conceived as a statistically uniform physical system. The order three for the transition derives from the additional assumption that all the chains crystallizing within the same crystallite have their end-to-end vectors parallel to the direction of the crystalline stems, according to the criterion of highest probability for each separate chain. If this assumption is relaxed, we conclude that the transition should be of order no less than three.

Experimental support to the above conclusions, although of a semi-quantitative nature, is provided by a series of careful X-ray investigations on cis-poly-1,4-butadiene stretched rubberlike samples[3,11]: the arcs observed on the diffraction photographs, showing the angular range of crystallite orientation, tend to become extremely narrow just before complete disappearance, in agreement with the prediction of the present model (i.e., with a transition of order three at least). Also in agreement with some experimental results[12], the predicted transition does not imply any discontinuity either in the equilibrium retractile force $f(\lambda)$ or in its derivative, at constant temperature[3]. It should be noted that a second-order transition would give a first-derivative discontinuity.

We have also shown that the lower the entropy of melting s_f of the smallest conformationally independent chain portion - identifiable with Kuhn's statistical segment - the better is the rubber's technical performance both at $T > T_m^o$ and $T < T_m^o$, provided T_m^o is in the proper range. In turn, a small value of s_f should generally be related with some conformational disorder within the crystallites. This remark may represent a useful guideline to prepare new rubberlike materials; existence of a reasonably wide class of crystallizable copolymers[13,14,15], which are likely to display some amount of conformational dis-

order, appears to be an encouraging pre-requisite for successful attempts along this direction. Perhaps a word of caution
should be added in this connection. It seems reasonable to
expect that, if the crystalline disorder is too large, i.e.,
if both s_f and h_f are too small, the crystallites should not
act as effective additional crosslinks, with little contribution to the sample's ultimate properties.

The author gratefully acknowledges useful discussions
with Prof. Sergio Brückner and Dr. Stefano V. Meille.

REFERENCES

1. P. J. Flory, Thermodynamics of crystallization in high po-
    lymers. I. Crystallization induced by stretching, J.
    Chem. Phys. 15:397 (1947).
2. G. Allegra and M. Bruzzone, Effect of entropy on strain-
    induced crystallization, Macromolecules 16:1167 (1983).
3. G. Allegra, The angular distribution of crystallinity in a
    stretched rubberlike polymer, Makromol. Chem. 181:1127
    (1980).
4. W. L. Wu, A Thermodynamic approach to the stress-induced
    crystallization in cross-linked rubbers, J. Polym. Sci.
    Polym. Phys. Ed. 16:1671 (1978).
5. K. J. Smith, Jr., Crystallization of stretched networks
    and associated elasticity, ACS Symp. Ser. 193 (Elasto-
    mers Rubber Elasticity): 293 (1982).
6. D. E. Roberts and L. Mandelkern, Thermodynamics of crystal-
    lization in high polymers: natural rubber, J. Am. Chem.
    Soc. 77:781 (1955).
7. M. Berger and D. J. Buckley, Structure effects and related
    polymer properties in Polybutadiene. I. Preparation and
    characterization, J. Polymer Sci. Pt.A·1:2945 (1963).
8. P. Corradini, Conformation of polymer molecules and entropy
    of melting, J.Polym. Sci., Polym. Symp. 50:327 (1975).
9. M. Bruzzone, A. Carbonaro and L. Gargani, Crystallizable
    trans-butadiene-piperylene elastomers, Rubber Chem.
    Technol. 51:907 (1978).
10. P. Corradini, Chain conformation of the high-temperature
    polymorph of trans-1,4-polybutadiene, J. Polym. Sci.,
    Part B 7:211 (1969).
11. G. Perego and M. Cesari, ENIRICERCHE, S. Donato Milanese
    (Italy), private communication.
12. K. J. Smith, Jr., Crystallization of networks under stress,
    Polym. Eng. Sci. 16:168 (1976).

313

13. G. Natta, P. Corradini, D.Sianesi and D. Morero, Isomorphism phenomena in macromolecules, J. Polymer Sci. 51: 527 (1961).
14. G. Allegra and I. W. Bassi, Isomorphism in synthetic macromolecular systems, Adv. Polymer Sci. 6:549 (1969).
15. B. Wunderlich, Isomorphism, in "Macromolecular Physics. 1. Crystal Structure, Morphology, Defects", p.147, Acad. Press, N.Y. (1973).

STRESS-TEMPERATURE BEHAVIOR OF STRETCHED TRANS-POLYISOPRENE NETWORKS

IN THE CRYSTALLIZATION REGION

Der Gun Chou and K.J. Smith, Jr.

Department of Chemistry
SUNY College of Environmental Science and Forestry
Syracuse, New York   13210

INTRODUCTION

Stretched networks of trans-polyisoprene (or gutta percha) exhibit interesting stress-temperature profiles as temperature passes through the crystallization zone.[1] Upon cooling, stress turns sharply downward, then climbs back to a high level as temperature continues to fall , Subsequent reheating causes the stress to fall precipitously to a negative value, then after an interval of several degrees ascend steeply back into the positive domain as melting is completed. It is generally believed that the heating phase of the cycle produces a stress-temperature profile more indicative of equilibrium conditions.

Gent[1] suggested that stress regeneration with increasing crystallization in trans-polyisoprene networks might be attributed to volume contraction accompaning crystallization. In this regard he rejected explanations involving a correlation between fibrillar-lamellar crystalline morphological transformations and stress regeneration. Recent calculations[2,3,4] lead to a similar conclusion that morphological transformations are unnecessary for stress regeneration. But contrary to Gent's suggestion, volume contraction upon crystallization is not endorsed as being of primary importance, except perhaps at quite low deformations. Instead, stress regeneration is a normal consequence of crystallite growth, apparently little influenced by morphological detail except perhaps quantitatively. At the molecular level two characteristics of polymer network crystallization are important: First, crystallites grow laterally (transverse to the chain axis) by accreting additional chains as they are encountered in the melt without large scale sub-chain rearrangements, and second, crosslinks continuously relocate to their most probable positions as crystallization progresses or declines, thereby eliminating disparate chain tensions generated by crystallization. The tension generated by a chain depends upon its end-to-end distance. Early stages of crystallization shorten end-to-end distances of the amorphous chains, on average, and decrease tensions, but high crystallization forces the amorphous chains into more extended configurations of elevated tensions.

This study of trans-polyisoprene networks was undertaken within the general guidelines established by the new theory. We wanted to determine if any consequences stemming from the theory are contradicted by experimental behavior to such as extent as to force abandonment of the theory. Although this study is far from complete, and in some instances inconclusive, nothing has been found that outright invalidates the theory. However, some refinements and modifications of the theory may be necessary to improve quantitative agreement with experimental results.

The theory yields three fundamental equations describing the melting temperature, T_m, the degree of crystallization, ω, and the retractive force, f, of the stretched network as a function of the uniaxial deformation ratio, α, and temperature, T. These equations are:

$$\frac{1}{T_m} = \frac{1}{T_m^{\,\circ}} - \frac{k}{2N\,\Delta H_u}\left((6\sigma-1)\alpha^2 - \frac{2}{\alpha}\right) \tag{1}$$

$$\omega = 1 - \left(B/(A+B)\right)^{1/2} \tag{2}$$

$$A = \frac{2N\,\Delta H_u}{k}\left(\frac{1}{T} - \frac{1}{T_m}\right)$$

$$B = (3\sigma-1)^2\alpha^2 + 2/\alpha$$

$$f = \frac{GkT}{L_i}\left(\frac{1}{1-\omega}\right)\left[(1-3\sigma\omega)^2\alpha - \frac{1}{\alpha^2}\right] \tag{3}$$

The parameters and constants are:

$T_m^{\,\circ}$ —   melting temperature of uncrosslinked polymer

k —   Boltzmann constant

N —   average number of monomer units between crosslinks

ΔH_u —   heat of fusion of a monomer unit

G —   total number of network chains

L_i —   length of unstretched sample

σ —   number of statistical units required for the nucleation of a chain for lateral crystallite growth

These equations are restricted to the region of high deformation, which insures that the crystallites are oriented along the stretch direction.

The quantity σ is the number of links (statistical) required for secondary nucleation of a new chain onto a growing crystallite – its value is expected to be small, at most only two or three statistical links and perhaps smaller still for more extended polymers. The theory as expressed by equations (1,2,3) treats σ as a constant, but it is recognized that σ may depend upon deformation, extent of crosslinking, and the temperature of crystallization, which must be determined by experimentation. Variations of σ, once established, can be taken into account if necessary.

Our general plan of investigation was to measure T_m as a function of α to determine the value of σ via Eq. (1) for several networks of different crosslinking (N). Amounts of crystallinity might then be calculated with Eq. (2). The stress-temperature profiles, calculated with Eq. (3), could then be compared with those obtained experimentally. Several questions are of particular interest:

Are the values of σ reasonable? Are they constant?
Are the theoretical values of ω reasonable?
Are the theoretical stress-temperature profiles consistent with experimental ones, particularly with regard to shape and valley width?

An experimental technique suitable for our study is that of crystallization under stress, by which a network is stretched at a high temperature (insuring amorphism during deformation) to a predetermined length, cooled at constant length (after a suitable period for stress stabilization) to generate crystallization, then reheated in a controlled fashion to obtain the stress-temperature profile as temperature ascends. The cooling program is an important factor, particularly for trans-polyisoprene networks: in addition to the connection between rate of cooling and equilibrium crystallization, these networks also display different crystallite modifications depending upon the program of cooling[5,6]. One form, _alpha_, is favored by high stress and high crystallization temperatures (>40°C). The other form, _beta_, predominates if stress and crystallization temperatures are low. Therefore, preparation of networks containing a great abundance of beta relative to alpha requires quick cooling to temperatures below 40°C, and this necessarily displaces the specimen away from conditions more conducive for equilibrium crystallization.

EXPERIMENTAL

Trans-polyisoprene was obtained from Polysciences, Inc., (Lot #0527). The polymer was dissolved in benzene and filtered. From 0.5% to 2.0%, by weight, of Dicumyl Peroxide (DCP) was added for crosslinking. This solution was cast on a teflon surface to evaporate the solvent and the resulting film was pressed at 150°C for one hour. (thickness was controlled by using a mold of two metal plates separated by a brass rectangular gasket). The amount of DCP added determined the degree of crosslinking. Experimental samples were cut from these films and washed respectively in toluene and benzene for 3 days to remove uncrosslinked material. Antioxidant was added by swelling in toluene containing 0.5 percent of antioxidant 2246 (American Cyanamid Company, N-phenyl-B-napthylamine) and drying. The degree of crosslinking was determined from the modulus of the stress-strain curve of the amorphous networks via the equations of the kinetic theory of elasticity. This information is reported in Table A.

The apparatus for measuring force has been previously described[8]. Film strips are clamped, one end to an analytical balance and the other to a movable rod. The sample is thermostated by water-jacketed cell. Length is measured with a cathetometer. All samples were stretched at 85°C to the desired elongation and allowed 1-1/2 hours for stress relaxation. The temperature was varied according to three programs H_1, H_2, H_3 given in the Appendix. Force was usually recorded for the reheating phase at a rate of 3°C/10 minutes.

Table A: Crosslinking data and molecular characteristics for gutta percha networks.

Network	Dicumyl Peroxide (wt.%)	Molecular Weight (Mc)	No. Monomer Units (N)
A	0.5	28,355	405
B	1.0	18,480	264
C	1.25	11,813	169
D	2.0	7,603	109
E	2.25	5,550	79

Crystallization was determined by density measurements in a density gradient column at room temperature. For conversion to crystallinity, amorphous density[9] is 0.905 g/cm^3 and crystalline density (beta)[6] is 1.02 g/cm^3.

The heat of fusion of the beta form is reported[5] to be 8 ± 2 KJ/mole (~2000 cal/mole). Mukherji and Woodward[10] report 8 KJ/mole.

Melting temperatures of the stretched networks (at constant length) were estimated by simply observing where on a force-temperature plot force begins to deviate from the f-T behavior expected of amorphous networks. These points are indicated in the figures by arrows and are at best only very crude estimates, especially at higher deformations where the curvature is very gradual relative to the abruptness displayed at lower deformations. Little use is made of these temperatures except to obtain a rough estimation of the parameter σ and whether or not σ depends upon the degree of crosslinking. A more reliable way to determine σ is to calculate its value with eq.(2), using measured crystallinity. The two methods give qualitative agreement with each other.

RESULTS AND DISCUSSION

Gutta percha may have three crystal structures: alpha, beta, gamma.[6,7] The gamma form is stable at low temperatures in the absence of stress, but heating destroys it, causing the formation of beta in its stead. Beta, being metastable, should revert to gamma upon cooling but this transformation is so extremely slow that recently heated and cooled samples contain only beta. The alpha form appears in stressed samples only. Accordingly, it is expected that specimens used in our study are free of gamma. On the other hand, no one other than Fisher[7] has ever presented experimental evidence for more than two crystalline forms of gutta percha, usually denoted alpha and beta or respectively, high melting form and low melting form. Specifically, Takahashi[12] argues that the diffraction patterns observed by Fisher do not result from three different crystallite structures- alpha, beta, gamma - but from only two which he calls alpha and beta. The beta form is the same but the alpha and gamma forms of Fisher become the single alpha form of Takahashi. No matter which investigation is correct our study of stretched networks involves only two structures, alpha and beta. Indeed, we have established the presence of both structures in our samples using x-ray diffraction. In general, we have found with x-ray diffraction that slowly cooled networks (and networks crystallized at high temperatures) crystallized in two forms (alpha and beta), whereas rapid cooling to low temperatures produced the beta form almost exclusively. Furthermore, the higher the deformation the greater the proportion of alpha, as indicated by intensities. In other words, we find that high temperature and high stress encourages the formation of alpha. A detailed report on this work will be forthcoming in a separate manuscript. Since the simultaneous presence of alpha and beta seems an impediment for verifying the theory, an effort was made to isolate one form or the other by varying temperature and stress at crystallization.

Typical results obtained with the temperature program H_1 are shown in Figure 1. Deformation ratios are indicated by the symbol α. The figure reveals a significant difference between the incipient crystallization temperature Tc (determined by cooling), and the melting temperature T_m (determined by heating), indicated by the abrupt slope changes in the higher temperature region. The stress-temperature profiles tend to be U or V shaped depending upon strain and the phase (heating or cooling) of the cycle. Such large undercooling required to initiate crystallization demonstrates that equilibrium is more closely approached by the heating phase. Accordingly, only heating phases are discussed in the remainder of this report.

To favor alpha crystallization, the temperature program H_2 was followed (crystallization at 40°C for 20 hours). The results shown in Figure 2 indicate two melting points about 10° apart and suggest that both alpha and beta are abundant even at the highest deformation. At low deformations the presence of alpha appears quite limited. Two features of the curves should be noted. First, their shapes change considerably as deformation increases, so that the curve for $\alpha=3.04$ fails to fall to zero stress as do those at lower deformations. Second, the width of the null-stress zone is of the order of 5 or 6 degrees, which, as will be seen, is a narrow interval.

Here and throughout this work we attribute the two apparent melting points displayed on the force-temperature diagrams to the alpha and beta crystalline forms, the higher melting form being alpha. This is done in light of our diffraction studies that show clearly the presence of both crystalline forms in samples prepared by slow cooling or by crystallization at temperatures generally in excess of 40°C. Rapid cooling to low temperature produced only beta. We did not obtain the diffraction patterns for the actual samples used in force-temperature studies nor attempt to precisely duplicate cooling programs. Nevertheless, it is difficult to doubt the presence of alpha and beta crystalline forms in all slowly cooled samples. However, the question arises if the apparent melting temperatures indicated on the force-temperature diagrams are in fact due to the melting of alpha and beta crystallites. It seems reasonable to us that this is the case. After all, if alpha and beta are both present, they must melt somewhere, and melting should be indicated by the force-temperature behavior of stretched networks. But it is well known that a polymer crystalline form can of itself exhibit more than a single melting point. The melting temperature depends upon the size (thickness) of the crystallite; consequently, a network containing crystallites of two sizes will show two melting temperatures[13]. Or a distribution of crystallite sizes will show a distribution of melting temperatures, which also creates a zone of melting and recrystallization. No doubt some distribution of crystallite sizes is present in our networks. A bimodal distribution of a single crystallite form might account for our observations of two apparent melting temperatures several degrees apart, but if that were the case, how would we account for the melting (or lack of melting) of the second crystallite form that we know is also present? For the time being, the simplest explanation is that the apparent melting temperatures correspond to alpha and beta crystallites. Further investigation of this question is underway.

Other temperature programs allowing increased annealing yielded the same results. For example: showly cooling from 85° to 50°C (20 hours)

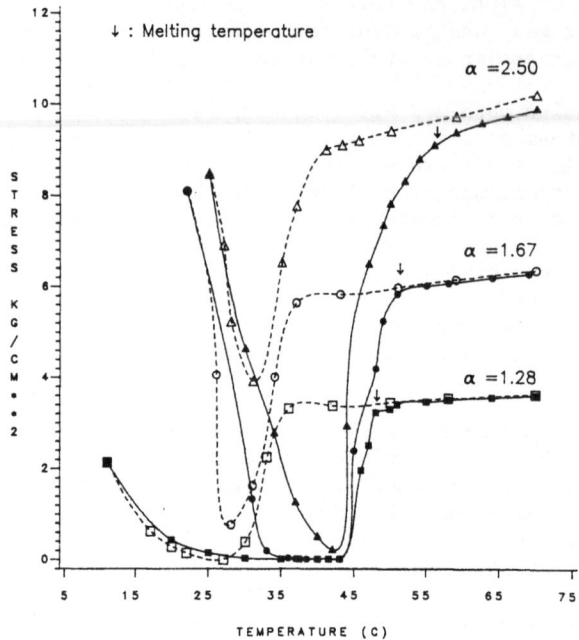

Fig. 1. Experimental stress-temperature profile of sample E. (dotted line is cooling curve, solid line is heating curve)

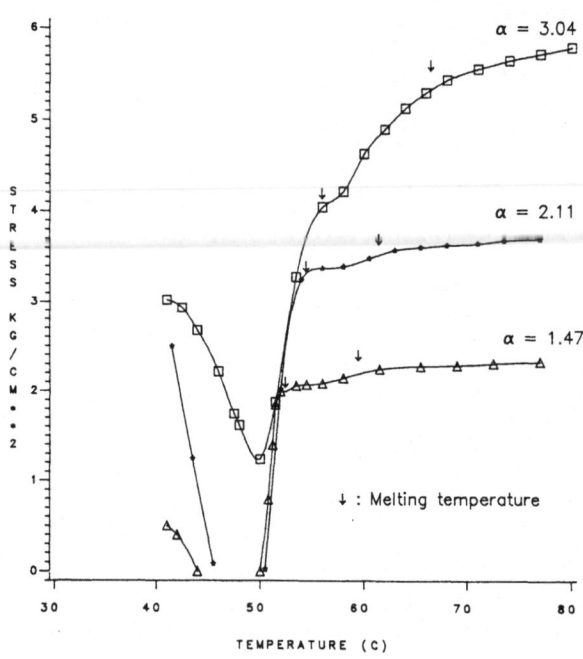

Fig. 2. Experimental stress-temperature profile of sample C crystallized at 40°C

to 40°C (20 hours) to room temperature (20 hours) produced the results in Figure 3. An interesting aspect of this figure is that the low temperature stresses greatly exceed the corresponding stresses of the amorphous networks (high temperatures), contrary to the observation reported by Gent[1] of a substantial diminution of the upturn stress as deformation increases. We found no indication that the upturn stress is diminished at high deformations.

Figures 2 and 3 show that the alpha crystalline form cannot be isolated from beta by crystallizing above 40°C under stress. Accordingly, we tried to isolate beta by quick cooling amorphous networks to 16-18°C (temperature program H₃). Results are shown in Figure 4 for a lightly crosslinked network (A) and in Figure 5 for a highly crosslinked sample (D). A network of intermediate crosslinking (C) is shown in Figure 6. Deformation ratios are indicated in the figures by α. Inspection of these figures reveals that valley widths of the profiles decrease as crosslinking increases; the null-stress zones of sample A are at least twice as wide as sample D. Curve irregularities also increase with crosslinking. The profiles of sample A (Figure 4) are quite regular relative to those of sample D, which suggests a greater abundance of beta crystallization in A. This is also suggested by the lower stress levels at comparable deformations of sample A; high stress favors the formation of alpha. And it might be noted that X-ray diffraction revealed only beta crystallization in networks of crosslinking and deformations similar to those of sample A in Figure 4. Approximate melting temperatures, indicated by arrows, are quite imprecise, particularly for samples C and D for which the transitions for crystalline to amorphous appear irregular and less abrupt than for sample A.

The uncertainty of the melting temperatures notwithstanding, the temperatures indicated in the figures and for duplicate samples if plotted

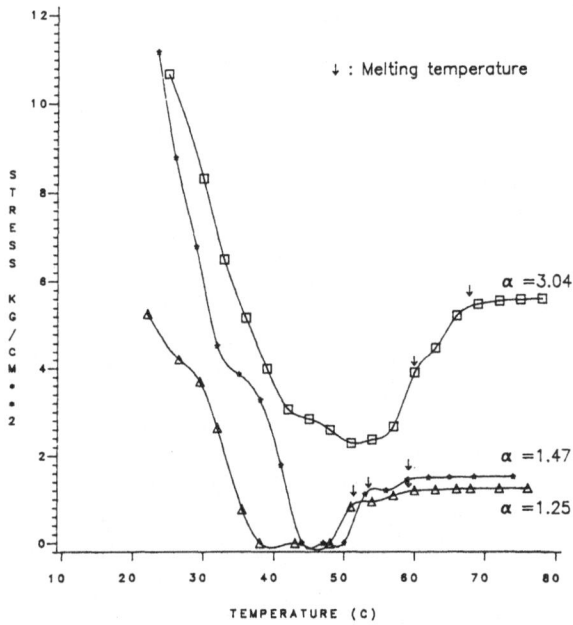

Fig. 3. Experimental stress–temperature profile of Sample C crystallized at 50-40-25°C

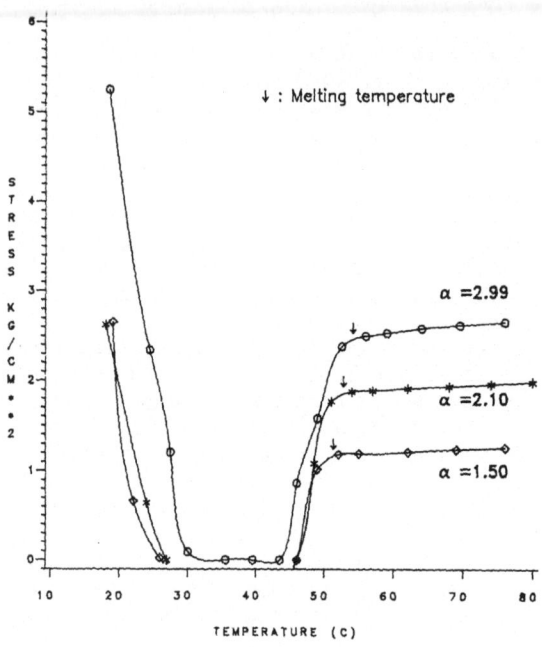

Fig. 4. Experimental stress-temperature profile of sample A

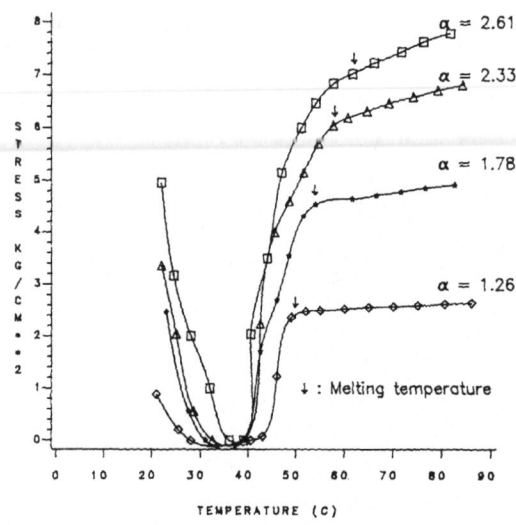

Fig. 5. Experimental stress-temperature profile of sample D

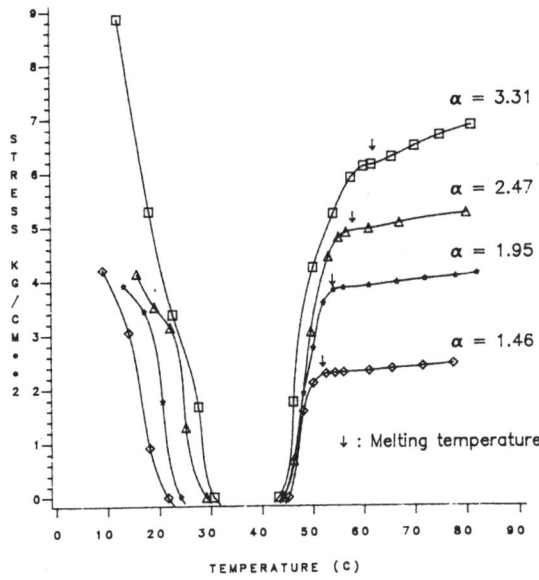

Fig. 6. Experimental stress-temperature profile of sample C

in accordance with equation (1) (assuming the term $2/\alpha$ can be neglected as a first approximation) provide a crude estimate of σ. The heat of fusion of the beta crystalline form is believed to be 2000 cal/mole of repeating units[5]. Such values of σ are shown in Figure 7 plotted against $N^{1/2}$. Estimates obtained from Gent's work[1] are also shown. These values of σ are not unreasonable, bearing in mind that the units of σ are <u>statistical</u> links (1 statistical link approximates 5.5 monomer repeating units[1]). But σ might be in error perhaps as much as 100%.

If σ is known the degree of crystallization and retractive force can be calculated with equations (2) and (3). Such calculations for sample C appear in Figures 8 and 9. Figure 8 shows the retractive force at $\alpha=3.31$ with $\sigma=0.75$ (from Figure 7); the theoretical melting temperature is shifted about 5° downward to give a better fit to the somewhat irregularly shaped experimental curve. Also shown in the figure for comparative purposes is the stress-temperature variation as predicted by Flory's original theory[11], which does not account for the stress elevation at lower temperatures and gives quite poor agreement above 30°C. In other words, Flory's theory does not fit the data anywhere. Figure 9 shows the theoretical crystallinities, via equation (2), for three deformations with $\sigma=0.75$. The calculated crystallinities are much higher than observed values; our measured crystallinities at room temperature (using a density gradient column) scattered between 36% and 42%. But a value of $\sigma=1.2$ brings the calculated crystallinity at $\sigma=3.31$ down to 38%. Recognizing the large uncertainty of σ estimated from melting point determinations, evaluations based upon crystallinity are expected to be considerably more reliable.

Density measurements at several elongations failed to reveal any significant variations of crystallinity with deformation, contrary to the theoretical prediction that equilibrium crystallization should decrease

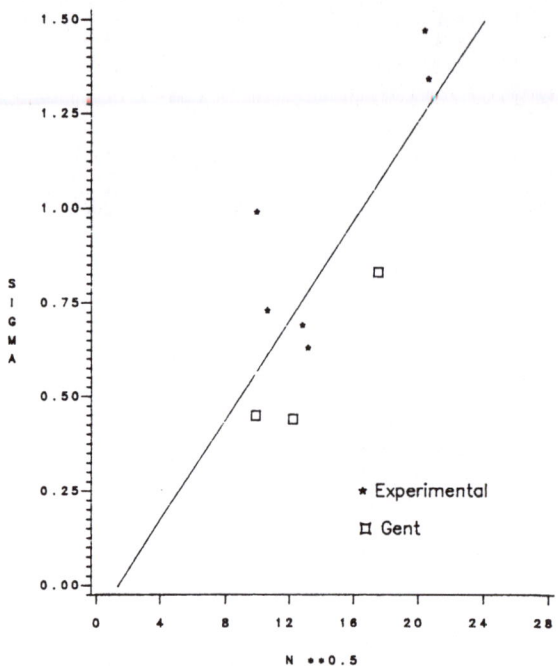

Fig. 7. Sigma vs. degree of crosslinking density

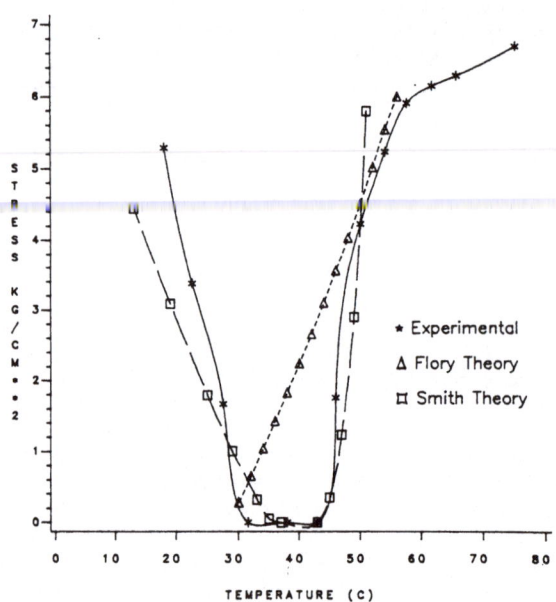

Fig. 8. Stress-temperature profile of sample C at α = 3.31 (sigma = 0.75 and 5.5 monomer units/statistical link)

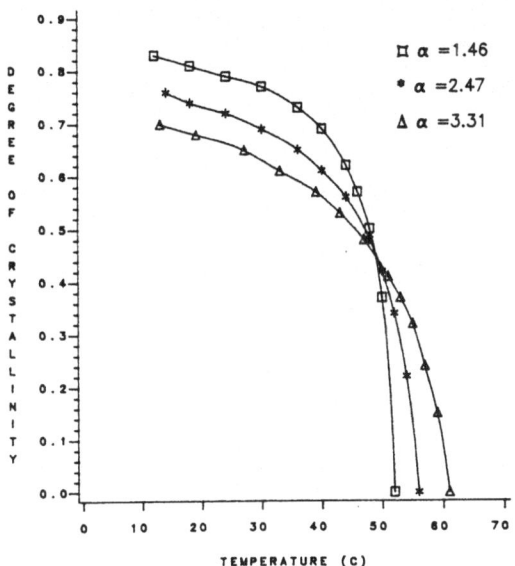

Fig. 9. Theoretical degree of crystallinity vs. temperature of sample C. (at constant sigma = 0.75)

with elongation, as shown in Figure 9. This can only mean that σ decreases with deformation, a reasonable conclusion since increasing uniaxial stress stabilizes nuclei and crystallites and necessitates fewer units for secondary nucleation for lateral crystallite growth. As an example, sample C requires σ to increase from 1.2 at α=3.31 to 1.8 at α=1.5 if crystallization is to remain at 38%. Consequently, we deduce that σ is a function of deformation as well as of crosslinking; i.e., σ=σ(N,α).

A variation of σ with deformation could alter the stress-temperature profile in a major way. In Figure 10 theoretical stress-temperature profiles for sample A are shown for three deformations with σ=1.60. This value of σ is determined by equation (2) for crystallinity of 38% and α=2.99. The value of T_m^o was estimated with DSC to be 42°C. A comparison of these curves with those in Figure 4 reveals several differences. The valley widths of the theoretical curves are much more narrow than those in Figure 4, and they tend to increase with increasing deformation in contrast to the experimental curves. These discrepancies can be eliminated by letting σ decrease with increasing deformation, but the results are sensitive to the exact form of the analytical relationship. For example, the function σ ∝ 1/α½ yields nearly constant crytallization as deformation changes but utterly fails to improve the stress-temperature profiles. The function σ ∝ exp (1/α), however, allows but a small change in crystallization with deformation (42% at α=1 and 38% at α=3) and produces stress-temperature profiles, shown in Figure 11, more in accord with experimental results.

Other functions could be sought in order to obtain better agreement with our experimental results, but at this stage such an effort might not be productive, the reason being that equilibrium conditions may not have been achieved. It is expected that quick cooling to produce beta

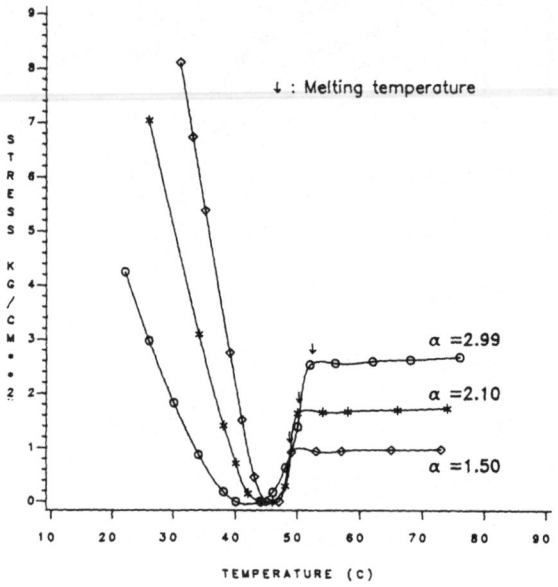

Fig. 10. Theoretical stress-temperature profile of sample A. (at constant
sigma = 1.60)

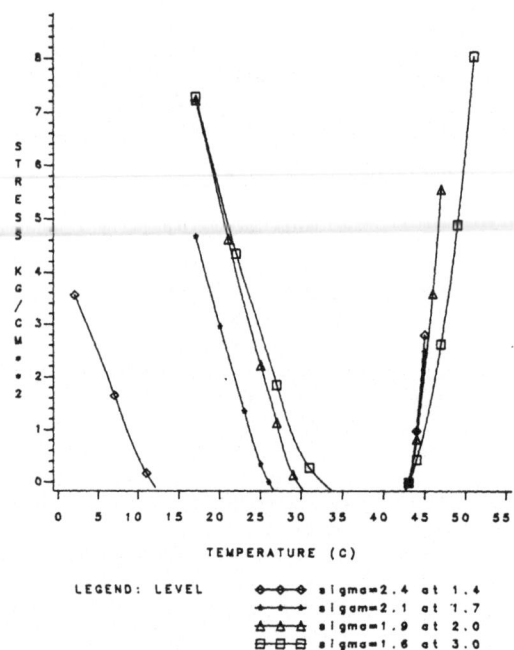

Fig. 11. Theoretical stress-temperature profile of sample A. (sigma is
varied with elongation)

crystallization generates non-equilibrium conditions. Consequently, the valley widths of our experimental curves could be considerably wider than they would be if equilibrium could be obtained, but we are unable to speculate at this time how serious the change might be or if other changes might occur. Therefore, further improvement of the theory must be postponed until this question is resolved. It should also be pointed out that the theory is invalid at low deformations in any instance because of the restrictions that crystallites be perfectly aligned along the stretch direction. Strictly, the theory should not be expected to hold below deformation of $\alpha=2.5$ or 3, but it may still give useful approximations a little below this range.

We have found that σ decreases with increasing crosslinking and increasing uniaxial deformation. Is this reasonable? If σ is a nucleation parameter for secondary nucleation (lateral crystallite growth) as suggested in the original theory[3], this indeed might be the case. σ is the least number of units (statistical links) deposited onto a crystallite face necessary to prevent dissolution of the chain back into the amorphous domain by configurational fluctuations. The cohesive forces between σ units and the crystallite surface exactly suffice to prevent such fluctuations, allowing additional links along the chain to crystallize. Anything that enhances the stability of the crystalline phase also enhances the stability of the secondary nuclei, and consequently provides for a smaller value of the critical parameter σ. It is reasonable to expect tension (uniaxial deformation) to be a stabilizing factor. Tension of a stretched network is increased if the deformation (α) is increased, and for a given deformation a network of higher crosslinking experiences a greater tension than one of lesser crosslinking. It then follows that the same factors that govern tension of stretched networks (α, N) also determine the value of σ, which agrees with our experimental findings. This argument indicates that we should not be surprised to find that σ behaves as it does.

The stress-temperature profile is intimately tied to σ and also to crystallinity ω, which too is controlled by σ. This makes it difficult to speculate about stress-temperature profiles in general. As a nucleation parameter for lateral growth, σ is expected to be unique for the polymer in question. Consequently we cannot draw conclusions about the expected behavior of other polymers (say polychloroprene) from what we have learned about trans-polyisoprene. To do so requires a detailed understanding of all factors affecting σ, which we lack at this time. Further progress in this area therefore requires not only a thorough continuation of the studies reported herein, but an expansion of the program to include additional polymers. We are moving in that direction.

REFERENCES

1. A.N. Gent, J. Polymer Sci., A2(4), 447 (1966).
2. K.J. Smith, J. Polymer Sci., 21, 45 (1983).
3. K.J. Smith, J. Polymer Sci., 21, 55 (1983).
4. K.J. Smith, ACS Syposium Series 193 Elastomers and Rubber Elasticity, J. Mark and J. Lal, Eds., 1982.
5. L. Mandelkern, F. Quinn, and D. Roberts, J. Am. Chem. Soc., 78, 926 (1956).
6. C.W. Bunn, Proc. Roy. Soc. (London), A180, 40 (1942).
7. D. Fisher, Proc. Roy. Soc. (London), B66, 7 (1953).
8. A. Britton, J. Sullivan, and K. J. Smith, J. Polymer Sci., 17, 1281 (1979).
9. W. Cooper and G. Vaughan, Polymer, 4, 329 (1963).
10. S. Mukherji and A. Woodward, J. Polymer Sci., 22, 793 (1984).

11. P. J. Flory, J. Chem., Phys., <u>15</u>, 397 (1947).

12. Y. Takahashi, T. Sato, and H. Tadokoro, J. Polymer Sci.
     (Phys.), <u>11</u>, 233 (1973).

13. E.G. Lowering and D.C. Wooden, J. Polymer Sci., A2 <u>7</u>,
     1639 (1969).

APPENDIX: Temperature Programs

   H_1 – Temperature was lowered 3°C/20 minutes then reheated to
        85°C at 3°C/10 minutes.

   H_2 – Temperature was gradually lowered to 40°C and held there
        for 20 hours, then reheated.

   H_3 – Films were quenched to 18°C with cold water, annealed at room
        temperature for 4 hours, then cooled slowly to 18°C overnight.

INTRAMOLECULAR REACTION AND NETWORK PROPERTIES

R.F.T. Stepto

Department of Polymer Science and Technology
The University of Manchester Institute of Science
 and Technology
Manchester
M60 1QD
U.K.

INTRODUCTION

There has been much activity in recent years relating theoretical and experimental values of the static shear moduli of polymer networks using model systems based on end-linking reactions; essentially non-linear polymerisations[1-11]. Almost all the work has assumed that, with stoichiometric reaction mixtures, chemically perfect networks are formed at complete reaction. The number of elastic chains per unit volume is taken to be defined precisely by the molar masses of the monomers and/or prepolymers reacted and the density of the network formed. Attention has then been focussed on whether affine or phantom chain behaviour occurs[2,4,5,7], and whether chain entanglements or interactions contribute additionally to the moduli measured[1-3,7]. However, a knowledge of imperfections in the network structure is essential if absolute values of moduli are to be interpreted meaningfully. Thus, it has been shown that in some siloxane systems[1,6] incomplete reaction gives inelastic free chain ends and reduces the modulus whilst entanglements cause it to increase, with the result that apparent perfect, affine behaviour can be observed.

A further type of imperfection which occurs naturally in end-linking polymerisation is inelastic loop formation[8-11], again causing a reduction in modulus. Such loops arise from intramolecular reaction. It is the aim of this paper to investigate, using theoretical arguments and modulus measurements, the extent of inelastic loop formation in model networks.

CONSIDERATIONS OF INTRAMOLECULAR REACTION AND NETWORK FORMATION

Intramolecular reaction in linear polymerisations leads to cyclic species and the concentrations of such species have been studied for systems in chemical equilibrium and for irreversible polymerisations[12]. In the former case, individual species have been isolated but in the latter only overall number fractions of rings have been measured. In both cases experimental data are interpreted satisfactorily by theory. For ring-chain equilibrates, Jacobson-Stockmayer theory may be used and for irreversible polymerisations, cascade, kinetics or rate theory may be employed[12]. Most polymer preparations, including those forming networks, use kinetically controlled, reversible reactions or irreversible reactions to obtain high molar mass products. Under these conditions and in bulk it has been shown[12] that the amount of ring formation will usually be negligible. Thus, given the equal reactivities of like functional groups, Flory's most probable distribution of molecular species results. However, a sharp contrast should be drawn with non-linear polymerisations. Even in the pre-gel regime, data show[9,12] that intramolecular reaction can rarely be neglected. Hence, for a linear polyurethane (PU)-forming system in bulk the number fraction of ring species (N_r) was about 0.02 at 90% reaction, whilst for a bulk tri-functional PU-forming system N_r was about 0.30 at the gel point. This difference occurred in spite of the fact that the probability of forming a ring structure per opportunity for so doing was higher in the linear polymerisation studied[9]. The difference arose naturally from the much larger number of reactive groups per molecule in a non-linear polymerisation. This number increases as a polymerisation proceeds and becomes infinite at the gel point and beyond.

The formation of a perfect network from an end-linking polymerisation requires that pre-gel intramolecular reaction is negligible and that post-gel intramolecular reaction, which must occur, always leads to elastically active chains. These requirements are unlikely to be met. First, pre-gel intramolecular reaction cannot be neglected and, of the ring structures formed, it is known from chain-statistical considerations and from experiment that the smallest is the most numerous. The smallest ring structure in a $RA_2 + RB_3$ polymerisation is illustrated in fig 1. In general, such structures in non-linear $RA_2 + RB_f$ polymerisations must give inelastic loops in the networks formed at complete reaction. (Larger loops may give elastically active chains at complete reaction.) Although the polymerisation

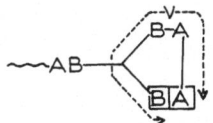

Fig 1: Smallest ring structure in an $RA_2 + RB_3$ polymerisation. ν is the number of bonds in the chain which can form the smallest ring. The chain has mean-square end-to-end distance $\langle r^2 \rangle = \nu b^2$.

statistics of the pre-gel and post-gel regimes are different, intramolecular reaction leading to the smallest ring structure must occur. Thus, inelastic loops will arise from pre-gel intramolecular reaction <u>and</u> post-gel intra-molecular reaction.

PRE-GEL INTRAMOLECULAR REACTION

A satisfactory, quantitative description of pre-gel intramolecular reaction in non-linear polymerisation has yet to be made. Cascade theory and rate theory supply only approximate descriptions at present[13-15]. However, the general statistical considerations which have been shown to apply to linear polymerisations will also hold for non-linear polymerisations.

Referring to fig 1, ν is the number of bonds in the chain which can form the smallest ring structure. Alternatively, ν can be considered the number of bonds in the linear repeating structure of the chains formed by the polymerisation [12,16]. Because the mean-square end-to-end distance of the chain ($\langle r^2 \rangle$) increases as ν increases, intramolecular reaction is less for higher values of ν, i.e. higher reactant molar masses. Intramolecular reaction is also less for stiffer chains, which have larger effective bond lengths(b). In a linear polymerisation, there are only two reactive ends on each molecule. In non-linear polymerisations the number of reactive groups per molecule is not fixed. However, at a given extent of reaction in $RA_2 + RB_f$ polymerisations it increases with the functionality of the branch species(f). Thus[9-12,17], intramolecular reaction increases with f. Finally, intramolecular reaction also increases if reaction systems are diluted with solvent[9-12,16,17]. The probability of reaction between groups on different molecules (intermolecular reaction) is then reduced, although intramolecular reaction itself is not affected, except in cases where chain expansion occurs. Expansion effects will usually be small unless extreme dilutions are used. Even in bulk, the reactive groups in a polymerising mixture are diluted by the inert portions of the chains between the groups. Only in the limit of zero molar masses of reactants is there zero dilution of reactive groups. Of course, in that limit ν is also zero so that, apart from considerations of ring strain, intramolecular reaction will still occur.

The importance of the factors enunciated, reactant molar mass (ν), chain flexibility (b) and functionality (f), and dilution, on amount of pre-gel intramolecular reaction and resulting delays in the gel point have been discussed in several previous publications [9-12,16,17]. They are perhaps most easily summarised in terms of the approximate gel-point equation of Ahmad and Stepto[16] for $RA_2 + RB_f$ polymerisations, namely

$$(f - 1)\alpha_c (1 - \lambda_{ab})^2 = 1 . \tag{1}$$

α_c is the product of extents of reaction of A and B groups and λ_{ab} is a ring forming parameter. Eqn(1) may be rearranged to give

$$\lambda'_{ab} = (f - 1)^{\frac{1}{2}} \alpha_c^{\frac{1}{2}} - 1 \tag{2}$$

where

$$\lambda'_{ab} = \lambda_{ab}/(1 - \lambda_{ab}) \tag{3}$$

and

$$\lambda'_{ab} = \frac{c_{int}}{c_{ext}} = \frac{(f - 2)(3/(2\pi\nu\,b^2))^{3/2}.\phi(1,3/2)}{N.(c'_a + c'_b)} \tag{4}$$

Eqn(2) may be used to evaluate λ'_{ab} from experimental values of α_c with λ'_{ab} interpreted according to eqn(4). In eqn(4), $c_{ext} = (c'_a + c'_b)$ is an average external concentration of A and B groups around a given molecule, that is, an average concentration from all other molecules. It has to be chosen arbitrarily and may be equated to the initial molar concentration of reactive groups, $(c_{ao} + c_{bo})$, and the gel-point concentration, $(c_{ac} + c_{bc})$, as extreme values. $(c'_a + c'_b)^{-1}$ represents the dilution of reactive groups discussed previously. c_{int} is the (internal) concentration of B groups around an A group on the same molecule, or vice versa. In the expression for c_{int} in eqn(4), $(f - 2)$ is the number of opportunities for intramolecular reaction for each size of ring, $(3/(2\pi\nu b^2))^{3/2}$ is the probability (assuming Gaussian statistics) that the groups at the ends of the sub-chain of ν bonds are coincident; $\phi(1,3/2) = \sum_{i=1}^{\infty} 1^i i^{-3/2} = 2.612$ is the sum due to ring structures of all sizes and N is the Avogadro constant. It can be seen that c_{int} reflects the dependence of intramolecular reaction on ν, b and f.

EXPERIMENTAL DATA – CORRELATION BETWEEN GEL POINT AND NETWORK PROPERTIES

The experimental data to be considered in the present paper have been published previously[9-11]. They refer to hexamethylene diisocyante (HDI) reacting with polyoxypropylene (POP) triols and tetrols in bulk and in nitrobenzene solution at various dilutions, that is, to series of $RA_2 + RB_f$ polymerisations. All reactions were stoichiometric, i.e.$[NCO]_o/[OH]_o = r=1$. The distinct feature of the data is that extents of reaction at the gel points (α_c) as well as the moduli of the networks at complete reaction were measured. As one is not yet able to predict network structure and therefore properties directly from reactant structures and reaction conditions, α_c is a good indicator of the propensity of a polymerisation mixture for intramolecular reaction. It serves as a reference point – one that is too often neglected in preparations of model networks.

The experimental data are shown in fig.2, where M_c/M_c^o, the molar mass of chains between elastically effective junction points relative to that of the perfect network, is plotted versus $p_{r,c}$, the extent of intramolecular reaction at gelation. M_c was evaluated from accurate, small-strain compression measurements on swollen and dry samples for each network. The averaging of values of M_c from swollen and dry networks introduces about 10% uncertainty in each point[8,9]. M_c^o was calculated directly from the molar masses and functionalities of the reactants and is essentially the molar mass of a chain of ν bonds (see fig 1 and ref 9-11). M_c/M_c^o is equal to the proportional reduction in modulus below that expected for the perfect, dry network.

For reactions at r = 1, $p_{r,c}$ is defined by the equation

$$p_{r,c} = (\alpha_c)^{\frac{1}{2}} - (f - 1)^{-\frac{1}{2}} \qquad (5)$$

where $(f - 1)^{-\frac{1}{2}}$ is $(\alpha_c^o)^{\frac{1}{2}}$, with α_c^o the value of α_c at the ideal, Flory-

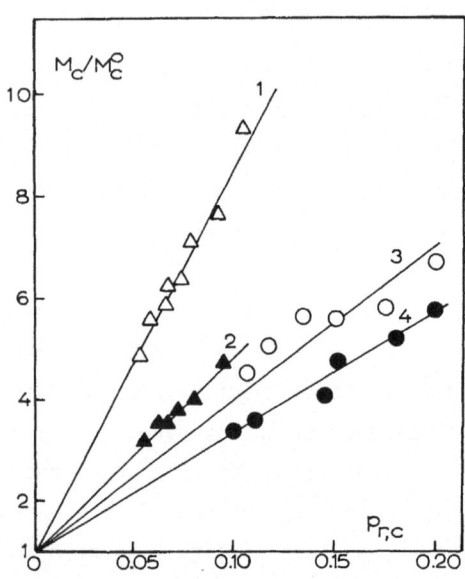

Fig 2: Molar mass between elastically effective junction points(M_c) relative to that for the perfect network(M_c^o) versus extent of intramolecular reaction at gelation($p_{r,c}$).

Polyurethane networks from hexamethylene diisocyanate(HDI) reacting with polyoxypropylene(POP) triols and tetrols at 80°C in bulk and in nitrobenzene solution[9-11].

Systems 1 and 2: HDI/POP triols; ν_1 = 33, ν_2 = 61. Systems 3 and 4: HDI/POP tetrols; ν_3 = 29, ν_4 = 33.

Stockmayer gel point. From eqn(2) an alternative definition of $p_{r,c}$ is

$$p_{r,c} = \lambda'_{ab}/(f - 1)^{\frac{1}{2}}. \tag{6}$$

$p_{r,c}$ was varied by carrying out polymerisations at various initial dilutions in nitrobenzene. For the four groups of polymerisations, the lowest values of $p_{r,c}$ come from polymerisations in bulk.

In fig 2, systems 1 and 2 refer to HDI/POP triol polymerisations using two triols of different molar masses giving the values of ν indicated in the legend. Systems 3 and 4 refer to HDI/POP tetrol polymerisations using two different tetrols. The values of M_c/M_c^o are all greater than unity. Thus, on an affine basis (see later) marked reductions in modulus occur, even for bulk reaction systems, for which the excess reaction at gel($p_{r,c}$) is about 5% for the triol polymerisations and about 10% for the tetrol polymerisations. Bulk system 1 shows a 5-fold reduction in modulus and bulk system 4 a 3-fold reduction.

The slopes of the lines in fig 2 relate to proportions of inelastic loops formed for given extents of pre-gel intramolecular reaction. Thus, for a given value of $p_{r,c}$ more inelastic loops are formed in the triol networks (systems 1 and 2) compared with the tetrol networks (systems 3 and 4). In addition, for a given functionality, as ν increases the proportion of inelastic loops decreases (compare systems 1 and 2 and systems 3 and 4). Similar results have been obtained from polyester-forming systems using POP triols and diacid chlorides[8].

The data show clearly that M_c/M_c^o increases with $p_{r,c}$. However, the linear relationships between M_c/M_c^o and $p_{r,c}$ are purely empirical[8-11]. The lines have been extrapolated to $M_c/M_c^o = 1$ at $p_{r,c} = 0$, predicting that perfect, affine networks are formed by ideal gelling systems. As discussed previously, this prediction is unlikely to be true, as there is no reason to assume that some smallest loops are not formed post-gel. The validity of this prediction is examined more closely in the following sections.

REINTERPRETATION OF GEL POINT AND MODULUS CORRELATIONS

The simplest approximation that can be made concerning the network defects is that all are due to the smallest loops[15]. The occurrence of such loops in $RA_2 + RB_3$ and $RA_2 + RB_4$ polymerisations is shown in fig 3 and 4. Structures 3(c), 3(d) and 4(b) refer to sol fraction species. In the

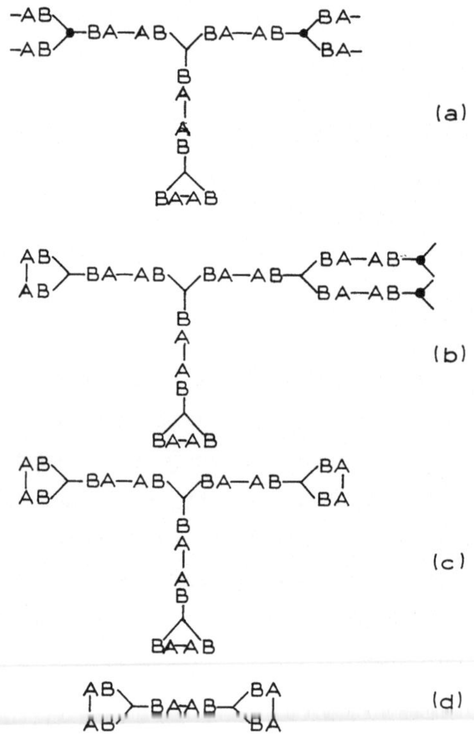

Fig 3: The occurrence of the smallest loops at complete reaction in a RA_2 + RB_3 polymerisation. Structures (a) and (b) introduce inelastic chains, and (c) and (d) contribute to the sol fraction. ⤙ denotes an elastically effective junction point.

experiments carried out, the amount of sol fraction formed was usually
negligible and it was in any case removed by swelling the networks to
equilibrium in several changes of nitrobenzene. Structures 3(a) and 3(b)
show that for trifunctional networks every smallest loop removes two junction
points, whilst structure 4(a) shows that for tetrafunctional networks only
one junction point is lost per loop. Notwithstanding that more complex ring

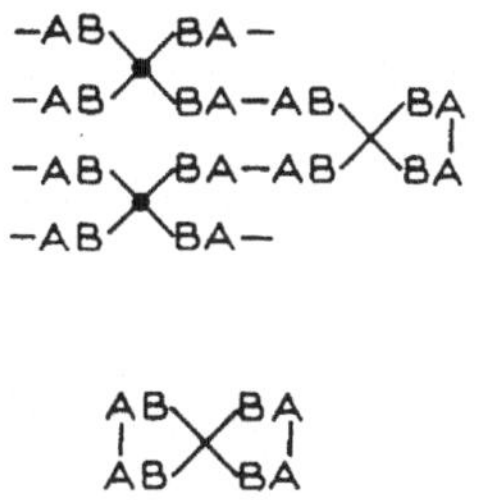

Fig 4: The occurrence of the smallest loops at complete reaction in an RA_2
+ RB_4 polymerisation. Structure (a) introduces inelastic chains, and (b)
contributes to the sol fraction. $\prec$ denotes an elastically effective
junction point.

structures will occur, this difference is the basic reason why trifunctional
networks are more sensitive to loop defects than tetrafunctional ones
(cf fig 2).

The equations used for the evaluation of M_c were[8]

$$\sigma = G(\Lambda - \Lambda^{-2}), \tag{7}$$

and

$$G = ART\rho\phi_2^{1/3}(V_u/V_F)^{2/3}/M_c. \tag{8}$$

Eqn(7) is the Gaussian stress-strain function, with σ the force per unit
area of the undeformed network, G the shear modulus and Λ the deformation
ratio. In eqn(8), A is the so-called front factor, ρ the density of the

dry network, ϕ_2 the polymer volume fraction during measurement, V_u the value of the dry, underformed network, and V_F the volume of the network at formation, taken to be its strain-free reference state. ϕ_2 is equal to 1 for dry networks. A is equal to 1 and $(1 - 2/f)$, respectively, for the extreme, idealised affine and phantom chain behaviours[2,4,5].

Under the small deformations by which the present data were obtained, A is expected to be unity, and was assumed to be so in fig 2. However, in general, M_c/A is determined from measurements of G. ρ/M_c is the number of elastic chains per unit volume(n), and for a perfect network

$$\frac{\rho}{M_c^o} = n^o = \frac{f}{2}.N_B , \qquad (9)$$

where N_B is concentration of RB_f units in the polymerisation. Fig 3 shows that for $f = 3$ the number of elastically effective junction points is

$$\frac{\rho}{M_c} = n = \frac{f}{2}.N_B(1 - 2fp_{1,e}) , \qquad (10)$$

where $p_{1,e}$ is the extent of intramolecular reaction at the end of the reaction, with $N_B.f.p_{1,e}$ the number of intramolecularly reacted B(or A) groups. Thus,

$$M_c/M_c^o = 1/(1 - 6p_{1,e}). \qquad (11)$$

Eqn(11) may be rearranged to allow evaluation of $p_{1,e}$ from measured values of M_c/A with

$$p_{1,e} = \frac{1}{6}(1 - \frac{1}{A}.(AM_c^o/M_c)) . \qquad (12)$$

Similar arguments for f = 4 give

$$p_{1,e} = \frac{1}{4}(1 - \frac{1}{A}.(AM_c^o/M_c)) , \qquad (13)$$

as only one junction point is lost per smallest loop. Although the detailed arguments leading to eqn(12) and (13) are based on smallest loops, the actual equations themselves are valid provided each inelastic loop causes the loss of two junctions points when f = 3 and one junction point when f = 4, irrespective of the size of the loop.

To evaluate $p_{1,e}$, a value of A has to be assumed, and used together with the values of M_c/AM_c^o, which are numerically equal to these of M_c/M_c^o in fig 2. Accordingly, fig 5 and 6 show $p_{1,e}$ plotted versus $p_{r,c}$ for affine (A = 1) and phantom (A = 1 - 2/f) chain behaviour. One condition that must be obeyed in the plots is that $p_{1,e} > p_{r,c}$, as $p_{1,e}$ includes pre-gel and post-gel reaction. The lines $p_{1,e} = p_{r,c}$ are indicated in both figures and it is apparent that this condition is met only for affine behaviour (fig 5).

In keeping with the preceding discussion on post-gel intramolecular reaction, the points in fig 5 do not show a tendency for $p_{1,e}$ to tend to zero as $p_{r,c}$ tends to zero. That is, even in the limit of a perfect gelling system inelastic loops are formed post-gel. Extrapolation to $p_{r,c} = 0$ gives $p_{1,e}^o$, the extent of reaction leading to inelastic loops at complete reaction in the perfect gelling system. The values of $p_{1,e}^o$ range from about 9% to 18% for the systems studied. As expected from considerations of pre-gel intramolecular reaction, the values of $p_{1,e}^o$ are smaller for f = 3 compared with f = 4 and they increase as ν decreases.

The derived values of $p_{1,e}^o$ may be reconverted to values of M_c/M_c^o for $p_{r,c} = 0$ using eqn(11), and fig 7 shows M_c/M_c^o versus $p_{r,c}$, as in fig 2, but with the values of M_c/M_c^o at $p_{r,c} = 0$ consistent with those of $p_{1,e}^o$ from fig 5. The curves give just as satisfactory a fit to the data as the straight lines in fig 2. Because it is not possible to have concentrations of reactive groups higher than those in bulk, points at lower values of $p_{r,c}$ than those shown cannot be obtained for the particular systems studied. Thus, there are uncertainties in the values of $p_{1,e}^o$ from fig 5 and in the intercepts shown in fig 7. However, merely the existence of non-zero intercepts shows that post-gel intramolecular reaction indeed leads to network imperfections. At their present stage of development, theories of

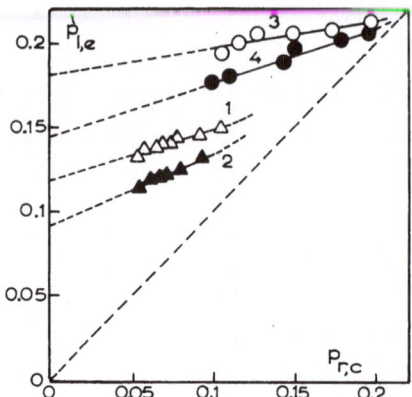

Fig 5: Extent of intramolecular reaction at complete reaction leading to inelastic loops ($p_{1,e}$) versus $p_{r,c}$. Values of $p_{1,e}$ derived from modulus measurements on the basis of smallest loops and affine chain behaviour. Intercepts at $p_{r,c} = 0$ define values of $p_{1,e}$ for ideal gelling systems, denoted $p_{1,e}^{o}$.

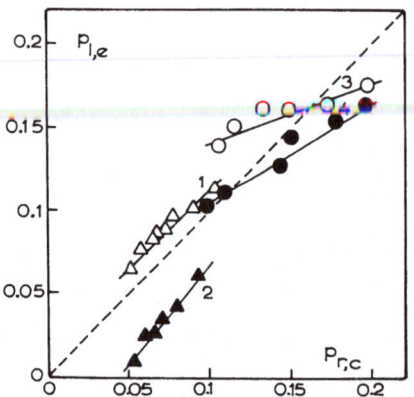

Fig 6: $p_{1,e}$ versus $p_{r,c}$. As fig 5 but values of $p_{1,e}$ derived assuming phantom chain behaviour.

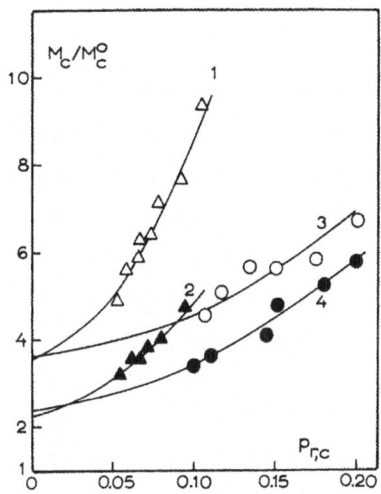

Fig 7: M_c/M_c^0 versus $p_{r,c}$ with curves drawn consistent with values of $p_{1,e}^0$ from fig 5.

polymerisation including intramolecular reaction[15,18,19] cannot treat this phenomenon. They all predict that $p_{1,e} = 0$ when $p_{r,c} = 0$.

UNIVERSAL OCCURRENCE OF IMPERFECTIONS

The reaction systems used do not have particularly large values of ν. Hence, it may be argued that, with triols and tetrols of higher molar mass, networks with $M_c = M_c^o$ would have been obtained, at least for bulk reaction systems. The decrease of $p_{1,e}^o$ for a given functionality as ν increases (fig 5) indicates a trend in this direction. However, for $M_c = M_c^o$, $p_{1,e}^o$ must be equal to zero. That such a value is unlikely to be obtained, irrespective of ν, is emphasised by using the present values $p_{1,e}^o$ to predict a value for infinite ν. Thus fig 8 shows $p_{1,e}^o$ versus $(f - 2)/(\nu b^2)^{3/2}$. The abscissa is proportional to c_{int} of eqn(4) and relates to the probability of pre-gel intramolecular reaction. The values of b required for the reactions studied have been evaluated previously[9-11] from the gel points, i.e. λ_{ab}', using eqn(4) and $(c_a' + c_b') = (c_{ao} + c_{bo})$. If $(c_a' + c_b') = (c_{ac} + c_{bc})$ is used then different values of b are obtained, but again linear behaviour is found with approximately the same intercept at $(f - 2)/(\nu b^2)^{3/2} = 0$. The positive intercept indicates that post-gel (inelastic) loop formation is influenced by the same factors as pre-gel intramolecular reaction but is not determined solely by them. Imperfections still occur in the limit of infinite reactant molar mass or very stiff chains ($\nu b^2 \to \infty$). The relationship is indicated to be a universal one, irrespective of ν, b and f, and shows that about 8% of the overall reaction between groups leads to inelastic loops. Such reaction occurs post-gel because of the unlimited number of groups per molecule in the gel fraction. It will be important to explore the universality of the relationship by measuring gel points and networks properties using other types of reaction systems.

CONCLUSIONS

The assumption that end-linking polymerisations lead to perfect networks implies that the intramolecular reaction which occurs post-gel never leads to inelastic loops. This assumption is unlikely to be true and has been examined critically in the present paper.

Previous correlations of shear moduli and extents of reaction at the gel point (fig 2) have shown that the network imperfections increase with pre-gel intramolecular reaction and thus depend to some extent on the same parameters, namely, ν, b and f. A more detailed analysis shows clearly that,

as expected for the small strain measurements made, affine rather than
phantom chain behaviour occurs (fig 5 and 6). The analysis is subject to
the assumption that each loop removes two junction points for f = 3 and
one junction point for f = 4 reactions. The assumption is strictly correct
if only the smallest loops are formed. Such loops are by far the most
frequent in pre-gel intramolecular reaction[20].

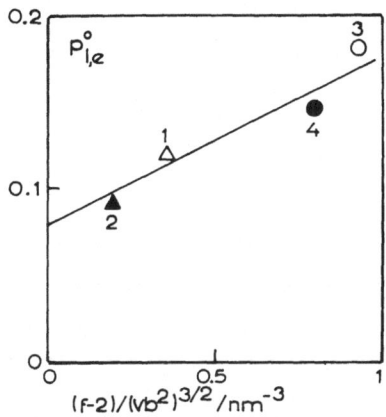

Fig 8: Dependence of the post-gel reaction leading to inelastic loops
($p_{1,e}^{o}$) on the parameters affecting pre-gel intramolecular reaction
$((f - 2)/(\nu b^{2})^{3/2})$.

Extrapolation to the limit of perfect gelling systems shows that
between 9 and 18% of reaction leads to inelastic loops (fig 5), the amount
($p_{1,e}^{o}$) increasing with f and increasing with decrease in ν, as for pre-gel
intramolecular reaction. Correlation with ν, f and b predicts that, even
for reactants of infinite molar mass, network imperfections persist with
$p_{1,e}^{o} \simeq 0.08$ (fig 8). Subject to the present analysis, this may be a
universal value and needs to be checked with other systems. Loop imperfections are a natural (law of mass action) consequence of the reactions of
gel molecules which contain unlimited numbers of reactive groups. Thus,
although pre-gel intramolecular reaction may be in principle be suppressed

or made negligible, post-gel intramolecular reaction must occur and naturally leads to some inelastic loops.

The occurrence of network imperfections can most clearly be investigated by forming networks from reactions carried out at different initial dilutions and by using the gel point as measure of the propensity of a system to form loops. Even for reactions in bulk, there is a dilution of reactive groups and more inelastic loops are formed than would be expected for perfect gelling systems. Inelastic loops have marked effects on the moduli of the networks formed at complete reaction, with values of M_c for the present systems being 3 to 5 times higher than M_c^o; the value often assumed in model network preparations. Further even the limiting value of $p_{1,e}^o$ of approximately 0.08 predicted for reactants of infinite molar mass leads, from eqn(12) and (13), to $M_c = 1.92 \, M_c^o$ for f = 3 and $M_c = 1.47 \, M_c^o$ for f = 4. Thus, gross errors are introduced if it is assumed that $M_c = M_c^o$ in interpretations of measured moduli of networks formed from end-linking polymerisations.

REFERENCES

1. M.Gottlieb, C.W.Macosko, G.S.Benjamin, K.O.Meyers and E.W.Merrill, Macromolecules, 14 1039(1981).

2. J.E.Mark, Adv. Polym. Sci., 44,1(1982).

3. W.Batsberg and O.Kramer, Chain Entangling in Elastomers, in: "Elastomers and Rubber Elasticity," J.E.Mark and J.Lal, eds, Amer. Chem. Soc. Symposium Series 193, Washington, D.C.(1982).

4. P.J.Flory, Br. Polymer J., 17,96(1985).

5. B.Erman, Br. Polymer J., 17,140(1985).

6. C.W.Macosko and J.C.Saam, Polymer Preprints, 26,48(1985).

7. R.W.Brotzman and P.J.Flory, Polymer Preprints, 26,51(1985).

8. A.B.Fasina and R.F.T.Stepto, Makromol. Chem., 182,2479(1981).

9. J.L.Stanford and R.F.T.Stepto, Experimental Studies of the Formation and Properties of Polymer Networks, in: "Elastomers and Rubber Elasticity," J.E.Mark and J.Lal, eds., Amer.Chem.Soc. Symposium Series 193, Washington, D.C.(1982).

10. J.L.Stanford, R.F.T.Stepto and R.H.Still, Studies of the Formation and Properties of Polyurethanes suitable for Reaction Injection Moulding, in "Reaction Injection Molding and Fast Polymerization Reactions," J.E.Kresta, ed., Plenum Publishing Corp., New York(1982).

11. J.L.Stanford, R.F.T.Stepto and R.H.Still, Formation and Properties of Polymer Networks, in: "Characterization of Highly Cross-Linked Polymers,"

S.S.Labana and R.A.Dickie, eds., Amer.Chem.Soc. Symposium Series 243, Washington D.C.(1984).

12. R.F.T.Stepto, Intramolecular Reaction and Gelation in Condensation or Random Polymerisation, in: "Developments in Polymerisation - 3," R.N.Haward, ed., Applied Science Publishers, Ltd., London(1982).

13. J.L.Cawse, J.L.Stanford and R.F.T.Stepto, Proc. 26th IUPAC International Symposium on Macromolecules, Mainz, 1979, p.693.

14. V.Askitopoulos and R.F.T.Stepto, to be published; see V.Askitopoulos, M.Sc.Thesis, University of Manchester (1981).

15. A.C.Lloyd and R.F.T.Stepto, Br. Polymer J., 17,190(1985).

16. Z.Ahmad and R.F.T.Stepto, Colloid and Polymer Sci., 258,663(1980).

17. Z.Ahmad, R.F.T.Stepto and R.H.Still, Br.Polymer J.,17,205(1985).

18. M.Gordon and W.B.Temple, Makromol. Chem., 160,263(1972).

19. M.Adam, M.Delsanti, D.Durand, G.Hild and J.P.Munch, Pure and Appl. Chem, 53,1489(1981).

20. R.F.T.Stepto, Faraday Disc. Chem.Soc., 57,69(1974).

THE HYDROSILYLATION CURE OF POLYISOBUTENE

C. W. Macosko

Unv. of Minnesota
Dept. of Chemical
Engr.& Materials Science
Minneapolis, MN

J. C. Saam

Dow Corning Corp.
Midland, MI 48686

ABSTRACT

A liquid polyisobutene oligomer with unsaturated chain ends undergoes hydrosilylation with $HMe_2SiOMe_2SiOMe_2SiH$ or $Si(OMe_2SiH)_4$ to give higher molecular weight polymers or elastomers. A major side reaction consumes SiH to give redistributed siloxane in the resulting polymers and gaseous silanes and siloxanes as by-products. A second side reaction results in loss of reactivity in the oligomer due to a shift of the terminal double bond to an internal position. If the side reactions are taken into account, it is possible to forecast quantitatively molecular weight, gel point and modulus from the conversions of $\rightarrow$SiH, $>C=CH_2$ and the chain entanglement concentration reported for polyisobutene in the literature.

INTRODUCTION

Hydrosilylation (reaction 1) has been used extensively in cross linking studies of elastomer networks because of its presumed straight forward nature and freedom from side reactions.

$$(1) \quad \rightarrow SiH + -CH=CH_2 \longrightarrow \rightarrow Si\overset{|}{C}HCH_2-$$

For example, hydrosilylation was employed for end linking poly-dimethylsiloxane networks to verify a statistical method of describing elastomer network structures and how they related to macroscopic properties where the required empirical information was the degree of conversion of the reactive groups (1). Subsequent work on the same system revealed, however, an extraneous side reaction which consumed $\rightarrow$SiH without forming end links (2). The side reaction was originally presumed

347

to be a platinum-catalyzed reaction of the silane with moisture in the reagents to produce silanol.

The present investigation extends these studies by applying hydrosilylation to end-linking a terminally unsaturated liquid polyisobutene (B_2) described by Kennedy and coworkers (3).

$$CH_2=\overset{\underset{\displaystyle CH_3}{|}}{C}-CH_2 \left[\overset{\underset{\displaystyle CH_3}{|}}{\underset{\underset{\displaystyle CH_3}{|}}{C}}-CH_2\right]_n \overset{\underset{\displaystyle CH_3}{|}}{C} \underset{\underset{\displaystyle CH_3}{|}}{\bigcirc} \overset{\underset{\displaystyle CH_3}{|}}{C}-\left[CH_2\overset{\underset{\displaystyle CH_3}{|}}{\underset{\underset{\displaystyle CH_3}{|}}{C}}\right]_m -CH_2-\overset{\underset{\displaystyle CH_3}{|}}{C}=CH_2$$

$$(B_2)$$

This material, because of its hydrocarbon character, can be easily maintained in a dry state and interactions of SiH with moisture during end-linking will be minimized. Further, any new siloxane structures formed during end-linking can be readily detected by NMR. Bifunctional (A_2) or tetrafunctional (A_4) endlinkers are used with a dry catalyst, cis-$[(C_2H_5)_2S]_2PtCl_2$ dissolved in toluene solutions.

$$HMe_2SiOMe_2SiOSiMe_2H \qquad\qquad Si(OSiMe_2H)_4$$

$$(A_2) \qquad\qquad\qquad\qquad (A_4)$$

The aim of this work is to demonstrate an alternative to the sulfur cure for unsaturated polyisobutene as well as to show how the recursive method of statistically calculating molecular parameters in stepwise polymerization or cross-linking can be applied to yet another polymer system (4). Further light will also be shed on side reactions occurring during end linking via hydrosilylation.

EXPERIMENTAL

Siloxanes A_2 and A_4 were obtained from Petrarch Chemicals and B_2, $\overline{M}_n$=1400, was obtained from Prof. J. P. Kennedy, Univ. of Akron, Akron, OH. The catalyst was used as a 0.1% stock solution in toluene. Rheological measurements during end-linking were made with a Rheometrics System IV Mechanical Spectrometer. NMR data were obtained with the Nicolet 293A 300 MHz spectrometer. FTIR spectra were obtained on a Nicolet 60SX using 64 scans.

Previously dried B_2 that was devolitalized and contained 2.7 x 10^{-5}g of catalyst/g. of B_2 was mixed at room temperature with enough A_2 to obtain the desired initial ratio (r) of $\geq$SiH to $>C=CH_2$. Excessive amounts of gas often formed during end linking with A_4 and it was necessary first to conduct part of the end-linking experiment with a large excess of A_4. This was then diluted with more B_2 to give the

appropriate value of r and the process was continued in the fixture of
the rheometer. In this fashion foaming and bubbles were minimized.
Results from some runs are shown in Figures 1 and 2. The process was
considered complete when viscosity, η, or storage modulus, G', achieved
a constant value. It was presumed that G' at $\omega=10$ rad sec.$^{-1}$ and at 5%
peak strain was an equilibrium value under these conditions since little
or no change was seen when ω was varied from 10^2 to 10^{-1} rad sec.$^{-1}$ at
the end of a run.

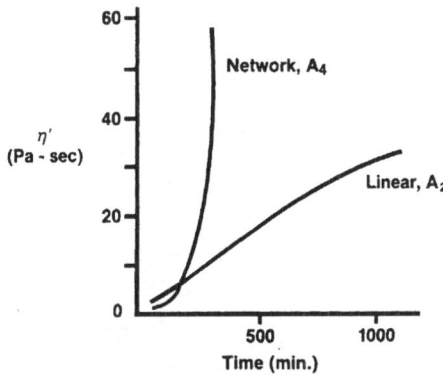

Figure 1. Plots of viscosity, η, vs. polymerization time where r=1.0 for
the hydrosilylation of B_2 with A_2 at 85°C and of B_2 and A_4 at
51°C.

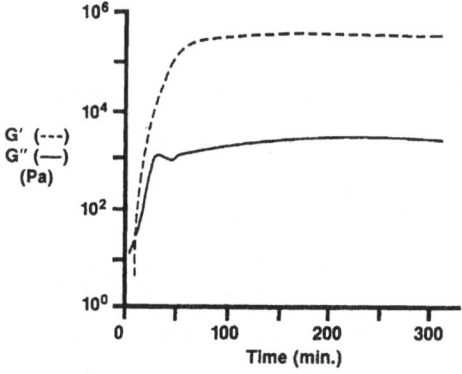

Figure 2. Dynamic shear moduli vs. time for the cure of B_2 with A_4 at
85°C, r=1.0. Measurements at 5% strain at $\omega=10$ rad sec.$^{-1}$.

Conversions of $\rightarrow$SiH, P_A, and $>$C=CH$_2$, P_B, were determined from the ^{1}HNMR spectrum obtained in CCl$_4$ solutions of polymers made from A$_2$. The area of the signals (SiH at 4.70 ppm and $>$C=CH$_2$ at 4.80 and 4.63 ppm, relative to SiMe$_4$) were referenced to the aromatic peak in B$_2$ (6.95–7.05 ppm) which remained unchanged during the process. A similar technique could be applied to gels made from A$_4$ when the end-linking process was terminated near the gel point. The resulting gels were swollen in CCl$_4$ containing about 1%S to prevent further reaction. It was necessary, however, to degrade the more highly cured elastomers made from A$_4$ to render them soluble in CCl$_4$ and amenable to quantitative analysis by ^{1}HNMR. The siloxane cross links were cleaved by first swelling the elastomers in CCl$_4$ and then equilibrating with (Me$_3$Si)$_2$O and CF$_3$SO$_3$H catalyst at room temperature. An identical procedure with previously characterized soluble polymers made from A$_2$ and B$_2$ established that these conditions did not alter the content of SiH but caused some shift of residual $>$C=CH$_2$ to an internal position of B$_2$.

Data on chemical shifts in the SiCH$_3$ region were obtained with the model compounds given in Table I. The series RMe$_2$SiO(Me$_2$SiO)$_n$Me$_2$SiR (where R=CH$_3$CH$_2$CH$_3$CHCH$_2$, n=0,1,2,3,4) was obtained by hydrosilylation with excess CH$_3$CH$_2$CH$_3$C=CH$_2$ of HMe$_2$SiO(Me$_2$SiO)$_n$Me$_2$SiH. The latter intermediates were >98% pure by GC. Products of the hydrosilylation were purified by distillation and structures were confirmed by both ^{1}HNMR and ^{29}SiNMR spectroscopy.

Table I

Chemical Shift Data for RMe$_2$SiO(Me$_2$SiO)$_n$Me$_2$SiR

where R=CH$_2$CH$_3$CHCH$_2$CH$_3$

Structure	n	^{1}HNMR[a] at position[b]			^{29}SiNMR[c] at position[b]		
		1	2	3	1	2	3
I	0	0.051	–	–	6.73	–	–
II	1	0.072	−0.003	–	6.97	−21.84	–
III	2	0.082	0.036	–	7.07	−22.23	–
IV	3	0.074	0.033	0.053	7.15	−21.93	−22.51
V	4	0.073	0.033	0.057	7.16	−21.89	−22.26

[a] p.p.m. at 200.1 MHz in CCl$_4$ relative to SiMe$_4$. [b] Silicon atoms are numbered from the terminal position attached to the R group, RMe$_2$SiO(Me$_2$SiO)$_n$Me$_2$SiR. [c] p.p.m. at 39.75 MHz in CCl$_4$ using

$\begin{matrix} 1 & 2 \text{ etc.} & 1 \end{matrix}$

SiMe$_4$ as an internal standard.

RESULTS

Clear polymers soluble in common solvents formed from A_2 and B_2. Number average molecular weights ranged up to $M_n=50,000$ and weight average to $M_w=90,000$ depending on r. Insoluble dry elastomers or tacky gels, depending the extent of reaction, formed from A_4. The ^{1}HNMR spectra of the polymers and gels resembled those of B_2 (3) with greatly diminished intensities for $>C=CH_2$ and new signals for $OSi(CH_3)_2$. Figure 3 shows that the signal for $OSi(CH_3)_2H$ (0.17 ppm) from A_2 in the original mixture disappeared. A strong new signal, A, corresponding to $R'(CH_3)_2SiO$ (0.095 to 0.085 ppm) appeared. Signal, C, at -0.009 ppm was assigned to the central $(CH_3)_2SiO$ unit in the corresponding structure II, Table I, which was anticipated from reaction 1. However, a strong spurious signal, B, for $SiCH_3$ also appeared at 0.028 ppm and was assigned to $(CH_3)_2SiO$ at position 2 in the corresponding structure III, Table I. The example in Figure 3 was run in high vacuum to eliminate the possibility of reaction of SiH with oxygen or moisture in the air. Identical chemical shift patterns, however, were obtained on samples run in air.

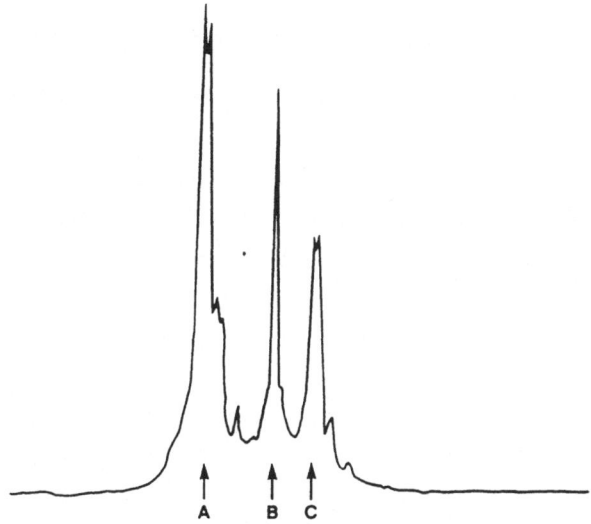

Figure 3. ^{1}HNMR spectrum of in the $SiCH_3$ region of the polymer obtained from the polymerization A_2 with B_2 (r = 1.0) in high vacuum (10^{-6} torr) using out-gassed reactants. Peak assignments based on data given in Table I. Peak A (0.094 ppm, relative area 1.00) was assigned the structure $R'(CH_3)_2SiO-$. Peak B (0.028 ppm, area 0.30) was assigned the structure $-O[(CH_3)_2SiO]_2$. C (-0.009 ppm area 0.36) was was assigned the structure $(CH_3)_2SiO$. Peak B was not anticipated from reaction 1.

The ^{29}SiNMR spectra of the same polymer gave major signals at 6.39ppm for $R'Me_2SiO$ and at -21.99ppm for Me_2SiO both of which were anticipated for structure II. Again a spurious separate signal was detected close to the peak corresponding to $R'Me_2SiO$ at 6.47 ppm. The later was also attributed to the presence of a higher siloxane homolog based on the chemical shift trends in Table I. The ^{29}SiNMR showed no signals in the region of -11 to -12 ppm which might be attributed to Me_2SiOH. This was further confirmed by the absence of any bands for $\rightarrow$SiOH in the region 3150 to 3600 cm^{-1} in the FTIR spectra in either the above polymer or in polymers made by polymerization of B_2 with A_2 in air.

Therefore, although end-linking proceeded via reaction 1, a second process was occurring which, from the relative area intensities in the NMR spectra, gave about 20 to 30% (based on end linker) of a structure similar to III in Table I. This was proposed to form via the desproportionation reactions 2 and 3.

(2) $\quad 2R'Me_2SiOMe_2SiOMe_2SiH \xrightarrow{\text{Pt}} R'Me_2SiO(Me_2SiO)_2Me_2SiR' + (HMe_2Si)_2O$

and

(3) $\quad (n+1)(HMe_2Si)_2O \xrightarrow{\text{Pt}} HMe_2SiO(Me_2SiO)_nMe_2SiH + n\ Me_2SiH_2$

$\quad$ (R'= H or the polymeric residue from B_2; n = 1 or 2)

The two processes, separately or in combination, amount to formation of new endlinker and simultaneous loss of SiH since by-produced disiloxane (reaction 2) or silane (reaction 3) are well over their boiling points and readily lost from the system under the present conditions. Similar processes are proposed when B_2 is cross-linked with A_4. Siloxane - $SiCH_3$ bond redistribution under conditions of hydrosilylation have been reported by Speier and Stober (5) and siloxane - SiH bond redistribution by Andrianov and coworkers (6). More recently Curtis and coworkers have shown that such processes were ubiquitous and readily took place with a variety of siloxanes bearing $-OMe_2SiH$. Complexes of Ir, Rh, Pt and Pd were catalysts for the process with varying degrees of effectiveness (7). The present system proved to be no exception.

Reactions 2 and 3 are considered to occur simultaneously with hydrosilylation, reaction 1. Figure 4 traces the disappearance of $\rightarrow$SiH, $>C=CH_2$ and the formation of products during polymerization of B_2 with an excess of A_2. This shows that more $\rightarrow$SiH is consumed than anticipated from reaction 1 and that the discrepancy developed while the adduct, $RSi(CH_3)_2O-$, formed. Structure III formed only during generation of adduct and its final amount compensated for the discrepancy between the observed consumption of >SiH and that expected from simple hydrosilylation.

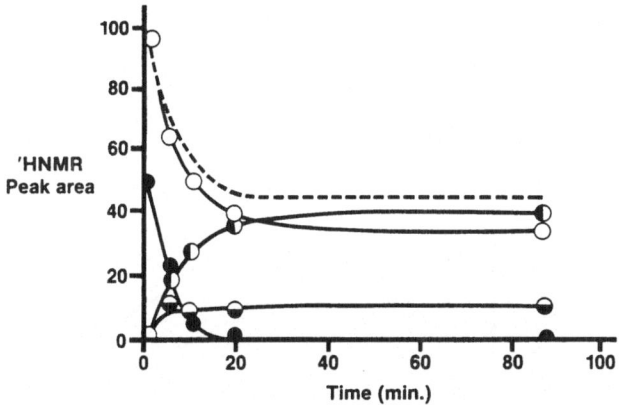

Figure 4: Peak areas relative to the aromatic peak in B_2 in the ^{1}HNMR vs. time during reaction at 80°C of with A_2 (r = 1.89). —◯—: Si$\underline{H}$, —●—: C=C$\underline{H}_2$/2, —◖—: R'Si(C$\underline{H}_3$)$_2$/6, corresponding to peak A, Fig. 3; —◓—: [(C$\underline{H}_3$)$_2$SiO]$_2$/12 corresponding to peak B, Figure 3; ---anticipated disappearance of SiH based on >C=CH$_2$

The consumption of >C=CH$_2$ in B_2 paralleled the appearance of product and there was no extraneous loss of this functionality. In other runs taken to the gel point with A_4, however, about 15% of the terminal double bond in B_2 shifted to an internal position as evidenced by the appearance of a signal at 5.10 ppm in the ^{1}HNMR. Similar rearrangements in structurally analogous low molecular weight olefins have been documented elsewhere when H_2PtCl_6 was the catalyst (8). Also, a single example of curing B_2 with A_4 with H_2PtCl_6 was documented but no evidence was given for the above shift of double bond or the reactions 2 and 3 (9).

Molecular weight averages in polymerization with A_2 and G' in cross-linking with A_4 reached maximum values as the initial ratio of $\Rightarrow$SiH to $>$C=CH$_2$ (r) was varied (Figures 5 and 6). However, the maxima at r=1.2 for A_2 and 1.3 for A_4 deviated significantly from the anticipated value of r=1 in an end-linking process based on reaction 1. The discrepancies can only be explained in terms of the side reactions 2 and 3 which lead to loss of $\Rightarrow$SiH and occur either prior to or during network formation.

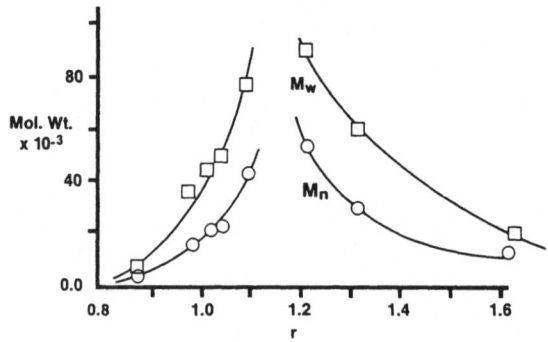

Figure 5: Effect of r on number average, M_n, and weight average, M_w, molecular weight for the polymerization of B_2 and A_2 at 85°C.

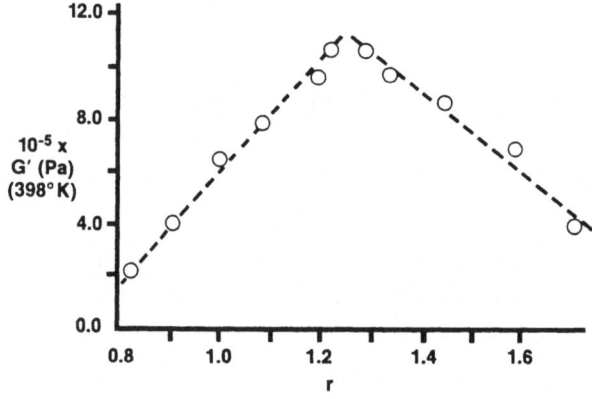

Figure 6: Effect of r on the final storage modulus in cures of B_2 with A_4 at 125°C.

354

DISCUSSION

Simultaneous occurrence of side reactions 2 and 3 during hydrosilylation results in loss of SiH without forming end links to change the value of r via volatilization of the by produced silanes. The value of r can also be altered by loss of the terminal double bond to an unreactive internal position in the polymer by a second side reaction. The net effect of both side reactions will be a lower than anticipated molecular weight with A_2 or modulus with A_4. A schematic representation of the over-all process can then be given by

$$
\begin{array}{l}
A + B \longrightarrow AB \\
\quad\quad \longrightarrow Z \\
\quad\quad \longrightarrow Y
\end{array}
$$

(4)

Where A represents SiH, B $>C=CH_2$, AB represents endlinks formed by hydrosilylation with A_2, A_4 or the non-volatile species formed in reactions 2 or 3, Y represents an equivalent of SiH lost as a gaseous silane in reaction 2 or 3 and Z represents an unreactive double bond in B_2. If A_o and B_o are the initial amounts of A and B respectively, $r = A_o/B_o$ and r' is the effective ratio of A to B in forming end links, the observed conversions in this scheme will be given by

$$
(5) \quad P_a \equiv \frac{A_o - A}{A_o} = \frac{AB + Y}{A_o} \qquad P_y = Y/A_o
$$

$$
P_b \equiv \frac{B_o - B}{B_o} = \frac{AB + Z}{B_o} \qquad P_z = Z/B_o
$$

and the effective conversions of $\rightarrow$SiH, P_a', or of $>C=CH_2$, P_b', to endlinks will be given by

$$
(6) \quad P_a' \equiv \frac{AB}{A_o - Y} = \frac{P_a - P_y}{1 - P_y} = \frac{P_b - P_z}{r(1-P_a)+(P_b - P_z)}
$$

$$
P_b' \equiv \frac{AB}{B_o - Z} = \frac{P_b - P_z}{1 - P_z} = \frac{r(P_a - P_y)}{1 - P_b + r(P_a - P_y)}
$$

$$
r' \equiv \frac{A_o - Y}{B_o - Z} = r\,\frac{1 - P_y}{1 - P_z} = P_b'/P_a'
$$

Thus, knowing either P_y or P_z, the effective conversions to end links can be calculated from the observed conversions.

Number and weight average molecular weights were predicted for the runs with A_2 using the expressions below and the effective conversions described in equation 6.

$$(7) \qquad M_n = \frac{r'M_A + M_B}{r'(1+r')}$$

$$(8) \qquad M_w = \frac{r'(1+r')M_A^2 + 1/2(3+r')M_B^2 + 4M_A M_B}{(r'M_A + M_B)(1+r')}$$

Where M_A and M_B are the number average molecular weights of A_2 and B_2 respectively. Equations 7 and 8 are special cases of the general equations for M_n and M_w given in reference 4 where either P_a' or P_b' are unity (the case in runs with A_2) and where $M_w/M_n = 1.5$ for B_2. Little loss of $>C=CH_2$ to rearrangement was seen in polymerization with A_2, hence $P_z = 0$. Good agreement between observed and predicted molecular weights was seen, Figure 7, with one exception at the highest molecular weight and conversion.

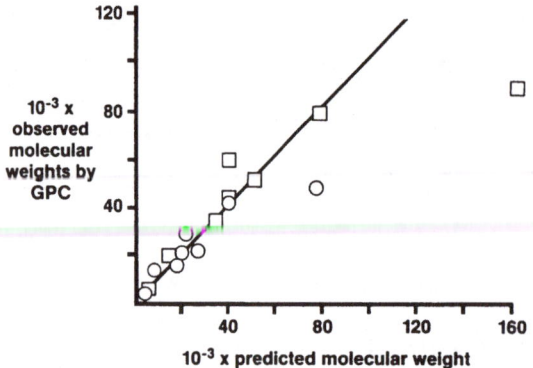

Figure 7: Observed vs. predicted molecular weights in reactions of A_2 and B_2. $\bigcirc:M_n$ $\square:M_w$. Solid line indicates perfect agreement with theory.

The scheme in reaction 4 is therefore presumed to represent the process within the present limits of accuracy. It is also presumed that an analogous process occurs in cures with A_4. This was illustrated in runs with A_4 taken to the gel point and then quenched by cooling. The gels could readily be analyzed directly by ^{1}HNMR when they were swollen in CCl_4 which contained 0.5 to 1% sulfur to suppress further reaction. The observed conversions and those defined by equation 6 could then be compared in the expression for critical conversions at the gel point.

$$(10) \qquad P_a'P_b' \text{ gel} = \frac{1}{(f_a-1)(f_b-1)}$$

Where f_a is the weighted average functionality of end-linker and f_b that of B_2 (4).

Coupling of A_4 via processes analogous to reactions 2 or 3 would form volatile species indentical with those from A_2 but changes in functionality might be anticipated. A hexafunctional non-volatile species, A_6, for example, could also form. Thus, along with a loss of SiH an accompanying increase in f_a would be expected. Table II shows that the critical conversions at the gel point with A_4 are in agreement with theory when the sequence in scheme 4 is presumed and conversions are defined by equation 6. The data also suggests that the side reactions tended to occur to a lesser extent at lower temperatures. The experiment was insensitive, however, to any changes in functionality that might have occurred (footnotes c and d, Table II).

Table II

Conversions at the Gel Point for the

Reaction of A_4 and B_2 $(r=1)^{a)}$

t°C	P_a	P_b	P_z	P_aP_b	P_a'	P_b'	$P_a'P_b'$
65	0.70	0.55	--	0.38	0.65	0.55	0.36
80	0.73	0.58	0.18	0.42	0.60	0.49	0.29
95	0.79	0.55	0.17	0.43	0.65	0.46	0.30
110	0.76	0.59	0.17	0.45	0.64	0.51	0.32
(Predicted)[b)]	0.58	0.58	--	0.33	--	--	--
(Predicted)[c)]	0.73	0.60	--	0.44	0.62	0.52	0.32
(Predicted)[d)]	0.74	0.61	--	0.45	0.62	0.53	0.33

a) P_a: conversion of SiH, P_b: conversion of $C=CH_2$ based on the relative ^{1}HNMR peak area in gels swollen in CCl_4, P_a' and P_b' are defined in equation 6 and P_z estimated from the area of the peak at 5.10 ppm in the ^{1}HNMR. b) Assume equation 10 and no side reactions $f_a = 4.0$ and $f_b = 2.0$. c) Assume equation 10 and side reactions in scheme 4 so 0.17 of B_2 forms a monofunctional species B_1 and 0.30 of A_4 couples via reaction 2 and 3 to form 0.15 of a hexafunctional species, A_6. $f_a=4.42$ and $f_b = 1.91$. 0.15 of disiloxane or silane are also formed which are volitile and lost. d) Same assumption as in c, but $f_a = 4.0$ and $f_b = 2.0$.

The equation for ideal elastic behavior can be expressed as equations 11 or 12, depending on the importance of topological chain entanglements. These can then be compared in predicting of equilibrium shear modulus, G, in the fully cured elastomers.

$$(11) \qquad G/RT = g[A_4]_o P(X_F)$$

or

$$(12) \quad G/RT = g[A_4]_o P(X_F) + G_N^o/RT(T_E)$$

Where g is a constant converting concentrations of cross links to effective concentrations of network chains and g = 1 for A_4; $[A_4]_o$ the initial concentration of A_4; $P(X_F)$ the probability that a cross link is connected to the network; G_N^o/RT the entanglement concentration of uncured PIB based on the measured plateau modulus (10), and T_E the probability that an entanglement is connected to the network.

The probabilities $P(X_F)$ and T_E were estimated by the recursive statistical method for stepwise polymerization which requires only knowledge of conversions of the reactive groups (4). In the present case the conversions P_a' and P_b' were obtained from ^{1}HNMR measurements on elastomers which had been solubilized by the procedure described in the experimental section. It was not possible to estimate P_z directly by this procedure but it was assumed to be the same as that seen in the runs with A_4 terminated at the gel point. It was also approximated that the weighted average functionality of A_4 was effectively 4 regardless of the side reactions (see note d, Table V). The results of these calculations are shown in Figure 8. Best agreement is seen when equation 12 which includes the effect of entanglements is used to compute G/RT.

CONCLUSIONS

Hydrosilylation proved to be an effective method of end-linking this type of elastomer but a side reaction which consumes SiH functionality and generates gaseous by-products caused complications. Among these were a tendency to foam or bubble and a need to use an excess of SiH functional end-linker for optimum properties. The use of ^{1}HNMR proved to be an effective tool to trace end linking and side reactions. When the side reactions were taken into account the recursive method was effective in predicting molecular weights, gel

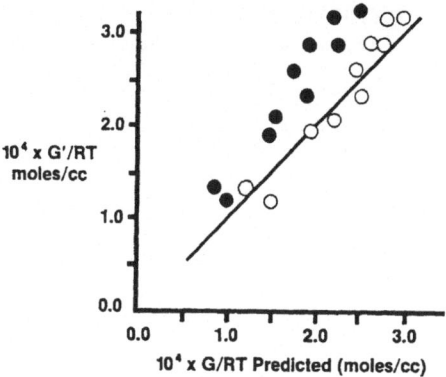

Figure 8: Predicted G/RT at 125°C Vs. that obtained from modulus
measurements in cures of B_2 with A_4. Solid line indicates
perfect agreement with theory. ●: Based on equation 11 where
entanglements are not taken into account. ○: Based on
equation 12 where entanglements are included.

point and the small strain shear modulus in the fully cured elastomers
if the contributions from the concentrations of entrapped topological
chain entanglements were included in the calculations. In view of these
findings, previous conclusions about network structure and behavior
based on "model networks" which assume end linking by simple
hydrosilylation should be carefully reappraised.

ACKNOWLEDGEMENT

The authors wish to acknowledge the Dow Corning Corporation for
granting J. Saam leave so he could conduct the major part of this work
at the University of Minnesota.

REFERENCES

1). [a]E. M. Valles and C. W. Macosko, Macromolecules, 12, 521
    (1979) and 12, 673 (1979).
    [b]M. Gottlieb, C.W. Macosko, G.S. Benjamin, K.D. Meyers
      and E. W. Merrill, Macromolecules, 14, 1039 (1981) and
      leading references therein.
2. C. W. Macosko and G. S. Benjamin, Pure and Appl, Chem. 1505 (1981).
3. J. P. Kennedy, V.S.C. Chang, R. A. Smith and B. Ivan, Polymer
    Bulletin 1, 575 (1979).
4. D. R. Miller, E. M. Valles, and C. W. Macosko, Polym. Eng. and
    Sci., 19(4), 272, (1979) and leading references therein.
5. M. R. Stober, M.C. Musolf, and J. L. Speier, J. Org. Chem. 30, 1651
    (1965).
6. K. A. Andrianov, B. G. Zavin, G. F. Sablina, L. A. Leites, B. D.
    Lavrukhin and A. M. Endokomov, Vÿsokomol. Soedin., Ser.
    B16(5), 330 (1974).

7. W. A. Gustavson, P. S. Epstein and M. D. Curtis, J. Organometall. Chem. 238, 87 (1982).

8. J. Saam and J. Speier, J. Am. Chem. Soc. 83, 1351 (1961).

9. P. H. Sung, S. J. Pan, J.E. Mark, J. E. Lackey, and J. P. Kennedy, Polymer Bulletin 9, 375 (1983).

10. J. W. Ferry "Viscoelastic Properties of Elastomers" John Wiley and Sons, Inc., N.Y. (1980), p. 606.

THE ELASTIC BEHAVIOR OF CIS-1,4-POLYBUTADIENE

R. W. Brotzman

Dept. of Chemistry, CUNY
Staten Island, NY 10301

ABSTRACT

This article presents uniaxial extension measurements on cis-1,4-polybutadiene networks of known junction functionality. The observed values of the reduced force from uniaxial extension measurements conform to the constrained junction theory of Flory. The reduced force intercept at $1/\alpha = 0$ is fully comprehensible in terms of the cycle rank of the network, and can be calculated from chemical considerations. This holds even though the polybutadiene melt has a high plateau modulus. Therefore, discrete topological entanglements do not contribute perceptibly to the equilibrium modulus of polybutadiene networks.

INTRODUCTION

The reduced force is a convenient measure of the elastic response of a polymer network to an applied stress. It is defined by

$$[f^*] = f^*(V/V^o)^{-1/3}(\alpha - \alpha^{-2})^{-1} \qquad (1)$$

where f^* is the tensile force per unit area in the reference state, V^o is the volume of the reference state, V is the system volume at measurement, and α is the extension ratio relative to the length of the sample when isotropic at the same volume V. Recent theory reformulates $[f^*]$ of a real polymer network as[1-4]

$$[f^*] = [f_{ph}^*](1 + f_c/f_{ph}) \qquad (2)$$

where f_{ph} represents the force which would be exerted by a topologically equivalent phantom network, f_c is the contribution of the force from the effects of constraints on the fluctuations of network junctions, and

$[f_{ph}{}^*]$ is the reduced force for the equivalent phantom network (i.e., a hypothetical network devoid of material properties; its constituent chains can transect one another, no chain excludes others from the volume it occupies, and the fluctuations of cross-links are not constrained by surrounding chains). The reduced force for a perfect phantom network is given, according to theory[2,5], by

$$[f_{ph}{}^*] = \xi kT/V^o = (\phi - 2)\, \mu_J/2V^o \tag{3}$$

where ξ is the cycle rank of the network, μ_J the number of network junctions, and ϕ the junction functionality. The ratio f_c/f_{ph} depends on two parameters κ and ζ; κ is of primary importance and is defined as the ratio of the mean-square radius of the fluctuations of the junctions in the phantom network to the mean-square radius of the Gaussian domain of constraint in the undistorted network while the secondary parameter ζ is believed to reflect inhomogeneities on the network topology. It is given by eq. 43 in ref. 3. In the limit of high extension and/or dilution, f_c/f_{ph} vanishes according to the theory.

Thus elastic behavior in the limit of high extension and/or dilution is fully comprehensible from a consideration of the covalent structure of the network. Equation 3 implicitly excludes contributions to the reduced force from discrete entanglements[3] of one network chain with another. If a contribution should be included for permanently "trapped entanglements" of this nature, as often contended to be necessary, their contribution would be reflected in apparent values of ξ, or of μ_J, exceeding those deduced from the chemical constitution of the network alone. Some authors have attempted to attribute such an alleged contribution to the plateau modulus of the uncross-linked polymer melt.[6,7] Equilibrium swelling data provide additional confirmation of the relationship between the reduced force of a phantom network and the cycle rank density determined from chemical considerations. Theoretical treatment of swelling data in the phantom network limits yields[8-10]

$$(\xi/V^o) = -[\ln(1 - v_{2,s}) + v_{2,s} + \chi\, v_{2,s}{}^2](v_{2,s}{}^{-1/3}/V_1) \tag{4}$$

where $v_{2,s}$ is the volume fraction of polymer at swelling equilibrium, χ is the familiar interaction parameter, and V_1 is the molar volume of the diluent.

This study presents uniaxial extension measurements on cis-1,4-polybutadiene (PBD) networks of known junction functionality. Primary PBD chains were cross-linked by a hydrosilylation reaction between the PBD double bonds and 1,1,3,3,5,5,7,7,-octamethyltetrasiloxane (OMTS)

$$H - Si(Me)_2-O-Si(Me)_2-O-Si(Me)_2-O-Si(Me)_2 - H$$

to produce a network which contains tetrafunctional junctions. The value of $[f_{ph}^*]$ determined from uniaxial extension measurements is compared with the stoichiometric value calculated from eq. 3. The polybutadiene melt has a high plateau modulus (1.2 MPa)[11] and thus provides a model system well suited for assessment of whether or not discrete topological entanglements contribute to the equilibrium modulus of an elastomer network.

Previous investigators have studied PBD networks formed by the action of free radicals induced photochemically, by peroxide, or by high energy radiation. In a dicumyl peroxide charged system, Van der Hoff[12] estimated the cross-linking efficiency to be approximately 12 for each pair of free radicals released by the peroxide in an antioxidant-free material. Infrared analysis showed that free radical propagation proceeds via the unsaturated double bonds in both the main chain (1,4) and in vinyl side groups (1,2), a process denoted by "cross polymerization". Thus, the polybutadienyl radical, once initiated, is capable of successively joining double bonds in neighboring polymer chains. Large junctions of high functionality (with $\phi > 20$ in some instances) may be formed through cross polymerization in this manner. Other investigators[13,14] have studied the peroxide cross-linking of PBD by measuring the sol-gel ratio of a series of samples charged with varying amounts of dicumyl peroxide. They concluded that the free radicals have large kinetic chain lengths (3-8) which result in network junctions of 4-8 functionality. Through the intervention of chain transfer the functionality of these junctions may be somewhat less than the kinetic chain length of the radical. In any event, the structures of networks formed through the action of free radicals are difficult to quantify.

EXPERIMENTAL

Solvents. Reagent grade toluene and 95% ethanol were used for fractionation without further purification. Spectroscopic grade benzene was used as the solvent in sample-deposition solutions. Inasmuch as the OTMS reacts rapidly with moisture, all traces of water should be removed from the benzene prior to its use for this purpose. To this end, the benzene used for dissolution of samples C, CDS-B-4, D-1, D-2, and D-3 was first refluxed over calcium hydride and distilled to remove any residual water. A less stringent procedure was employed for samples A and B.

Benzene, practical grade decane, and hexadecane (Aldrich 99%) were used as swelling diluents. The density of hexadecane is calculated to be 0.7866 g/cm^3 at 33.6° C[15].

Polymer. The PBD used in the stress-strain measurements and for one of the samples in the sol-gel study (see below) was Goodyear Budene 1207, having a cis-1,4 content of 98%. The samples were prepared for cross-linking by fractionation from a 2-2.5% toluene solution using ethanol as the precipitant. Throughout this and subsequent dissolution procedures an antioxidant (p-benzoquinone) concentration of approximately .05% was maintained. The goal of fractionation was to confine somewhat the molecular weight distribution of PBD; narrow fractions were not desired. Four batches of polymer were fractionated; the molecular weight characteristics of each were determined by size exclusion chromatography (Waters Model 150°C liq/gel Permeation Chromatograph with a refractive index detector) and are given in Table I. After fractionation, the polymer was dried under vacuum and stored in the dark at 0°C.

The second sample for the sol-gel study (see below) was prepared from Goodyear - Batch No. CDS-B-4 polybutadiene. This polymer has a molecular weight of 22000, a polydispersity index of about 1.0, and a microstructure of: 40% cis 1,4; 52% trans 1,4; and 8% vinyl 1,2. It was used without further purification.

PBD, the cross-linking agent OMTS (Petrach Systems) at the concentrations given in Table I, and 1.0-1.5 ppm of a platinum catalyst (chloroplatinic acid - Petrach Systems) were dissolved in benzene. The solution was deposited in a Teflon mold. Solution deposition and all procedures which follow were conducted under a flowing, dry nitrogen atmosphere at room temperature. After thorough evaporation of solvent a section of the sample was dissolved in toluene to check for incipient gelation; in no case was evidence of gelation detected. The sample was transferred to a stainless steel mold which was pressed between two stainless steel plates covered with Teflon sheets in order to obtain samples of uniform thickness. The samples were allowed to relax in the mold for four days prior to the execution of cross-linking in bulk at 140°C for 3 h through hydrosilylation[16] of PBD double bonds. It was imperative to exclude all water from the sample to prevent the hydrolysis[17] of OMTS.

The cross-linked sheets were extracted with toluene for 30 h at room temperature. Soluble constituents removed during this treatment yielded the percentages of gel given in Table I. Exercise of extreme care was necessary during the filtration of the supernatant of samples D-2 and CDS-B-4 used in gelation measurements. Toluene was gradually removed by deswelling with solutions of increasing ethanol concentration.

Drying was completed by subjecting the samples to vacuum for 24 h at room temperature. Visually the swelling of the gels was uniform, indicating good cross-linking uniformity and little if any residual orientation from the molding process.

The specific volume of PBD at 33.6°C interpolated from data of Barlow[18], 1.1216 cm^3/g, was used for reducing the measured masses to volume fractions. The density of OMTS is reported to be 0.8632 at 20°C.[16]

Swelling Measurements. Equilibrium degrees of swelling in benzene, decane, and hexadecane were determined at 35°C on rectangular specimens (ca. 0.3 x 0.5 cm^2). Changes in the linear dimensions (unswollen versus swollen) of two sides of the specimen were measured with a cathetometer and averaged.

Mechanical Measurements. Rectangular specimens of ca. 3 cm in length were cut with a stainless steel die and the cross-section of each specimen was measured with a cathetometer. The samples were mounted in light-weight grips ca. 2.5 cm apart, the lower grip being equipped with a hook for holding weights; the upper one was attached to a load cell (Straindyne Engineering Co. - model CFT12-70GR) situated above the chamber. The load cell was mounted on a temperature controlled block maintained at 35°C. Calibration of the cell verified that the relationship between applied stress and output voltage was linear throughout our experimental range. Reference marks were provided by two lengths of thin wire inserted near the extremities of the measurement area. In order to minimize damage to the sample, they were inserted while the specimen was swollen. A load was applied and the distance between the wires was measured at periodic intervals until the sample was judged to be at equilibrium (see below). The weight was then increased or decreased and a new equilibrium point was established. Data obtained during loading and unloading cycles showed no significant differences.

Stress-strain measurements on unswollen samples and those partially swollen with hexadecane were carried out at 33.6°C. Samples swollen to equilibrium in decane were submerged in the liquid (plus .05% antioxidant) in a cylindrical container at 35°C. Partially swollen samples were prepared by immersing specimens in the diluent for a limited time and allowing 48 h for uniform distribution of the diluent throughout the specimen. The amount of diluent was determined by weighing and the volume fraction of polymer calculated from the known specific volumes of polymer and liquid, additivity of volumes being assumed.

Although equilibration during stress-strain measurements was fairly rapid for swollen samples, a period of 2 h was allowed for attainment of equilibrium after each load. However, the question of whether equilibrium was established for bulk specimens is difficult to answer. In our experiments a quasi-steady-state reading was taken after 48 hr.

RESULTS

Gelation. The cross-linking of polybutadiene is analogous to random condensation of bifunctional diene units with ϕ-functional units. The following development is a review of classical gelation theory[19] and assumes the following: the primary PBD chains may be approximated by a most probable distribution; all functional groups are chemically equivalent; the reactivity of a given diene unit is independent of the size or structure of the molecule to which it is attached; and intramolecular cross-linking reactions may be neglected. In support of the first assumption, we note that the ratios M_w/M_n given in Table I approximate two, as for the most probable distribution. The departures, therefore, should not vitiate the following analysis significantly. Accordingly, we take the probability that a given functional group of a branch unit, leads via a chain of bifunctional units, to another branch unit to be

$$\alpha_b = p\varrho/[1 - p(1 - \varrho)] \qquad (5)$$

where ϱ is the cross-linking density defined as the ratio of interlinked units to the total number of diene units, and p corresponds to the extent of reaction in the analogous polycondensation. Thus, 1 − p is the probability of occurrance of an end of a primary molecule. For a most probable distribution of primary chain lengths, whose weight and number average degrees of polymerization are $\bar{y}_w$ and $\bar{y}_n$, respectively, p is given by $1 - 2/(\bar{y}_w + 1) = 1 - 1/\bar{y}_n$. The critical condition for incipience of infinite networks is

$$\alpha_b{}^c = 1/(\phi - 1) \qquad (6)$$

The foregoing relations and those that follow should apply to the networks of higher functionality ϕ formed by cross polymerization as well as to those with $\phi = 4$ formed by cross-linking pairs of units.

Using the complexity distribution approach[19,20] and assuming that the proportion of ϕ-functional units is small ($\varrho \ll 1$), the weight fraction of sol when $\alpha_b > \alpha_b{}^c$ is

$$w_s = (1 - \alpha_b)^2/(1 - \alpha_b')^2 \qquad (7)$$

where α_b' and α_b are, respectively, the lower and upper roots of

$$\beta = \alpha_b (1 - \alpha_b)^{\phi - 2} \tag{8}$$

The term β represents no physical quantity but serves only to quantify α_b and α_b'. The weight fraction of gel is

$$w_g = 1 - (1 - \alpha_b)^2 / (1 - \alpha_b')^2 \tag{9}$$

For the random interlinking of primary chains having a most probable distribution the corresponding definition of p may be substituted into eq. 5 to yield

$$\rho = 2\alpha_b / (1 - \alpha_b)(\bar{y}_w - 1) = \alpha_b / (1 - \alpha_b)(\bar{y}_n - 1) \tag{10}$$

Once the cross-linking density for a given α_b and $\bar{y}_w$ or $\bar{y}_n$ is obtained, the partitioning of cross-linked units between sol and gel in a ϕ-functional system is given by

$$\rho' = \rho w_s^{(\phi - 2)/2}(1 + (\phi/2)\rho [1 - w_s^{(\phi - 2)/2}]) \tag{11}$$

for the sol and

$$\rho'' = \rho (1 - w_s)^{-1}(1 - w_s^{\phi/2}[1 + (\phi/2)\rho (1 - w_s^{(\phi - 2)/2})]) \tag{12}$$

for the gel.

Calculations pertaining to interlinking of primary molecules through introduction of junctions with functionalities $\phi = 4$ and 24 are shown in Figure 1, where the weight fraction of gel and the densities ρ' and ρ'' of cross-linked units in the sol and gel, respectively, are plotted against the overall cross-link density ρ. (These cross-link densities retain the definition given above, namely, the ratio of interlinked units to the total number of diene units; they are related to the junction density by the factor $(2/\phi)$(polymer density/segment molecular weight).) The solid lines represent the designated tetra-functional behavior while the dashed line indicates the percent gel in a system containing junctions of 24-functionality. The difference in cross-link density between networks of tetra- and 24-functionality at a given fraction of gel is greatest at low gel fractions. However, this difference persists even at 90 percent gelation where it is approximately 10 percent. Although these results illustrate the consequence of cross-linking primary chains having a most probable distribution (with M_w = 358,000), the following analysis is general and elucidates the gelation behavior of all cross-linking systems with regard to the functionality of the junctions which are formed.

The effect of functionality on the gelation characteristics of a network are clearly illustrated in Figure 1. While distinctions between networks of differing junction functionality become less (in the gelation sense) at higher gel fractions, the effect of junction functionality on the equilibrium mechanical properties of a network are significant, as this study demonstrates.

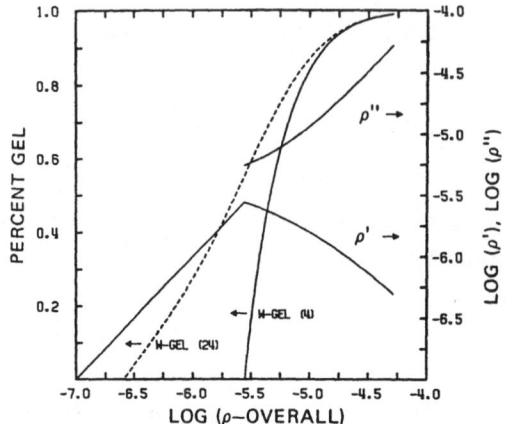

Fig. 1. Percent gel and distribution of cross-link density between sol and gel versus the log of the total cross-link density for the cross polymerization of primary chains having a most probable distribution. Two curves representing networks with junction functionalities of 4 (solid) and 24 (dashed) are shown with arrows pointing to the left. In each case M_w=358,000. The other solid lines represent the gelation behavior of a tetrafunctional system and have arrows pointing to the right.

Cross-linking Efficiency. The efficiency of the process of cross-linking with OTMS was tested by comparing observed weight fractions w_g of gel with those calculated according to theory. The theoretical values for the weight fraction of gel were determined by substituting the $\bar{y}_w$ and ϱ values from Table I into eq. 10 to give α_b and using this result in eq. 8 to obtain α_b'; α_b and α_b' were then substituted into eq. 9. The theoretical and experimental values of w_g for samples D-2 and CDS-B-4 are 50 versus 47 and 85 versus 81 percent gel, respectively. Considering the steep slope of the curve for tetrafunctional networks in Figure 1 deviations of 3 and 4 percent gel indicate good agreement with theory. This close agreement, coupled with the order of magnitude differences in molecular weight of the uncross-linked polymers used to prepare the samples, indicates that quantitative cross-linking occurs.

Table I

Polymer	$\underline{M}_w \times 10^{-3}$	$\underline{M}_w/\underline{M}_n$	mole % OTMS/monomer $10^2 (\rho/2)$	Expt. % gel	Expected from Flory Theory
A	358	2.5	.535	97.6	99.9
B	414	2.1	.263	100.	99.8
C	345	1.9	.253	99.8	99.7
D-1	350	2.1	1.50	100.	100.
D-2*	350	2.1	.0119	47.	44.7
D-3	350	2.1	.160	97.3	99.3
CDS-B-4*	24.7	1.1	.423	81.	85.6

*Sample used in gelation study (see text).

Stress-Strain Measurements. The stress-strain isotherms at 33.6°C for PBD networks of varying inter-junction chain molecular weight are given in Figures 2-6. In each figure the reduced force is plotted versus 1/ with the solid lines representing the best theoretical fits in the least-square sense to the experimental isotherms. The parameters used to calculate these theoretical lines are given in Table II. The lower dashed lines represent the cycle rank density calculated from the chemical constitution with the correction due to dangling chain ends included. This correction is given by

$$(\breve{\xi}/V^o)_{corrected} = (\breve{\xi}/V^o)_{chemical} - 1/(M_n v_{sp}) \tag{13}$$

where M_n is the number average molecular weight, v_{sp} is the specific volume of PBD, and the cycle rank density from chemical considerations is calculated from eq. 3. The upper dashed lines represent the

Table II

Figure	Sample	v_2	κ	$[f*]/_{RT} \times 10^4$ $1/\alpha = 0$	$1/\alpha = 1$
2	A	.256	.46	.680	.693
3	B	.191	5.06	.320	.359
4	C	.800	*	.395	.725
		.539	6.84		.551
		.339	6.16		.497
		.186	5.92		.469
6	D-3	.920	*	.241	.419
		.739	6.73		.358
		.411	4.82		.300
		.155	3.99		.266

$\zeta = 0$ for all fits.

* The κ values selected as best fits to the data in the least-squares sense are abnormally high. This, plus the knowledge that a wide range of κs, lower in magnitude, could have been used to fit the data equally well, undermines the relevance of these κ values.

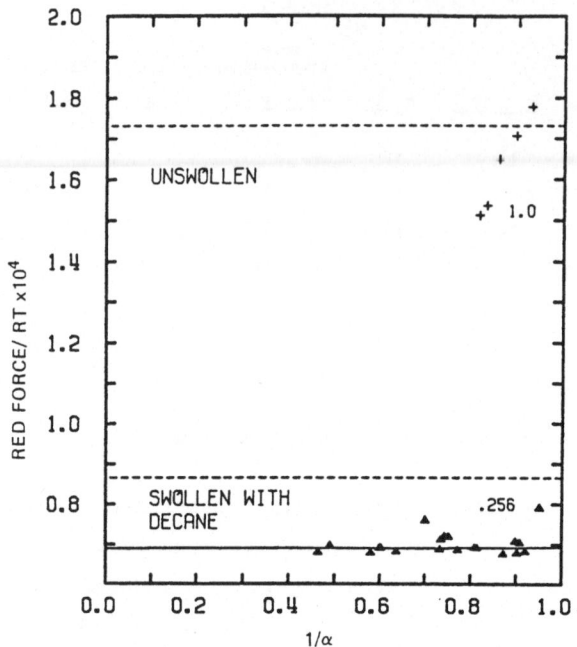

Fig. 2. Reduced force as functions of α^{-1} for
Network A. Data for bulk samples (+)
and samples swollen with decane (triangles)
are given and the polymer volume fractions
are indicated with each isotherm. The
solid curves were calculated according
to theory with parameters given in Table
III. The lower dashed line represents the
cycle rank density calculated from the
chemical constitution of Newtwork A with
corrections due to dangling chain ends
included. The upperdashed line represents
the affine limit for a perfect network
obtained with respect to the corrected
chemical value of Network A.

affine limit for a perfect network obtained with respect to the corrected
chemical value. The polymer volume fraction of the sample is given next
to each isotherm.

Stress-strain measurements, obtained for samples swollen to equilibrium with decane, have been corrected for the effect on the stress of
the small increase in swelling with extension[21]; the values of χ given
in Table III were employed in making this correction. No correction was
required for samples swollen with hexadecane because it is a non-volatile
solvent.

Figures 2 and 3 contain data for Networks A and B, respectively.
In each case the reduced force for samples swollen with decane (triangles)
fall below the value determined from "the chemistry" of each system by

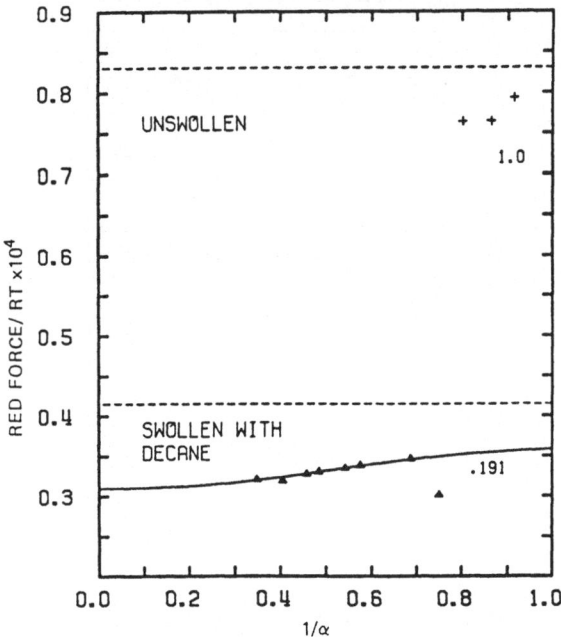

Fig. 3. Same as Fig. 2 but for Network B.

21 and 25%. The apparent chemical value is the reduced force intercept of the theoretical fit to stress-strain data at $1/\alpha = 0$; this was used in Table III to calculate χ values which are starred. In both figures bulk data are given by +; this data may not be true equilibrium data due to long relaxation times and the onset of strain-induced crystallization, which was observed to occur in bulk samples at higher elongations. Strain-induced crystallization was not observed for any of the samples swollen with diluent.

Table III

Sample	Diluent	$v_{2,eqm}$	χ	$(\xi/V^0) \times 10^4$ moles/cm^3
A	Benzene	.0995	.253*	.680*
	Decane	.256	.477*	
B	Benzene	.0679	.275*	.320*
	Decane	.191	.477*	
C	Benzene	.0800	.292	.395
	Decane	.186	.445	
	Hexadecane	.304	.545	
D-1	Benzene	.236		2.47
	Decane	.422		
	Hexadecane	.492		
D-3	Benzene	.0588	.280	.241
	Decane	.155	.453	
	Hexadecane	.256	.538	

*The values were determined using the cycle rank density deduced from elasticity measurements (see text).

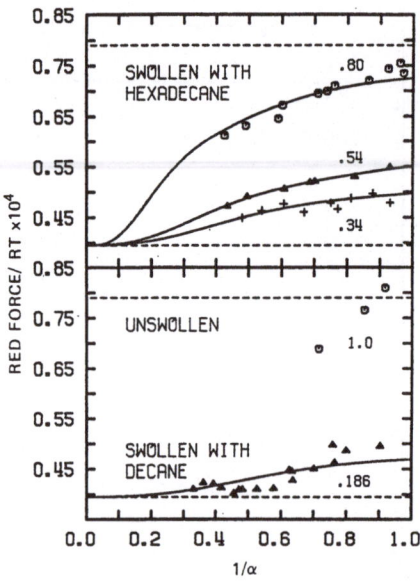

Fig. 4. Data for Network C. The lower portion presents data for bulk samples (o) and samples swollen with decane (triangles) while the upper portion contains data for samples swollen to varying degrees with hexadecane. The polymer volume fractions are indicated with each set of data and dashed lines are as indicated in Fig. 2.

In considering the discrepancy between experiment and theory for Networks A and B the effect of water during sample preparation became a concern. To correct this problem, the benzene was further purified (see experimental) and all mixing and deposition procedures occurred under a dry nitrogen atmosphere. The result of this modification to the experimental procedure is embodied in the strain-strain isotherms of Networks C, D-1, and D-3 in Figures 4, 5, and 6. The lower portions of the Figures 4 and 6 contain data for bulk sample (C only) and samples swollen to equilibrium with decane; the upper portion contains data for samples swollen to varying degrees with hexadecane. All the theoretical curves were generated with the cycle rank density from chemical data. There is good agreement between experiment and theory. Samples A and B are therefore deemed to be somewhat inferior because the cycle rank deduced from elastic measurements are more reliable owing to inefficiency of "the chemistry".

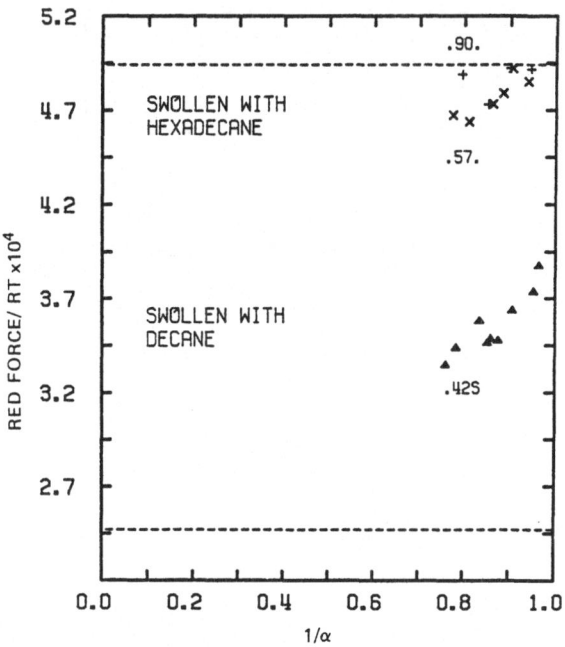

Fig. 5. Same as Fig. 2 but for Network D-1.

Equilibrium stress-strain data on bulk samples were difficult to obtain because of the aforementioned problems. Therefore, the behavior of PBD networks in which the fluctuations of the junctions are highly constrained was explored with Network D-1, which is highly cross-linked. Although the junctions in highly cross-linked networks are subject to smaller relative constraint from entanglements, these junctions will remain highly constrained. Measurements on samples swollen with hexadecane eliminate the problem of equilibrium and provide a measure of this limit; the experimental data exhibit large departures from phantom behavior. But these departures are within the framework of the theory; Network D-1 exhibits affine-like behavior.

Swelling. The polymer volume fractions at swelling equilibrium at 35°C are given in Table III. The χ values have been calculated from the cycle rank density ξ as determined from the molar ratios of OMTS, eq. 4 being used for this purpose under the assumption that the swollen network conforms to the phantom limit. This is a valid assumption. The χ for sample D-1 was not determined since its mean inter-junction chain molecular weight, 1800, is too low to validate the Gaussian approximation. The starred values for samples A and B were determined using the cycle rank density deduced from elasticity measurements. An examination of the apparent χ values reveals that those for benzene are more sensitive to the cycle rank density than the corresponding values for decane.

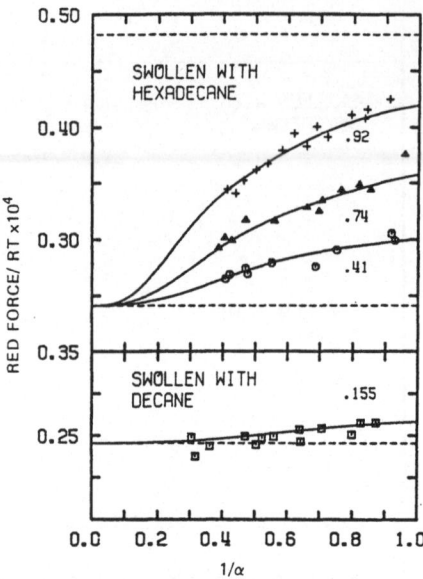

Fig. 6.   Same as Fig. 4 but for Network D-3.

Even so, for each diluent the χ for the networks agree within experimental error.   The χ values for the PBD and benzene system are contrasted with previously reported values of .325[22] (unspecified PBD microstructure) and .21[23] (98% cis-1,4 and ca. 34% cis-1,4; 54% trans-1,4; 12% vinyl-1,2) obtained by vapor sorption techniques.   From Table III the mean χ values for benzene, decane, and hexadecane are respectively .265, .463, and .542.

CONCLUSIONS AND DISCUSSION

The results of this study are in difference with those from previous investigators.   Mark[24] has reported stress-strain isotherms of networks prepared using peroxide, sulfur, and high energy radiation.   The reduced force behavior of these networks was compared in terms of the ratio of linear-fit parameters ($2C_2/2C_1 = R$) derived from $[f^*] = 2C_1 + 2C_2\alpha^{-1}$. Although this technique of fitting data linearly has now been thoroughly discredited[1-5], it did provide a means of examining earlier work.   The R values for peroxide (.35 $\pm$ .15) and radiation (1.69 $\pm$ .2) cross-linked networks are compared with that of the sulfur (1.30 $\pm$ .2) cross-linked networks.   The low value of R for peroxide cross-linked networks is comprehensible if junctions are considered to be of high functionality.

The higher value of R for radiation cross-linked networks is explicable by a spatially non-random distribution of functions of high functionality This conclusion has support from the nonequality of inter-junction chain molecular weights deduced from swelling and elasticity measurements[24,25]

Dossin and Graessley[6] (D-G) have examined polybutadiene networks in which cross-linking was induced by high-energy radiation. Their networks were assumed to be tetrafunctional; according to their analysis junction fluctuations were deemed to be completely suppressed. This is precisely the expected effect of junctions of high functionality. The fluctuations of these junctions are small and ineffectual in reducing the stress below that expected for an affine network. The results of Mark and Llorente[26] (M-L) demonstrates that the polymer studied by D-G (ca. 50% cis-1,4 and 50% trans-1,4) could have crystallites persisting "up to temperatures well above 60°C" and they are likely to exist in the unstretched state; the experiments of D-G were conducted at 25°C. The protracted relaxation of stress reported by D-G, from 1 to 3 days being required for equilibrium, also suggests the presence of crystalline aggregates. Colby and Graessley[20] (C-G) recently reported some observations on low molecular weight polybutadienes (M_n < 40,000) with this microstructure which indicates crystallization at temperatures below -26°C. C-G determined that the degree of crystallinity increases with decreasing molecular weight; samples with molecular weight ranges which approximate our interjunction molecular weights (3000-16000) had from 5.6 to 1.3 percent crystallinity. D-G reported reduced force values from elasticity measurements that were 2.5 to 6.5 times that from chemical considerations. M-L observed that swelling their networks with decalin decreased their reduced force measurements by six-fold. The presence of a diluent evidently suppressed crystallization or facilitated the attainment of true equilibrium, or both.

Recently, Macosko and Saam[27] demonstrated that side reactions consume approximately 25 percent of the SiH without forming end-links in the hydrosilylation cure of polyisobutene. This estimate of reaction inefficiencies is certainly large considering the low sol fractions reported in Table I for the samples subjected to uniaxial deformations and the close agreement between experiment and theory in our gelation study. In any event, the modulus should be reduced below the value expected from chemical considerations by the degree of reaction ineffi-ciency - purportedly 25 percent in this instance. Dossin and Graessley[6] reported the contribution to the modulus from "topologically trapped" entanglements to be between 250 and 650 percent greater than that deduced from the covalent structure of the network. The expected contribution to the equilibrium modulus from trapped entanglements is so great that the existence of small reaction inefficiencies in the cross-linking reaction will not alter the following conclusions.

Previous studies on PDB networks, which were reputed to be well-characterized, are therefore vitiated by the non-quantitative chemistry of the networks, the effect of strain-induced crystallization, and the question of equilibrium attainment during elastic measurements. Departures from theory can not be attributed to the highly speculative contributions from interchain entanglements. PBD of high cis-1,4 content is free from the problems of crystallization at our experimental conditions[26], and gelation studies establish that the chemistry of our system is quantitative. The elastic behavior of well-characterized polybutadiene networks conforms to the constrained junction theory. At large extension (and/or high dilution) phantom network behavior is approached. The reduced force intercept at $\alpha^{-1} = 0$ is fully comprehensible in terms of the cycle rank of the network and can in fact be calculated from the chemical constitution. This conclusion receives strong support from recent results on this system by Hsu and Mark[28]. Discrete topological entanglements therefore do not contribute perceptibly to the equilibrium modulus of polybutadiene networks.

ACKNOWLEDGEMENTS

This work was performed under the direction of Professor P.J. Flory at IBM Research Labs in San Jose. It details our presentation and the subsequent discussion at the 1986 ACS Meeting in Chicago.

We are grateful for the polybutadiene samples received from Dr. J. Lal of Goodyear. C. Cole of IBM Research Labs performed the SEC analysis and P. Cotts assisted with column calibration corrections for this analysis. R.W.B. benefited from an IBM Postdoctoral Fellowship. This work was supported by IBM and, in part, by a National Science Foundation Grant — NSF DMR 85-06669.

REFERENCES

1. P. J. Flory, The Polymer Journal, 17, 1 (1985).
2. P. J. Flory, J. Chem. Phys, 66, 5720 (1977).
3. P. J. Flory and B. Erman, Macromolecules, 15, 800 (1982).
4. B. Erman and P. J. Flory, Macromolecules, 15, 806 (1982).
5. P. J. Flory, Proc. R. Soc. Lond. A, 351, 351 (1976).
6. L. M. Dossin, W. W. Graessley, Macromolecules, 12, 123 (1979).
7. D. S. Pearson and W. W. Graessley, Macromolecules, 13, 1001 (1980).

8. P. J. Flory and J. Rehner, <u>J. Chem. Phys.</u>, 11, 521 (1943).
9. P. J. Flory, <u>J. Chem. Phys.</u>, 18, 108 (1950).
10. P. J. Flory, <u>Macromolecules</u>, 12, 119 (1979).
11. J. D. Ferry, <u>Viscoelastic Properties of Polymers</u>, 2nd Ed., Wiley, New York, (1970).
12. B. M. E. Van der Hoff, <u>Ind. Eng. Chem. Prod. Res. Develop.</u>, 2, 273 (1963).
13. M. Morton and J. C. West, <u>100th Rubber Division - ACS</u>, (1971).
14. L. O. Malotky and M. Morton, <u>Polymer Preprints</u>, 15, 714 (1974).
15. R. A. Orwoll and P. J. Flory, <u>J. Amer. Chem. Soc.</u>, 89, 6814 (1967).
16. S. S. Washburne, <u>Petrarch Systems, Inc.</u>, p. 10.
17. F. A. Cotton and G. Wilkenson, <u>Advanced Inorganic Chemistry</u>, J. Wiley, New York, N.Y., p. 334 (1972).
18. J. J. Barlow, <u>Polym. Eng. and Sci.</u>, 18, 238 (1978).
19. P. J. Flory, <u>Principles of Polymer Chemistry</u>, Cornell University Press, Ithaca, N.Y., Chapter IX (1953).
20. R. H. Colby and W. W. Graessley, <u>Macromolecules</u>, submitted.
21. B. Erman and P. J. Flory, <u>Macromolecules</u>, 16, 1607 (1983).
22. R. S. Jessup, <u>J. Res. Nat. Bur. Stand.</u>, 60, 47 (1958).
23. S. Saeki, J. C. Holste and D. C. Bonner, <u>J. Polym. Sci., Polym. Phys. Ed.</u>, 20, 794 (1982).
24. T. K. Su and J. E. Mark, <u>Macromolecules</u>, 10, 120 (1977).
25. M. A. Sharaf, J. E. Mark and S. Cesca, accepted in <u>Poly. Eng. and Sci.</u> (1985).
26. J. E. Mark and M. A. Llorente, <u>Polymer J.</u>, 13, 543 (1981).
27. C. W. Macoski and J. C. Saam, <u>Polymer Preprints</u>, 26(2), 48 (1985).
28. Y.-H. Hsu and J. E. Mark, <u>Polymer Preprints</u>, 27(1) (1986).

OPTICAL STUDIES OF NETWORK TOPOLOGY

Richard S. Stein, Vivek K. Soni, Hsinjin E. Yang

Polymer Research Institute

University of Massachusetts, Amherst, MA   01003

INTRODUCTION

   Polymer networks can usually sustain large recoverable deformations
due to the presence of chemical crosslinks which serve to bind long chains
into a permanent network structure. The elasticity of the network chains
is considered to originate primarily in terms of the entropy of the chains
(1). The elastic free energy of an elastomeric network is usually treated
as the sum of the contributions of its individual chains. Therefore, the
most important parameter in describing the properties of a network is the
molecular weight of the chain between crosslinks (M_c).

   Theories have been developed to describe the mechanical properties of
amorphous networks and their swelling behavior in terms of an average M_c
(1-3). Over the years there have been several modifications in the theories to account for the fluctuations of the junction points, the role of
network defects such as dangling chains and loops and the role of trapped
entanglements in determining the equilibrium elasticity of a network (4).
In recent years, with the development of model networks it has been possible to prepare networks of controlled M_c and junction functionality ϕ.
These are prepared by endlinking functionalized prepolymers with crosslinking agents of known functionality. Therefore, by choosing the appropriate molecular weight distribution of the prepolymers it is possible to
prepare unimodal and bimodal networks. Mark and coworkers (5-11) have
performed extensive studies on model networks to test the various theories
of rubber elasticity. In the case of unimodal networks they find that the
macroscopic properties such as stress or swelling ratios can be described
reasonably well by the Flory-Erman theory (12,13).

   Mark et al (9-11) have also studied bimodal poly(dimethylsiloxane)
(PDMS) networks. These were prepared by mixing a predetermined fraction
of short chains ($M_n \sim 1,000$) with long chains ($M_n \sim 20,000$). \For high
mole fractions of short chains, typically 90%, they found marked improvement in the ultimate mechanical properties i.e. high moduli at high elongations (9-11). This type of behavior would not be predicted on the basis
of affine deformation of gaussian chains. Affine deformation corresponds
to the situation in which the macroscopic and the molecular elongation
ratios are identical. Spatial segregation of the short chains did not
significantly affect the results, indicating that this was not a simple

379

filler type of effect. The short chains used are highly non-gaussian. This results in a non-affine deformation process with the long chains taking up most of the elongation. This work by Mark et al clearly indicates the importance of the distribution of the molecular weight between crosslinks in determining the characteristics of a network.

Recently, Picot, Bastide and coworkers (14-18) have used small angle neutron scattering (SANS) to measure the dimensions of a few labelled chains in a network. They find that the deformation of a single labelled chain between crosslinks is much less than the deformation predicted by the affine model. In particular, Bastide has observed very small changes in the radius of gyration of a labelled chain in a network which undergoes a considerable change in its macroscopic dimensions by osmotic deswelling. Bastide has suggested network deformation involves a rearrangement of the network chains such that the molecular deformation is less than affine (16-18). This implies that the network topology plays an important role in determining the overall properties.

Studies have been undertaken at the University of Massachusetts to understand the role of network topology, with special emphasis being given to the homogeneity of crosslinking. These studies have primarily involved scattering methods. In a scattering experiment one obtains information regarding spatial correlations in the sample and is therefore helpful for studying network topology. By labelling a few chains by deuteration, one can use SANS to study the molecular deformation in a network as has been done by Picot, Bastide and their coworkers. However, to observe correlations in the overall network structure it is necessary to have chains of similar scattering power. Usually there is not sufficient contrast in dry unlabelled amorphous polymers. Therefore, one has to swell the network with a diluent which has a scattering power significantly different from that of the polymer comprising the network. By choosing proper diluents it is possible to study the same network by light scattering (LS) and by SANS so as to study a wide range of spatial correlations.

Scattering: General Considerations

When a beam of a particular radiation is incident on a sample some of it gets scattered. The intensity of the scattered radiation depends on its type and on the composition of the sample. The commonly used types of radiation include light, x-rays and neutrons. The scattered intensity $I(q)$ can be expressed as

$$I(q) = K \sum \sum Z_i Z_j \exp (i\vec{q} \cdot \vec{r}_{ij}) \qquad (1)$$

Here one is concerned with interference between scattering elements located at points i and j and having scattering powers Z_i and Z_j respectively. $\vec{r}_{ij}$ is the vector connecting the points. $\vec{q}$ is the scattering vector, the magnitude of which is defined as

$$q = (4\pi/\lambda) \sin (\theta/2) \qquad (2)$$

where θ is the scattering angle and λ is the wavelength of the radiation used, K is a constant which will depend on the type of radiation used and the experimental conditions. It is convenient to express the scattered intensity in terms of a Rayleigh ratio $R(q)$ which is defined as

$$R(q) = \frac{I_s(q)S^2}{I_0 V} \qquad (3)$$

where the subscripts s and o refer to the scattered and incident beams. V is the scattering volume and S is the sample to detector distance.

The scattering power depends on the type of radiation used. In the case of neutron scattering ($\sim 5\text{Å}$) it is associated with the nature of the scattering nucleus and is called the scattering length. For x-ray scattering ($\lambda \sim 5,000\text{Å}$) it is the polarizability or the refractive index. Since these three properties differ for a given system for the different types of radiation, various structure elements will be weighted differently through their use. In this work, neutron and light scattering have been used.

Swollen networks are non-particulate systems and therefore it is convenient to use a statistical approach to describe the scattered intensity. The intensity is considered to arise due to fluctuations in the scattering power throughout the material.

If $\overline{Z}$ is the average scattering power and Z_i is the local scattering power, then the local fluctuation in the scattering power will be

$$\eta = Z_i - \overline{Z} \tag{4}$$

For random, isotropic and inhomogeneous materials, such as swollen networks, one has the Debye expression (19) for the Rayleigh ratio

$$R(q) = 4\pi K \overline{\eta^2} \int_0^\infty \gamma(r) \frac{\sin(qr)}{qr} r^2 dr \tag{5}$$

Here $\overline{\eta^2}$ is the mean squared fluctuation in the scattering power, $\gamma(r)$ is a correlation function which is defined as

$$\gamma(r) = \frac{\langle \eta_i \eta_j \rangle_r}{\eta^2} \tag{6}$$

where $\langle \ \rangle_r$ represents averaging over all scattering elements i and j separated by a distance r. In terms of material having a two phase structure, the correlation function $\gamma(r)$ represents the probability that a rod of given length r, will have both its ends in the same phase. For $r = 0$, $\gamma(0) = 1$, and for $r = \infty$, $\gamma(\infty) = 0$.

Therefore, such a function contains all the information regarding spatial correlations of scattering elements. The scattering is then completely defined in terms of the mean squared fluctuation in the scattering power, $\overline{\eta^2}$, and the correlation function $\gamma(r)$. For several systems it is found that $\gamma(r)$ can be expressed in an exponential form

$$\gamma(r) = \exp(-r/a_c) \tag{7}$$

where a_c is a correlation size and can be taken to represent the size of an inhomogeneity. Such a form for $\gamma(r)$ results in the familiar Debye-Bueche (19) equation for $R(q)$

$$R(q) = \frac{8\pi K \overline{\eta^2} a_c^3}{(1 + q^2 a_c^2)^2} \tag{8}$$

If the data is plotted as a Debye-Bueche plot, i.e. $R(q)^{-1/2}$ vs. q^2, it is possible to estimate a_c^2 from the value of (slope/intercept) and $\overline{\eta^2}$ from the intercept.

Light Scattering from Swollen Networks

In pure liquids, light scattering arises as a result of density fluctuations and is described in terms of the fluctuation approach developed by Einstein (20). The Rayleigh ratio arising from density fluctuations, R_d, is a function of the bulk compressibility of the material, β, the incident wavelength λ, the refractive index n of the material and the absolute temperature, T. For polymer solutions, in addition to R_d, there will be scattering arising from concentration fluctuations R_c. This term is given by

$$R_c = \frac{K_L RTC}{(\partial\Pi/\partial C)_T} \qquad (9)$$

where R is the gas constant, C is the polymer concentration, Π is the osmotic pressure and K_L is the light scattering constant.

For concentrated solutions, one can use the Flory-Huggins equation (3) for the osmotic pressure in equation (9) giving

$$R_c = \frac{K_L \overline{V}_1 v_2 \rho_2{}^2}{\dfrac{1}{1-v_2} + \dfrac{1}{X_w} - 1 - 2\chi v_2} \qquad (10)$$

where $\overline{V}_1$ is the partial molar volume of the solvent, v_2 is the volume fraction of the polymer, ρ_2 is the mass density of the polymer, X_w is the weight average degree of polymerization of the polymer and χ is the polymer-solvent interaction parameter.

For a polymer network at equilibrium swelling one can use the Flory-Rehner theory for the free energy of the network to obtain the following expression for R_c.

$$(R_c)_{id} = \frac{K_L \overline{V}_1 v_2 \rho_2{}^2}{\dfrac{1}{1-v_2} - 1 - 2\chi v_2 + \dfrac{\rho_2 V_1}{6M_c}(3 - 2v_2{}^{-2/3})} \qquad (11)$$

It should be noted that this expression corresponds to an ideal network since it does not take into account scattering due to crosslink inhomogeneities. In real networks, the scattered intensity is greater than that predicted by equation (11). The excess scattering is attributed to crosslink inhomogeneities and other topological features.

An inhomogeneously crosslinked network will swell inhomogeneously in the diluent solvent. If there is sufficient difference in the refractive index between the polymer and the diluent, the non-uniform distribution of the diluent can be studied by light scattering to characterize the inhomogeneity of crosslinking. Such an approach has been used by Stein (22), by Bueche (23) and by Wun and Prins (24). Using the Flory-Rehner theory for network swelling, Stein (24) developed expressions to estimate the mean squared fluctuation in the molecular weight between crosslinks, $\langle(\Delta M_c)^2\rangle$, from the excess scattering of light, $R_e(q)$.

$$R_e(q) = \frac{K_u \langle(\Delta M_c)^2\rangle}{M_c{}^2} \int_0^\infty \gamma(r) \frac{\sin(qr)}{qr} r^2 dr \qquad (12)$$

where K_u is a function of v_2, the refractive indices of the polymer and

382

the solvent and λ_0. Often the excess scattering data can be fitted with an exponential correlation function. Therefore, from the intercept of the Debye-Bueche plot, one can estimate $\langle(\Delta M_c)^2\rangle$. However, it should be noted that the excess scattering can arise from factors other than inhomogeneous crosslinking. One possibility is concentration fluctuations caused by topological rearrangements in a network. These may be of the nature of "network unfolding" processes which we shall discuss later. The excess scattering associated with such concentration fluctuations is difficult to estimate at this time.

Scattering: Experiments and Results

In the initial stages of this work, polybutadiene networks were studied. Various networks were prepared by crosslinking a high cis content polybutadiene with dicumyl peroxide at 140°C. The details of the sample characteristics are given elsewhere (25). These networks were swollen in benzene and studied by light scattering (25). Using deuterated benzene, these networks were also studied by SANS. In both the LS and the SANS experiments, it was observed that the swollen networks scattered considerably more than solutions of linear polymers at equivalent concentrations. Although it was not possible to analyze the excess scattering data in terms of correlation sizes it was possible to interpret these sizes in terms of the network structure. This was primarily due to uncertainties in network structure associated with dicumyl crosslinking of polybutadienes.

To have greater control of the chain topology, it was decided to use model networks. PDMS networks were chosen because of the extensive work by Mark and his coworkers on these systems. The networks were prepared by crosslinking divinyl terminated PDMS chains with a tetra silane in the presence of a platinum catalyst. All networks were prepared in bulk. By choosing the appropriate prepolymers, it was possible to prepare unimodal and bimodal networks. For the bimodal networks, the short chains had $M_n = 770$ and the long chains had $M_n = 22,500$. In both cases $M_w/M_n \sim 1.8$. The desired amounts of the two prepolymers were mixed together prior to the addition of the crosslinking agent to obtain random mixing of the chains.

Table 1 lists the swelling data for different bimodal PDMS networks swollen to equilibrium in benzene and in toluene at 25°C. x is the mole percent of short chains and M_c represents the average molecular weight between crosslinks.

$$M_c = xM_S + (1 - x)M_L \tag{13}$$

The subscripts S and L correspond to the short and long chains respectively.

The swelling results of these PDMS networks are consistent with the results of Mark and coworkers. This indicates that the macroscopic properties of networks such as the swelling behavior are reasonably well described by the current theories of rubber elasticity, such as the Flory-Erman theory.

SANS from PDMS Networks

SANS experiments were performed on swollen PDMS networks at the National Bureau of Standards, Washington, DC, in collaboration with Dr. Charles C. Han. The experimental details are given elsewhere (26). The networks were swollen with deuterated solvent to enhance the contrast for

Table 1

Swelling Data for Bimodal PDMS Networks at 25°C

x	M_c	$(v_2)B$	$(v_2)_T$
0	22500	0.200	0.161
25	17608	0.230	0.192
50	11635	0.276	0.251
75	6203	0.330	0.301
90	2943	0.380	0.339
95	1857	0.439	0.402
100	770	0.535	0.504

B: swelling in benzene
T: swelling in toluene

SANS. Polymer solutions containing the uncrosslinked polymer at the same concentration as the swollen network were also studied. Fig 1 shows the SANS data for a unimodal PDMS network ($M_n \sim 22,500$) swollen in C_6D_6 and the corresponding solution. In all cases the scattering from the networks is greater than that from the solutions. Fig 2 shows the excess scattering from the various swollen bimodal PDMS networks.

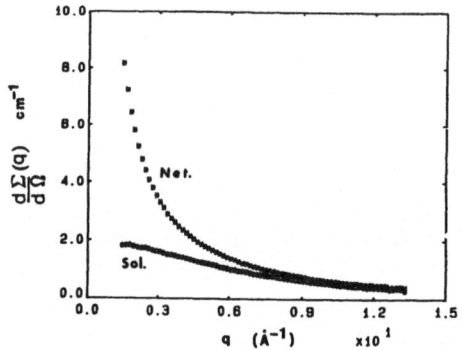

Fig 1 SANS data for swollen PDMS ($M_n = 22,500$) network and corresponding solution.

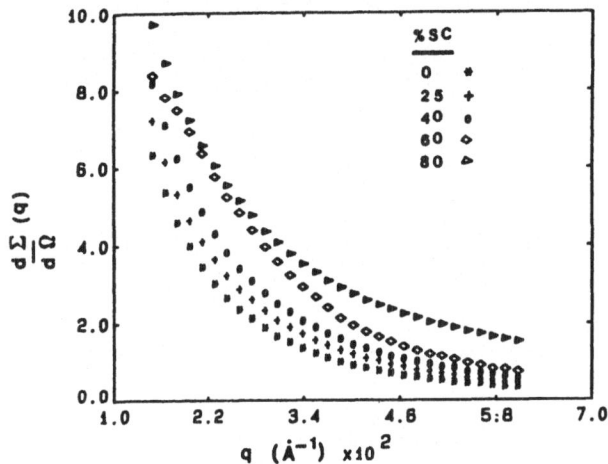

Fig 2   Excess SANS intensities for various
swollen bimodal PDMS networks.

At small scattering angles or q, the scattered intensity can be expressed in the form

$$R(q) = \frac{K_n}{1 + (q^2 \xi^2 / 3)} \qquad (14)$$

Here ξ is the correlation distance in the swollen network. Knowing the molecular weights of the prepolymers, one can estimate the average radius of gyration of a network chain in the dry or bulk state. Most of the SANS data to date on labelled chains in networks seems to indicate that the deformation from the dry to the swollen state is found to be close to deformation predicted by the phantom network model. Therefore, by using the elongation ratio predicted by this model, one can estimate the average radius of gyration for a chain between crosslinks in a swollen network. Let this be $(R_g)_{sw}$. Tables 2 and 3 lists the values of ξ and $(R_g)_{sw}$ for some swollen unimodal and bimodal networks. Q_c represents the ratio $\xi / (R_g)_{sw}$.

Table 2

SANS Results for C_6D_6 Swollen Unimodal
Tetrafunctional PDMS Networks at 25°C

M_c	v_2	$\xi(A)$	$(R_g)_{sw}(A)$	Q_c
22,500	0.19	131.8	74.2	0.78
5,200	0.30	74.1	30.0	2.47
770	0.54	16.1	12.2	1.32

Table 3

SANS Results for C_6D_6 Swollen Bimodal
PDMS Networks at 25°C

x	v_2	ξ(Å)	$(R_g)_{sw}$(Å)	Q_c
0	0.19	128	74.2	1.73
25	0.23	124	62.6	1.98
40	0.26	114	55.4	2.06
60	0.29	100	45.2	2.21
80	0.34	90	32.4	2.78
100	0.54	16	12.2	1.32

For the q range studied, the Q_c values for the bimodal networks are found to be similar to unimodal networks. The fact that the correlation size in a swollen network is much lager than in the corresponding semi-dilute solution is usually taken to be an indication of inhomogeneities in the network structure. However, the similarity of the results from the unimodal and bimodal networks seems to suggest that in the absence of non-random crosslinking, the distribution of crosslink densities is not the dominant factor in determining spatial correlations in a swollen network at the level of the network mesh size.

It has been suggested by Bastide (16-18) that the swelling of a network may involve the rearrangement of the conformations of the chains without significantly disturbing the end-to-end distance of a chain between crosslinks. Therefore, for such an unfolding mechanism to occur, correlations in a network at a size scale longer than just the dimension of a chain between crosslinks must occur. Even in an ideal or perfectly monodisperse unimodal network, correlations over distances greater than the radius of gyration of a single chain between crosslinks could be expected. As discussed later, topological variations associated with fluctuations in the types of crosslinks may occur over such extended distances. Our SANS results seem to be consistent with such a hypothesis.

Light Scattering from Swollen PDMS Networks

Bimodal PDMS networks were swollen in benzene and in toluene and studied by light scattering. To observe scattering at wide angles, the samples were in the form of thick flat films which were tilted with respect to the incident beam. As has been shown in the earlier work by Stein and Keane (26), sample tilting is necessary in order to cover an appreciable range of scattering angle, θ, as measured within the sample, when the scattered beam emerges from the front face of the sample. This is because of the refraction of the scattered ray in passing through the plane sample face which results in the scattering angle within the sample being less than that in air. Of course, procedures for correction for such refraction must be used (26). Great care was taken to ensure smooth surfaces and keep the samples free of dust. The details of the experiment and data reduction can be found elsewhere (27). The scattered intensities were converted to absolute units by using the scattering from toluene at 25°C for absolute calibration.

Fig 3 and 4 show the excess scattering data for the various bimodal networks swollen in benzene and in toluene at 25°C. The excess scattering data was analyzed using Debye-Bueche plots. Fig 5 shows one such typical plot.

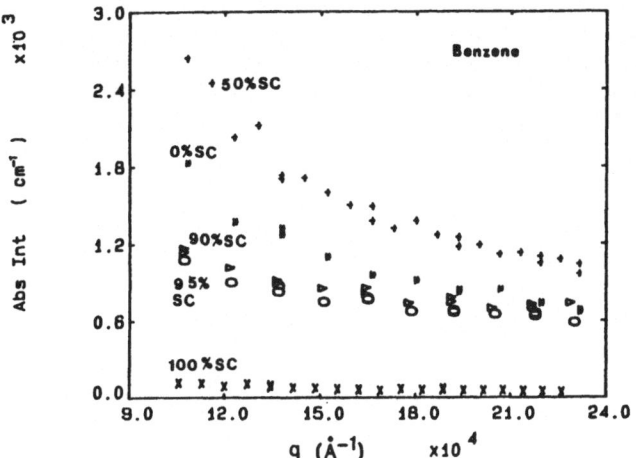

Fig 3 Corrected, absolute intensity light scattering
data for bimodal PDMS networks in benzene.

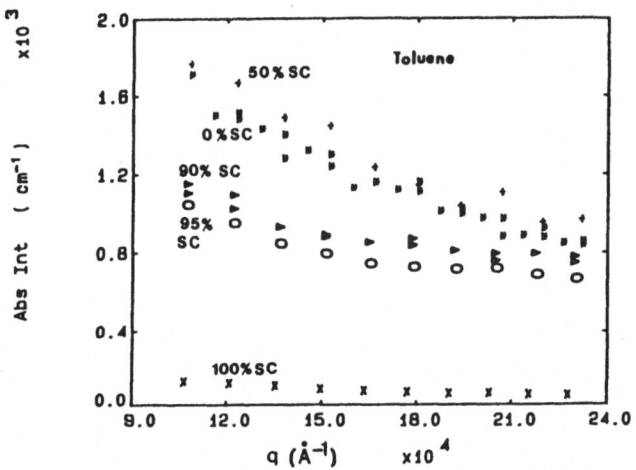

Fig 4 Corrected, absolute intensity light scattering
data for bimodal PDMS networks in toluene.

Table 4 lists the values of the mean-squared fluctuation in scattering power $\overline{\eta^2}$ which were estimated from the intercept of the Debye-Bueche plots for the various networks. The excess scattering determined from $\overline{\eta^2}$ consists of contributions resulting from crosslink inhomogeneities and other topological features. Using equation (12) it is possible to estimate the contribution to $\overline{\eta^2}$ if $\langle(\Delta M_c)^2\rangle$ is known. For the bimodal networks studied, the molecular weights of the prepolymers used are known and therefore it is possible to calculate $\langle(\Delta M_c)^2\rangle$. The simplest calculation is the macroscopic one, in which the short and long chains are weighted by their mole fractions.

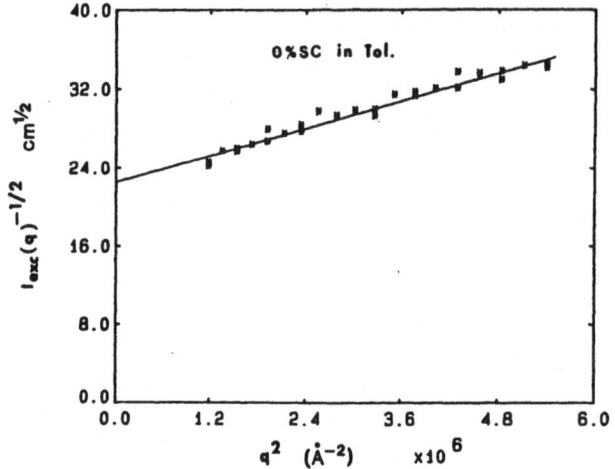

Fig 5 Debye-Bueche plot of light scattering data from PDMS swollen networks in toluene.

For monodisperse short and long chains one then has the following expression for $\langle(\Delta M_c)^2\rangle$

$$\langle(\Delta M_c)^2\rangle = x(1 - x)(M_L - M_S)^2 \tag{15}$$

Since the prepolymers are polydisperse, one has to include polydispersity into the calculation for $\langle(\Delta M_c)^2\rangle$ (27). Then, the mean-squared fluctuation in the scattering power corresponding to such a value of $\langle(\Delta M_c)^2\rangle$ can be estimated. These values are listed in Table 4.

Table 4

Comparison Values of $\overline{\eta^2}$ Obtained from
Debye-Bueche Plots with those Calculated Theoretically

Mole % of Short Chains	Solvent	$\eta^2 (cm^{-4})$ x 10^{-14} (exp.)	$\eta^2 (cm^{-4})$ x 10^{-14} (calc.)
0	toluene	23.8	8.75
50	"	18.7	17.68
90	"	11.3	51.93
95	"	12.0	70.20
100	"	1.2	21.86
0	benzene	15.2	7.55
50	"	16.6	13.94
90	"	12.2	36.95
95	"	12.9	52.53
100	"	1.1	16.80

In comparing the calculated values of $\overline{\eta^2}$ with the experimentally measured values it can be seen that the theoretical values greatly exceed the experimental. We believe that this difference is real in that most experimental errors would lead to too large an experimental value. This clearly suggests that the macroscopic calculation overestimates $\overline{\eta^2}$. Such an overestimation arises as a result of using macroscopic swelling theory to calculate microscopic swelling in a heterogeneous network.

The calculation of $\overline{\eta^2}$ from $\langle (\Delta M_c)^2 \rangle$ requires a knowledge of the variation of the local degree of swelling. This has been done by applying the conventional swelling theory of Flory (3) which relates the degree of swelling to the average molecular weight between crosslinking, M_c, and the polymer-solvent interaction parameter, χ, to relate the local degree of swelling to the value of $\overline{M_c}$ for a chain at that location. This is not correct in that the theory only applies to a chain which resided in a region occupied by other chains of the same $\overline{M_c}$.

The swelling of a network depends on its average $\overline{M_c}$. We have demonstrated that this is true with bimodal networks, and have shown that the swelling relates to $\overline{M_c}$ as calculated using Eq (13). The question is, in a heterogeneous network, how large is the region over which one takes this average? It is evident that this averaging process will lead to a lessening of the fluctuation on local degree of swelling over that calculated by direct use of the Flory theory. A theory of "micro-statistical mechanics" is required.

We propose the use of a cluster approach (28). The simplest cluster will contain four chains for a tetrafunctional network. Higher order clusters will contain larger numbers of chains with different contributions of short and long chains. Preliminary results based on such a cluster approach have shown that $\overline{\eta^2}$ decreases as one considers clusters with increasing number of chains. Such an analysis clearly indicates the necessity of taking into account the detailed network structure in describing some properties of a network.

The Flory theory also suffers in that it applies the affine hypothesis to the transformation of the average position of all crosslink points.

Its predictions appear to contradict the observations of changes in the radius of gyration, R_g, of labelled chains in deformed networks by SANS. As pointed out by Bastide, et al (15-18), the difficulty may lie in the lack of distinguishing between topological and geometric neighboring cross-links. Their concepts of "network unfolding" should be considered here.

Also, the theory does not directly address the role of entanglements, but treats them in the manner of Flory and Erman (12-13) as imposing restrictions on crosslink mobility. We regard it as an open question as to whether this approach is adequate for accounting for the effect of entanglements and its local variation on influencing local swelling.

Orientation of Network Chains

To understand the deformation of chains in a network it is useful to take into account the orientation of the network chains with respect to the stretching direction. The second order (Herman's) orientation function of the statistical segments with respect to the stretching direction is given by

$$P_2 = \frac{\langle 3\cos^2\theta_s - 1 \rangle}{2} \qquad (16)$$

where θ_s is the angle between the stretching direction and the statistical segment. For a gaussian network undergoing affine deformation it is predicted that (29)

$$P_2 = \frac{1}{5N} \left(\lambda^2 - \frac{1}{\lambda}\right) \qquad (17)$$

where λ is the deformation ratio and N is the number of statistical segments in the chain.

Birefringence is usually used to measure the orientation. It measures the orientation of all segments regardless of whether they belong to short or long chains. Infra-red (IR) dichroism, however, could be used to differentiate between different chemical species.

If D is the dichroic ratio, then

$$\frac{D - 1}{D + 2} = P_2 P_2^\circ \qquad (18)$$

where $D = A\|/A\perp$, where $A\|$ and $A\perp$ are the IR absorbancies for radiation polarized parallel and perpendicular to the machine direction respectively and P_2° is the orientation parameter for the transition moment of the segment.

Hrabowska and Stein (30) have studied the IR dichroism in bimodal PDMS on uniaxial deformation. In these networks both the short and long chains were hydrogeneous. The IR bands studied involved molecular vibrations of hydrogen containing groups and the measured orientation function was an average of all network segments. For the same extension ratio, affine deformation of gaussian chains would predict P_2 to increase with crosslink density (Eq 17). In the case of bimodal PDMS networks, they observed that at similar extension ratios, on increasing the fraction of short chains, P_2 was found to decrease. Such an observation clearly indicates that the deformation in bimodal networks is highly non-affine. This is in agreement with the hypothesis put forward by Mark and coworkers to account for the enhanced mechanical properties of bimodal PDMS networks.

In principle, if all the chains of one type (say long chains) are deuterated, one can simultaneously measure the orientation functions for the short and long chains in a bimodal network using IR dichroism. This would be done through observation of different IR bands involving vibrations of either hydrogen or deuterium containing groups. In addition to the IR measurements the materials could also be characterized by SANS so as to measure the actual molecular deformation ratio of the labelled chains. Currently, efforts are being undertaken to prepare model networks with deuterated prepolymers. One system under consideration is poly-(tetrahydrofuran) (PTHF) diol.

Strain Induced Crystallization

The PTHF prepolymers have an attractive feature that they crystallize around 40°C. These could be suitable for studying strain induced crystallization (SIC) of polymer networks. In randomly crosslinked systems like natural rubber, SIC is a very important means for self reinforcement. Use of model networks, especially bimodal networks, may be very helpful in studying this phenomenon. Due to the non-affine nature of the deformation in bimodal networks, it may be possible to have regions of local orientation greater than average, resulting in higher nucleation rates. The SIC phenomenon can be followed by measuring the stress and the birefringence as a function of temperature and elongation.

We anticipate quite different SIC behavior for unimodal and bimodal networks. The understanding of this phenomenon and its relationship to network micromechanics as affected by network homogeneity is of importance in understanding factors leading to crystallization reinforcement of elastomers.

Summary

Macroscopic or average properties of polymer networks seem to be reasonably well described by the current theories of rubber elasticity. However, these theories are unable to account for some of the local behavior in networks as observed by Bastide and coworkers. Our light scattering results on swollen networks clearly indicate that macroscopic swelling approaches cannot be used to account for the observed results. It is necessary to take into account the network structure to evaluate the contribution of neighboring chains in trying to describe local properties. The scattering results (LS and SANS) seem to be consistent with the network chain rearrangement for folding hypothesis proposed by Bastide. At this stage, the analysis in terms of topological rearrangement is qualitative and greater insight is required to understand the local behavior of networks.

Acknowledgment

This work was supported by the Center of University of Massachusetts and Industry Research in Polymers (CUMIRP). We are grateful to Dr. Charles C. Han, Professor K. Langley, Matt Bishop and Dr. Claude Picot for valuable discussion and generous assistance with the scattering experiments.

References

1. L.R.G. Trealor, "The Physics of Rubber Elasticity," 3rd Ed., Clarendon, Oxford (1975).
2. W. Kuhn, Kolloid Z., 76, 258 (1936); Angew. Chem., 51, 640 (1938).
3. P.J. Flory, "Principles of Polymer Chemistry," Cornell University Press, Ithaca, New York (1953).

4. J.P. Queslel and J.E. Mark, Adv. Polym. Sci., 65, 135, (1984).
5. J.E. Mark, Adv. Polym. Sci., 44, 1, (1982).
6. J.E. Mark and J.L. Sullwan, J. Chem. Phys., 66, 1006 (1977).
7. M.A. Llorente and J.E. Mark, J. Chem. hys., 71, 682 (1979).
8. M.A. Llorente and J.E. Mark, Macromolecules, 13, 681 (1980).
9. M.A. Llorente, A.L. Andrady and J.E. Mark, J. Polym. Sci., Polym. Phys. Ed., 19, 621 (1981).
10. M.A. Llorente, A.L. Andrady and J.E. Mark, Colloid Polym. Sci., 259, 1056 (1981).
11. J.E. Mark, in "Elastomers and Rubber Elasticity," J.E. Mark and J. Lal, Eds., Am. Chem. Soc., Washington, D.C. (1982).
12. P.J. Flory and B. Erman, Macromolecules, 15, 800 (1982).
13. B. Erman and P.J. Flory, Macromolecules, 15, 806 (1982).
14. M. Beltzung, C. Picot, P. Rempp and J. Herz, Macromolecules, 15, 1594 (1982).
15. M. Beltzung, J. Herz and C. Picot, ibid, 16, 580 (1983); ibid, 17, 663 (1984).
16. J. Bastide, C. Picot and S. Candau, J. Macromol. Sci., Phys. Ed., B19, 13 (1981).
17. S. Candau, J. Bastide and Delsanti, Adv. Polym. Sci., 44, 27 (1982).
18. J. Bastide, R. Duplessix, C. Picot and S. Candau, Macromolecules, 17, 83 (1984).
19. P. Debye and A.M. Bueche, J. Appl. Phys., 20, 518 (1949).
20. A. Einstein, Ann. de Physik, 33, 1275 (1910).
21. P.J. Flory, and J. Rehner, Jr., J. Chem. Phys., 11, 521 (1943).
22. R.S. Stein, J. Polym. Sci., Polym. Lett, 7, 657 (1969).
23. F. Bueche, J. Colloid Interface Sci., 33, 61 (1970).
24. K.L. Wun and W. Prins, J. Polym. Sci., Polym. Phys. Ed., 12, 533 (1974).
25. R.S. Stein, R.J. Farris, S. Kumar and V. Soni, in "Elastomer and Rubber Elasticity," J.E. Mark and J. Lal, Am. Chem. Soc., Washington, D.C. (1982).
26. R.S. Stein and J.J. Keane, J. Polym. Sci., 17, 21 (1955).
27. V. Soni, Ph.D. Dissertation, University of Massachusetts, Amherst (1986).
28. R.S. Stein, V. Soni, and H. Yang, in preparation.
29. W. Kuhn, F. Grun, Colloid Z., 101, 248 (1942).
30. J. Hrabowska and R.S. Stein, results to be published.

THEORY OF SEGMENTAL ORIENTATION IN AMORPHOUS POLYMER NETWORKS AND COMPAR-
ISON WITH EXPERIMENTAL DETERMINATION OF ORIENTATION IN POLYISOPRENE
NETWORKS BY FLUORESCENCE POLARIZATION

J.P. Queslel, B. Erman [*] and L. Monnerie

Laboratoire de Physicochimie Structurale et Macromoléculaire
associé au C.N.R.S. ESPCI, 10, rue Vauquelin 75231 Paris
Cedex 05 (France)
* School of Engineering, Bogazici University, Bebek,
Istanbul (Turkey)

In memory of Professor Paul J. Flory

ABSTRACT

Determination of segmental orientation in stretched rubbery networks
may serve to test the validity of elaborated models of rubberlike elasticity, in conjunction with stress-strain measurements.

As a preliminary, the recently proposed theory of segmental orientation of real networks with constraints on junctions is reviewed. Basic
principles of fluorescence polarization technique and its application to
investigation of molecular orientation of mobile, anisotropic networks are
then described. Illustrative data on elasticity and orientation behaviors
of cis-polyisoprene networks are consistent with theoretical predictions.

INTRODUCTION

Rubberlike elasticity is a unique property of long, flexible chains
with weak interchain interactions joined together by cross-linking points
to form tridimensionally stable networks. Early in the 1940's statistical
mechanics formalism has been developed to understand the molecular mechanism governing the behavior of the large ensemble of chains constituting
elastomeric networks [1, 2]. Two models of Gaussian networks, i.e. networks
with chains between junctions sufficiently long to have a Gaussian distribution of end-to-end distances, were proposed [3]. In the phantom model,
chain crossability and large junction fluctuations were allowed. In the
affine one, junction fluctuations were completely suppressed and the components of the vectorial length of each chain were changed by the deformation in the same ratio as the macroscopic deformation of the sample.
The reduced stress in uniaxial extension is conventionally defined as [4] :

$$[\tau] \equiv v_2^{-1/3}(\tau_x - \tau_y)/(\alpha^2 - \alpha^{-1}) \qquad (1)$$

Where $(\tau_x - \tau_y)$ is the difference between the principal components of
the equilibrium true stress along the longitudinal (x) and the lateral
(y) directions of deformation, α is the ratio of the swollen deformed
length to the swollen undeformed length, and v_2 is the volume fraction of
polymer in the swollen network.

For both the phantom and the affine networks, the reduced stress is calculated to be independent of deformation. However, stress-strain measurements carried out in uniaxial extension of dry and swollen networks have revealed departures from these predictions of simple models [5]. These observations then gave rise to phenomenological equations like the Mooney-Rivlin expression, i.e.

$$[\tau] = 2 c_1 + 2 c_2/\alpha \qquad (2)$$

Recently, new models were proposed which can indeed be used to characterize the structure of real elastomeric networks in view of their mechanical properties [6, 7]. However several parameters are necessary to describe the relationship between molecular and macroscopic deformations and therefore stress-strain measurements are generally not sufficient to conclude without any ambiguity on the validity of these elaborated theories. Another possible test consists in measuring molecular orientation in stretched rubbery networks. With this in view, the photoelastic properties of rubbers have been widely investigated [8 - 10]. However birefringence data suffer from the fact that segmental polarizability is affected by its immediate environment.

Recent developments in spectroscopic techniques allow for accurate measurement of orientations of structural units in deformed polymeric systems [11, 12]. Infrared dichroism(I.R.D), ^{2}H - N.M.R. spectroscopy, fluorescence polarization (F.P.) are three methods which find increasing use in experimental investigations. The experiments reported in the present paper concern effectively simultaneous measurements of orientation by fluorescence polarization and of stress-strain relationship [13]. It is worth noting that the transition moments of the label involved in fluorescence polarization are not sensitive to their immediate environment. Previous models are briefly reviewed and a new model of orientation of real networks is proposed. Then the fluorescence polarization technique is described and theoretical predictions are compared with measurements on polyisoprene vulcanizates.

REAL NETWORK MODEL OF ELASTICITY AND ORIENTATION

In real networks, diffusion of junctions about their mean positions may be severely restricted by neighboring chains sharing the same region of space. Real networks present strong chain overlapping. Intermolecular steric hindrances on chain motion, commonly termed as entanglements, have long been recognized to be the origin of departures from phantom and affine predictions. Different formalisms were proposed to include these intermolecular effects in the rubberlike elasticity analysis at thermodynamic equilibrium [2]. In the one due to Flory and Erman [7], entanglements are embodied as domains of constraints acting as restrictions of junction fluctuations. These domains can be initially represented as spheres. They are transformed to ellipsoids by the deformation. In uniaxially extension, the main axes of these ellipsoids are along the direction of stretching. Thus fluctuations increase in the direction along which the stress is measured. The behavior of real networks then tends to that of phantom networks with large fluctuations in the limit of infinite deformation. The C_2 effect, observed experimentally in uniaxially tensile tests of amorphous rubbers, results from the transformation of fluctuations with strain in the manner stated above and is predicted by this molecular model of constraints. The principal parameter κ of the theory is defined as the ratio of the mean square radius of the fluctuations of the junctions in the phantom network $< (\Delta R)^2_{ph} >$, to the mean square radius $< (\Delta s)^2 >_o$ of the Gaussian domains of constraints in the undistorted network, i.e.

$$\kappa = \langle (\Delta R)^2_{ph} \rangle / \langle (\Delta s)^2 \rangle_o \qquad (3)$$

The parameter κ has been postulated to depend on the degree of inter-penetration in the network, which is well supported by experiments [14, 15]. A second parameter ζ is believed to reflect the effect of inhomogeneities in the network topology which perturb the affine transformation of constraint domains with macroscopic deformation [16].

The molecular theory of elasticity outlined above predicts the following expression for the difference between the principal components of the true stress along the longitudinal (x) and the lateral (y) directions of deformation [7].

$$\tau_x - \tau_y = (\xi kT/V)\{\lambda_x^2 - \lambda_y^2 + (\mu/\xi)[\lambda_x^2 K(\lambda_x^2) - \lambda_y^2 K(\lambda_y^2)]\} \qquad (4)$$

where :

$$K(\lambda_t^2) = B_t[\dot{B}_t(B_t+1)^{-1} + g_t(\dot{g}_t B_t + g_t \dot{B}_t)(g_t B_t+1)^{-1}] \qquad (5)$$

$$t = x, y$$

with :

$$B_t = (\lambda_t-1)(\lambda_t+1 - \zeta\lambda_t^2)/(1+g_t)^2$$

$$g_t = \lambda_t^2[\kappa^{-1} + \zeta(\lambda_t-1)]$$

$$\dot{B}_t \equiv \partial B_t/\partial\lambda_t^2 = B_t\{[2\lambda_t(\lambda_t-1)]^{-1} + (1-2\zeta\lambda_t) \qquad (6)$$

$$[2\lambda_t(1+\lambda_t - \zeta\lambda_t^2)]^{-1} - 2\dot{g}_t(1+g_t)^{-1}\}$$

$$\dot{g}_t \equiv \partial g_t/\partial\lambda_t^2 = \kappa^{-1} - \zeta(1-3\lambda_t/2)$$

In Eq. 4, ξ/V represents the cycle rank (number of independent circuits in the network) density of the swollen network [3, 17], k the Boltzmann constant, T the absolute temperature, φ the junction functionality, and μ the number of junctions.

The deformation ratios λ_x and λ_y characterizing the deformation of the swollen network relative to the unswollen, undeformed state are related to α by the following equations :

$$\lambda_x = v_2^{-1/3}\alpha \qquad (7)$$

$$\lambda_y = v_2^{-1/3}\alpha^{-1/2}$$

The phantom network limit is reached when $\kappa \to 0$ and $\kappa\zeta \to 0$ [7]. Eq. 4 then simplifies to :

$$(\tau_x - \tau_y)_{ph} = (\xi kT/V)(\lambda_x^2 - \lambda_y^2) \qquad (8)$$

The affine limit corresponds to $\kappa \to \infty$ and $\zeta \to 0$.

The degree of molecular segmental orientation in a deformed amorphous polymeric network is obtained by the knowledge of the orientational distribution function. This distribution is represented by a function $f(\Theta_x)$, Θ_x being the angle between unit vectors affixed to chain segments and the laboratory fixed axis denoting the direction of stretch. The function

$f(\Theta_x)$ can be expanded in terms of Legendre polynomials in $\cos \Theta_x$ as follows :

$$f(\Theta_x) = \sum_\ell b_\ell P_\ell (\cos \Theta_x) \qquad (9)$$

with

$$b_\ell = [(1/2 \pi)(2 \ell + 1)/2] < P_\ell (\cos \Theta_x) > \qquad (10)$$

where $< P_\ell (\cos \Theta_x) >$ is the value of $P(\cos \Theta_x)$ averaged over the distribution. The average of the second Legendre polynomial is defined as :

$$S_x \equiv < P_2(\cos \Theta_x) > = (1/2) < 3 \cos^2 \Theta_x - 1 > \qquad (11)$$

$$< \cos^2 \Theta_x > = \int_0^\pi f(\Theta_x) \cos^2 \Theta_x \sin \Theta_x \, d\Theta_x \qquad (12)$$

It has been shown that for a uniaxial distribution the determination of $< P_2 >$ is sufficient in most cases. Previous treatments of segmental orientation were also based on the affine or phantom model [18,19]. The molecular theory of real networks presented above has been recently applied to the analysis of segmental orientation by Erman and Monnerie [20]. The expression for the second moment S_x of the orientation function for Gaussian networks is conveniently separated into two factors : a configurational factor D which represents the statistical properties of chains and a strain function $F(\lambda)$ which relates the orientation to macroscopic deformation :

$$S_x = D \, F(\lambda) \qquad (13)$$

For a long freely-jointed chain formed with N statistical segments

$$D = 1/5 \, N \qquad (14)$$

However, in real networks, chains are not freely jointed. Furthermore, chain segments are locally subject to orientational correlations from segments of the neighboring chains in the bulk state. These local intermolecular orientational correlations affecting local orientation of segments but not the overall chain configurations have been observed in birefringence [15,22] and fluorescence polarization [13,23] experiments and have been attributed to hindrance of rotations of neighboring segments about to each other due to intermolecular steric constraints (uncrossability of chains) and to alignment of neighboring chain segments with respect to each other due to anisotropic intermolecular dispersive forces [24]. In the case of weak local correlations, it can be shown [20] that the configurational factor for the chain in real network, D, is the sum of the configurational factor for the free chain, D_o, and an intermolecular contribution, D_{int}

$$D = D_o + D \text{ int} \qquad (15)$$

The configurational factor D_o for a free chain is given by Nagai's treatment [21].

$$D_o = (3 < r^2 \cos^2 \phi >_o /< r^2 >_o - 1)/10 \qquad (16)$$

where ϕ is the angle between a labeled segment and the chain end-to-end vector r. Brackets with the subscript zero denote the ensemble average for a free chain in the unperturbed state. The expression given by eq. (16) reduces to that given by eq. (14) for a freely jointed chain.

The value of D_o for a given chain may conveniently be evaluated by the rotational isomeric state formulation. Rational calculation of D_{int} in terms of molecular parameter awaits further work.

The strain function depends on the transformations of chain dimensions with deformation in real networks. According to the model proposed by Flory and Erman [7], the elements of the molecular deformation tensor Λ^2 can be expressed as the sum of contributions from the phantom network, Λ^2_{ph} and from the constraints, Λ^2_c

$$\Lambda^2_t = \Lambda^2_{t,ph} + \Lambda^2_{t,c}, t = x, y, z \qquad (17)$$

In addition to the deformation of the chains associated with the alteration of their chain vectors under strain, it is necessary to consider the action of the junctions on the constraint domains surrounding them. The junction and its domain are coupled elastic elements. Non-affine transformation of the fluctuation domains of junctions in the presence of chain connectivity creates an additional strain field in the medium around each junction. A domain deformation tensor $(H)^2$ can be introduced in analogy with Λ^2. $(H)^2$ relates the mean deformation (according to ensemble distribution) of the junction constraint domains in the deformed, connected network to that in the absence of connectivity. Consequently, the state of microscopic deformation in a real network is the sum of the molecular deformation tensor, Λ^2 and the junction domain deformation tensor, $(H)^2$.

The first contribution to the strain function is expressed from the molecular deformation tensor Λ^2.

$$F_1(\lambda) = \Lambda^2_x - (\Lambda^2_y + \Lambda^2_z)/2 \qquad (18)$$

The second contribution arises from the distorsion of the junction constraint domains. Depending on the extent of distorsion, a chain surrounding a given junction has to reorganize at a scale corresponding to the size of its constraint domain. This results in an additional orientation of the segment present in the domain proportional to the local domain deformation tensor $(H)^2$. In analogy with Eq. (18), the second contribution to the strain function may be expressed as :

$$F_2(\lambda) = e[(H)^2_x - ((H)^2_y + (H)^2_z)/2] \qquad (19)$$

where the parameter e denotes the strength of reaction of orienting chain segments to the distorsion of junction constraint domain. The resulting strain function is the sum of F_1 and F_2

$$F(\lambda) = F_1(\lambda) + F_2(\lambda) \qquad (20)$$

According to the theory, the components of the microscopic deformation tensors Λ^2 and $(H)^2$ are related to the macroscopic deformation tensor λ by

$$\Lambda^2_t = (1 - \frac{2}{\varphi})\lambda^2_t + \frac{2}{\varphi}(1 + B_t) \qquad (21)$$

$$(H)^2_t = 1 + g_t B_t$$

Using eqs. (21) in (19) and (20), and substituting in eq. (13) leads to the following expression for S_x

$$S_x = (1-2/\varphi)v_2^{-2/3} D\{\alpha^2-\alpha^{-1} + 2 v_2^{2/3}(\varphi-2)^{-1}[B_x-B_y + (\varphi e/2)(g_xB_x-g_yB_y)]\} \quad (22)$$

For a phantom network, the term in square brackets in Eq. 22 vanishes. Therefore

$$S_{x,ph} = (1-2/\varphi)v_2^{-2/3} D(\alpha^2 - \alpha^{-1}) \quad (22)$$

The difference between S_x and $S_{x,ph}$ reflects the effect of constraints on segmental orientation in networks.

In analogy with the reduced stress, the reduced orientation function in uniaxial extension may be defined as :

$$[S_x] \equiv v_2^{2/3} S_x/(1-2/\varphi)(\alpha^2 - \alpha^{-1}) \quad (23)$$

$$= D S_x/S_{x,ph}$$

The limit of $[S_x]$ in the infinite range of deformation is the configurational factor :

$$\lim_{\alpha^{-1} \to 0}[S_x] = D \quad (24)$$

The ratio Ψ of the orientation to the stress difference normalized with respect to the respective phantom contribution has the following form.

$$\Psi = (S_x/S_{x,ph})/\{(\tau_x - \tau_y)/(\tau_x - \tau_y)_{ph}\} \quad (25)$$

FLUORESCENCE POLARIZATION (F.P.) TECHNIQUE

Principles

Fluorescent molecules have the property of re-emitting in the form of visible light part of the energy acquired by the absorption of luminous radiation. After illuminating by a very short pulse at time t_o, the fluorescent light emitted at time t_o+u is proportional to $\exp(-u/\tau)$, where τ is the mean lifetime of the excited state (usually called the fluorescent lifetime). The most frequent values range from 10^{-9} to 10^{-7} s. When absorbing light of a suitable wavelength a molecule behaves as an electric dipole oscillator with a fixed orientation with respect to the geometry of the molecule. Such an equivalent oscillator is termed an absorption transition moment, M_o. In the same way, for the fluorescence emission we have an emission transition moment, M. When such a molecule receives an incident beam polarized along P direction, the absorption probability is proportional to $\cos^2 \gamma$, where γ is the angle between P and M_o. In the same way, the fluorescence intensity measured through an analyzer, A, is proportional to $\cos^2 \delta$, where δ is the angle between M and A. Thus, for the P and A directions of polarizer and analyzer the observed luminescence intensity is proportional to $\cos^2\gamma \cos^2 \delta$. Owing to the lack of phase correlation between excitation and emission lights, fluorescence emission can be described as resulting from three independent radiations respectively polarized along the X, Y, and Z laboratory axes with intensities I_x, I_y, I_z. The Curie symmetry principle, applied to excitation light polarized along Z, leads to $I_x = I_y$. The fluorescence polarization is characterized by the emission anisotropy :

$$r = (I_\parallel - I_\perp)/(I_\parallel + 2\, I_\perp) \qquad\qquad (26)$$

where $I_\parallel$ and $I_\perp$ correspond to the fluorescence intensity obtained with an analyzer direction parallel or perpendicular, respectively, to that of the polarizer.

When dealing with an isotropic set of fluorescence molecules for which the relaxation times of the molecular motions are in the range of the fluorescence lifetime, F.P. yields information on the mobility of the molecules. This high frequency mobility appears in polymers at temperature superior to Tg(DSC) + 50°C [25]

Orientation of uniaxially symmetric systems

For our present purpose we are mainly interested in the use of F.P. to look at the orientation distribution of fluorescent molecules. The transition moments in both absorption and emission are assumed to coincide with a molecular axis M of the molecule, the direction of which is specified by the spherical polar angle $\Omega = (\Theta, \beta)$ in the reference frame (Fig. 1). In the treatment [26] are then introduced the angular functions $N(\Omega_o, t_o)$, the orientation distribution of M at time t_o (M_o in Fig. 1), and $P(\Omega, t; \Omega_o, t_o)$ the conditional probability density of finding at position Ω at time t a vector M which was at position Ω_o at time t_o. After illuminating the samples by a linearly polarized short pulse of light at t_o, the intensity emitted at time ($t_o + u$) for the P and A directions of polarizer and analyzer is given by :

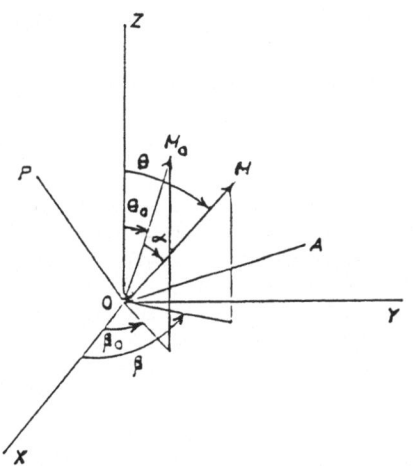

Fig. 1 - Illustration of the angles which define the orientation of molecular axis M_o at time t_o and M at time t_o+u with respect to the fixed frame OXYZ

$$i(P,A,t_o,u) = K\iint N(\Omega_o,t_o)P(\Omega,t_o + u; \Omega_o,t_o)$$

$$\times \cos^2(P,M_o)\cos^2(A,M) \exp(-u/\tau)d\Omega_o d\Omega \qquad (27)$$

Where K is an instrumental constant. In this expression t_o corresponds to the macroscopic evolution of the sample whereas u corresponds to a microscopic reorientational motion in the scale of the fluorescence lifetime τ . In most cases the t_o dependence of N and P can be ignored within the time τ (ca.10^{-8} s), and the fluorescence intensity emitted under continuous excitation is given by :

$$i(P,A,t_o,\tau) = \int_0^\infty i(P,A,t_o + u)du \qquad (28)$$

In the case of a uniaxially symmetric distribution of the molecular axes M, the intensities corresponding to the P and A directions lying along the fixed-frame axes (Z corresponds to the symmetry axis) can be conveniently expressed through the following quantities :

$$G_{20}^{(o)} = (1/2) < 3 \cos^2 \Theta_o - 1 >$$

$$G_{02}^{(o)} = (1/2) < 3 \cos^2 \Theta - 1 >$$

$$G_{22}^{(o)} = (1/4) < 3 \cos^2 \Theta_o - 1)(3 \cos^2 \Theta - 1) >$$

$$G_{22}^{(1)} = (9/16) < \sin \Theta_o \cos \Theta_o \sin \Theta \cos \Theta \cos(\beta-\beta_o) >$$

$$G_{22}^{(2)} = (9/64) < \sin^2 \Theta_o \sin^2 \Theta \cos 2(\beta-\beta_o) >$$

Thus, for example :

$$i(X,X) = (K/9)(1 - G_{20}^{(o)} - G_{02}^{(o)} - G_{22}^{(o)} + 8 G_{22}^{(2)})$$

$$i(Z,Z) = (K/9)(1 + 2G_{20}^{(o)} + 2G_{02}^{(o)} + 4G_{22}^{(o)})$$

$$i(Z,X) = (K/9)(1 + 2G_{20}^{(o)} - G_{02}^{(o)} - 2G_{22}^{(o)})$$

In uniaxial mobile systems, i.e. polymers at a temperature superior to Tg(DSC) + 50°C, both orientation and mobility contribute to the fluorescence polarization and the two effects have to be separated from the measured intensities. This is possible if we assume that during the fluorescence lifetime (ca.10^{-8} s) the orientation distribution doesn't change [26]. Nevertheless, special equipment is required and only $< P_2(\cos \Theta) >$ can be obtained from the measurement of fixed intensities, $i(P,A)$, corresponding to P and A directions which are not contained in the same plane. On the other hand, the mean amplitude of the motion performed during the fluorescence lifetime, $M(\tau)$, is available from the data [27].

$$\overline{M}(\tau) = \int_0^\infty M(u)e^{-u/\tau} du/\tau$$

with $\qquad\qquad\qquad\qquad\qquad\qquad\qquad\qquad\qquad$ (29)

$$M(u) = < (3 \cos \Theta(u) - 1)/2 >$$

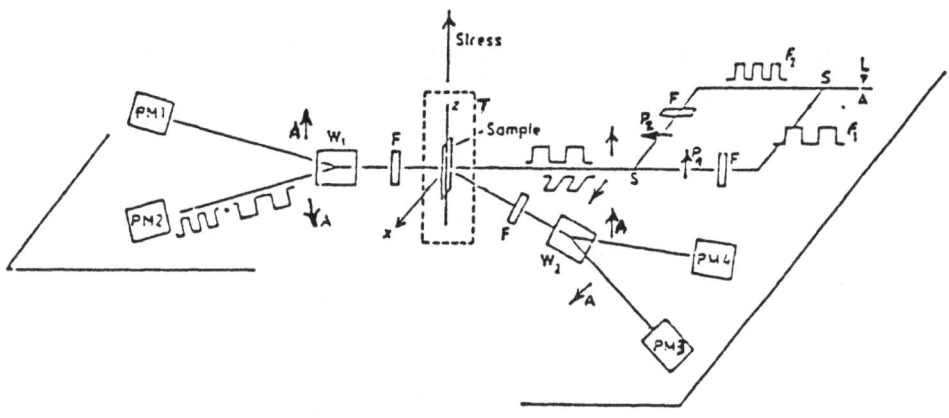

Fig. 2 - Optical equipment : L, mercury lamp; S, beam splitter;
f_1, F_2, modulation frequencies of the mechanical choppers;
F, optical filter; P_1, P_2 polarizers; W_1, W_2, wollaston prisms;
A, analyzing directions; PM, photomultiplier; T, temperature
chamber

where $\Theta(u)$ is the angle through which the M axis rotates between t_o and $t_o + u$.

An apparatus to measure simultaneously the convenient set of intensities i(P,A) has been developed [28] and is represented in Fig. 2. Values of the second moment S_x are obtained after correcting for the effect of the delocalization of the transition moments and for birefringence of the sample [26].

COMPARISON BETWEEN THEORY AND EXPERIMENT

Stress-strain and orientation-strain measurements have been carried out on polyisoprene networks [13] and have been compared with predictions of elasticity theories of real networks [29]. Some illustrative results are reported hereafter.

Materials and method

Networks formed from blends of 75 % by weight of an anionic commercial polyisoprene Shell IR 307 with a high cis-1,4 configuration (92 % cis, 5 % trans) (Tg(DSC) 60°C)   of high molecular weight (M_n = 46.3 x 10^4, M_w = 182 x 10^4 g.mole^{-1}) and 25 % of polyisoprene chains of low molecular weight (M_n = 50,000), the microstructure of which is similar to that of IR 307. These matrices (99 %) were mixed in solution with labeled anionic polyisoprene chains containing a dimethylanthracene (DMA) fluorescent group at their centers. The labeled polyisoprene had the same microstructure as IR 307 and its molecular weight was M_n = 60.10^4 g.mole^{-1}. Samples were molded and cross-linked with dicumyl peroxide. Measurements of stress and orientation were done during uniaxial stretching at constant cross-head speed, V = 50 mm/min, and at a temperature of 298°K.

Table I — Parameters giving the best agreement between stress and orientation experimental data and theoretical curves

Samples	$\xi kT/V$ N/mm^2	κ	ζ	$D \times 10^2$	e
IR1	0.121	7	0.025	1.16	1.4
IR2	0.063	10	0.025	0.76	1.6

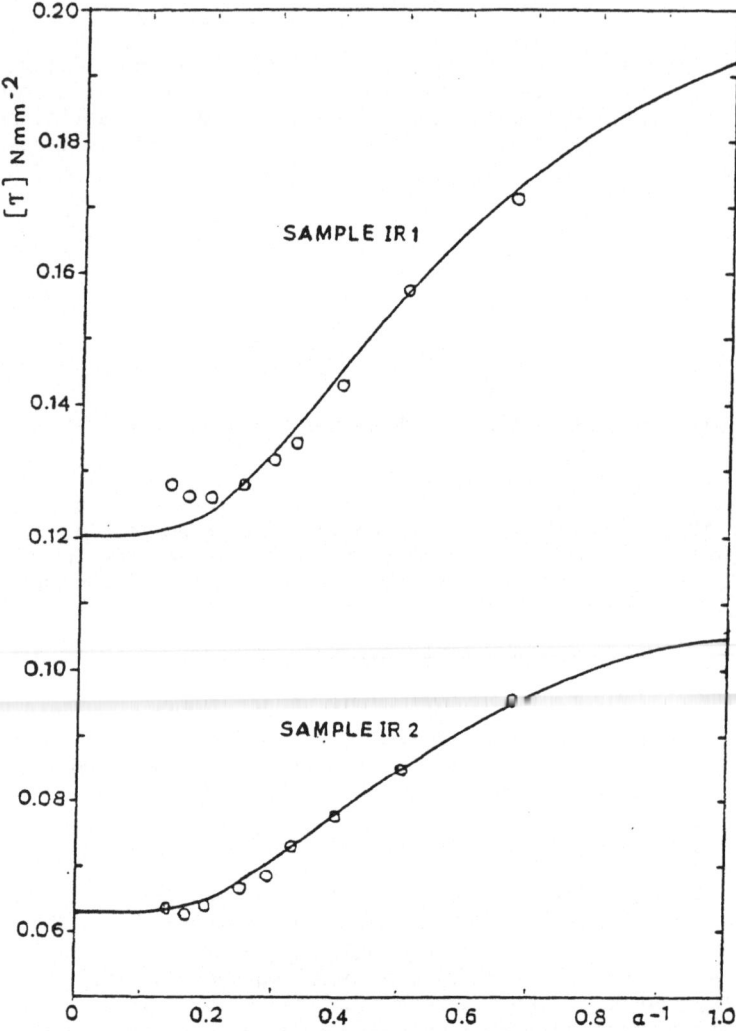

Fig. 3 — Dependence of reduced stress [τ] on inverse of deformation ratio α^{-1} for samples IR1 and IR2. Experimental data are represente by points and theoretical predictions by continuous curves.

Results and discussion

Experimental data for two samples IR 1 and IR 2 of different cross-linking density are represented in terms of reduced stress $[\tau]$, reduced orientation function $[S_x]$, and normalized ratio Ψ in Figs 3, 4 and 5, respectively. Continuous curves were calculated by making use of the equations derived in the preceding theoretical part with the parameters listed in Table I.

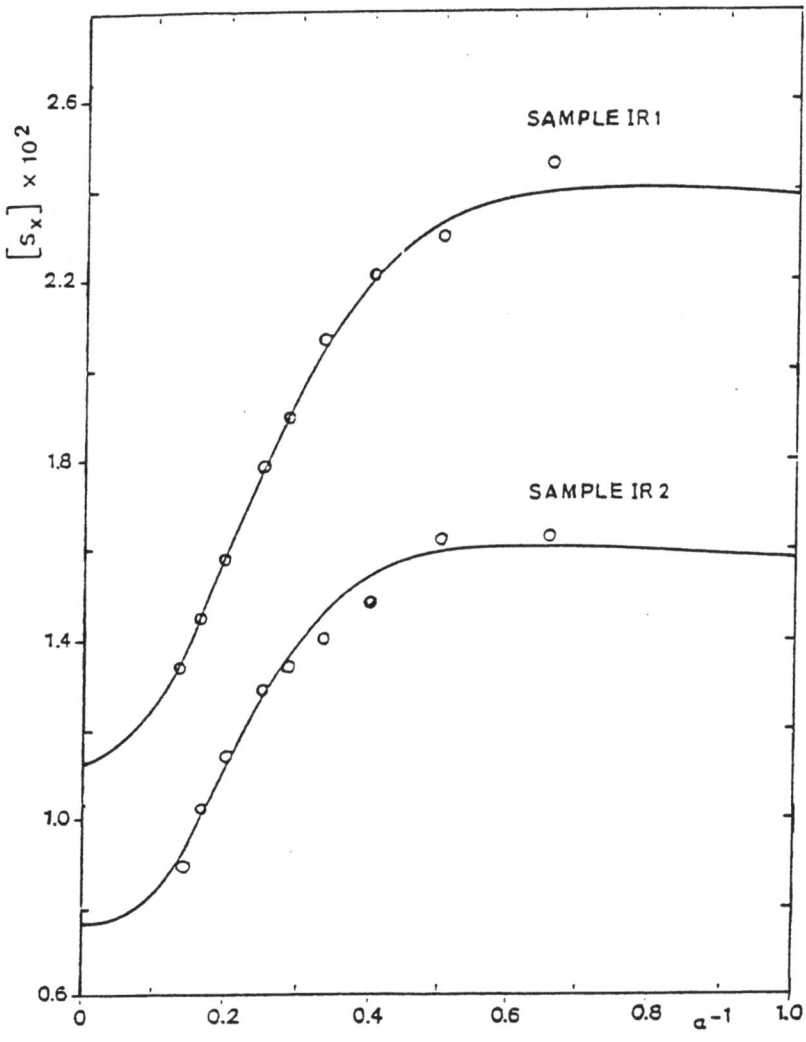

Fig. 4 – Dependence of reduced orientation function $[S_x]$ on α^{-1}. Same symbols as in Fig. 1.

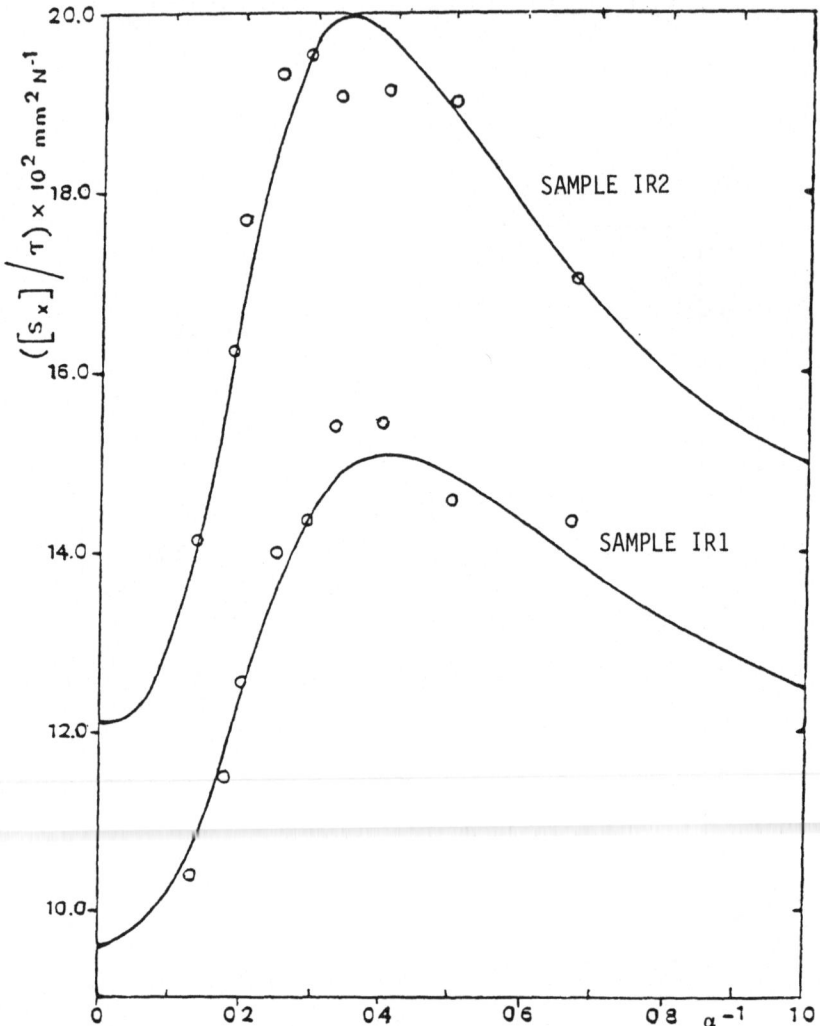

Fig. 5 – Dependence of normalized ratio Ψ on α^{-1}. Same symbols as in Fig. 1

Figure 3 is typical of elastomeric networks, showing the decrease of modulus with elongation, as already mentioned in the introduction. Simultaneous increase of cycle rank density and decrease of κ are also usual trends in real networks [14,30].

The decrease of reduced orientation with deformation observed in Fig.4 has often been reported and has led authors to propose a phenomenological equation similar to Mooney-Rivlin one in elasticity. However this equation is not able to predict the leveling of reduced orientation in the small deformation range. Our proposed theory gives good agreement over a large range of deformation.

More demanding is the comparison of data and theory in terms of the ratio in Fig. 5. Very good agreement is reached again. The maxima observed in the set of experimental points coincides with the respective maxima in the curves obtained from theory. The nonlinearity of the curves and data points in Fig.5 clearly indicate that orientation is not proportional to stress and that stress cannot be accepted as a measure of segmental orientation in amorphous networks.

Effects of swelling and temperature were also investigated in additional experiments. The observed lowering of values of $[S_x]$ upon swelling the networks can only be accounted for by considering a decrease in D, reflecting the gradual disappearance of local intermolecular orientational correlations with swelling. The effect of increasing temperature results in a linear decrease of D which may be attributed to the high sensitivity of D_{int} to temperature.

CONCLUSION

The expression of the orientation function derived on the basis of the real network model differs markedly from those obtained in previous treatments. The strain function includes the non-affine molecular deformation consecutive to restriction on junction fluctuations and distorsions of constraint domains. The configurational factor for the real chain accounts also for local intermolecular orientational correlations.

Illustrative experiments on polyisoprene networks were shown to be consistent with both stress and orientation theoretical predictions.

ACKNOWLEDGMENTS

We are grateful for the samples received from Manufacture Française des Pneumatiques MICHELIN (Clermont-Ferrand). The authors gratefully acknowledge financial support by the American Chemical Society (J.P. Queslel and B. Erman) and by MICHELIN (J.P. QUESLEL) for participation in the Chicago National ACS Meeting.

REFERENCES

1. P.J. Flory, "Principles of Polymer Chemistry", Cornell University Press, Ithaca, New-York, (1953)
2. B.E. Eichinger, Ann. Rev. Phys., 34:359 (1983)
3. P.J. Flory, Proc. Roy. Soc. London, A351:351 (1976)
4. M. Mooney, J. Appl. Phys., 11:582 (1940)
5. G. Gee, Trans. Farad. Soc., 42:585 (1946)
6. G. Ronca and G. Allegra, J. Chem. Phys., 63:4990 (1975)
7. P.J. Flory and B. Erman, Macromolecules, 15:800 (1982)
8. R.S. Stein, F.H. Holmes and A.V. Tobolsky, J. Polym. Sci., 14:443 (1954)

9. A.N. Gent, Macromolecules, 2:262 (1969)
10. T. Ishikawa and K. Nagai, J. Polym. Sci., A-2, 7:1123 (1969)
11. L. Monnerie, Faraday Symp. Chem. Soc., 18:57 (1983)
12. I.M. Ward, "Structure and Properties of Oriented Polymers", Applied
        Science Publ., London, (1975)
13. J.P. Queslel, Ph.D. Thesis, Paris (1982)
14. B. Erman and P.J. Flory, Macromolecules, 15:806 (1982)
15. B. Erman and P.J. Flory, Macromolecules, 16:1607 (1983)
16. J.P. Queslel, P. Thirion and L. Monnerie, submitted to Polymer (1986)
17. P.J. Flory, Brit. Polym. J., 17:1 (1985)
18. L.R.G. Treloar, "The Physics of Rubber Elasticity", Clarendon, London,
        (1975)
19. W. Kuhn and F. Grun, Kolloid-Z., 101:248 (1942)
20. B. Erman and L. Monnerie, Macromolecules, 18:1985 (1985)
21. K. Nagai, J. Chem. Phys., 40:2828 (1964)
22. M.H. Liberman, Y. Abe and P.J. Flory, Macromolecules, 5:550 (1972)
23. J.P. Jarry and L. Monnerie, J. Polym. Sci., Polym. Phys. Ed.,
        18:1879 (1980)
24. P.J. Flory, Rubber Chem. Technol., A8:513 (1975)
25. J.P. Jarry and L. Monnerie, Macromolecules, 12:927 (1979)
26. J.P. Jarry and L. Monnerie, J. Polym. Sci., Polym. Phys. Ed.,
        16:443 (1978)
27. L. Monnerie and J.P. Jarry, Ann. New-York Acad. Sci., 366:328 (1981)
28. J.P. Jarry, P. Sergot, C. Pambrun and L. Monnerie, J. Phys. E: Sci.
        Instrum., 11:702 (1978)
29. J.P. Queslel, E. Erman and L. Monnerie, Macromolecules, 18:1991 (1985)
30. J.P. Queslel and J.E. Mark, Adv. Polym. Sci., 65:135 (1984)

STUDY OF THE UNIAXIAL DEFORMATION OF RUBBER NETWORK

CHAINS BY SMALL ANGLE NEUTRON SCATTERING

Hyuk Yu*, Toshiaki Kitano[1a], Chung Yup Kim[1b], Eric J. Amis[1c],
Taihyun Chang[1d], Michael R. Landry[1e], and Jeffrey A. Wesson[1e]

Department of Chemistry, University of Wisconsin
Madison, Wisconsin 53706

Charles C. Han*, Timothy P. Lodge[1f], and Charles J. Glinka

Center for Materials Science, National Bureau of Standards
Gaithersburg, Maryland 20899

ABSTRACT

Small angle neutron scattering (SANS) measurements were performed on
poly(isoprene) networks at different uniaxial strains, i.e., $1.0 < \lambda$
(extension ratio) < 2.1. The networks were prepared from anionically
polymerized, α, ω-dihydroxy-poly(isoprene) precursors (H-chains) and the
corresponding poly(isoprene-d_8) isotopic counterparts (D-chains),
crosslinked in concentrated tetrahydrofuran solutions by trifunctional
crosslinkers, tri-isocyanates. The two components of the radius of
gyration of elastic strands, parallel and perpendicular to the strain
axis, were determined from the SANS data of the networks with 8% and 15%
D-chains. Two molecular weights of D-chains, 26,000 and 64,000,
crosslinked with approximately the same molecular weight H-chains (29,000
and 68,000 respectively) were examined for the deformation behaviors.
From the observed changes in the parallel and perpendicular components of
the radius of gyration relative to macroscopic extension ratio, after
appropriate correction for the dangling chain contributions, the chain
extensive deformation is found to follow a behavior intermediate between
the junction affine model and the phantom network model which allows
unrestricted fluctuations of network junctions. On the other hand, the
chain contractive deformation follows closely the chain affine model,
indicating an asymmetry between extensive and contractive chain
deformation. In either case, the deformation behavior is found to be the
same for the two molecular weights.

INTRODUCTION

Rubber elasticity of a polymer network is one of the most distinctive
features of long polymer chains. The elastic force of such a network is
mainly due to the change of conformational entropy of network strands
which are connected to other strands by chemical linkages or topological
constraints. The theoretical models to clarify the relationship between

macroscopic and microscopic deformations have been extensively examined for the past several decades.[2] One is to hypothesize that the macroscopic extent of deformation is affinely transformed to the microscopic deformation of a pair of junctions constraining an elastic strand,[3] which may be called the junction affine model. An alternative approach by James,[4a] and James and Guth[4b] is to take into account the thermal fluctuations of network junctions (phantom network). More recently, some reexamination of rubber elasticity based on an average of affine deformation with non-affine fluctuation has taken place.[5-8] A crucial issue is still how network chains deform under different macroscopic strains. With use of the SANS technique, the chain dimensions of polymer molecules have been determined not only in solutions but also in bulk state.[9-11] In the meantime, there have been several attempts to study network systems by SANS,[12-16] and the results are recently summarized by Ullman.[17] It has become apparent that after all these trials the data are still so scattered that the problem remains essentially unsolved. The main difficulties arise from network imperfection which results from a nonuniform molecular weight distribution of prepolymer chains, ambiguous functionality of crosslinkers and incomplete crosslinking reactions which lead to an inevitable formation of dangling chains as well as rings. The small extension ratios employed for most of the network experiments performed to date have also rendered the results inconclusive. Lately there has been steady progress in preparing uniform molecular weight prepolymer as well as more complete crosslinking reactions to produce an ideal network. It should also be noted that we can improve the strength of the network sample by making the molecular weight of the network strands larger in order to accommodate larger extension ratios. In this study, we have prepared α,ω-dihydroxypoly(isoprene) prepolymers with narrow molecular weight distributions by anionic polymerization (with bifunctional initiator) and crosslinked to make networks with a trifunctional crosslinking reagent following the synthetic procedures similar to Rubio[18] and Morton et al.[19] The network samples contain 8 and 15 wt % of perdeuterated poly(isoprene) with nearly the same molecular weights as those of the undeuterated host poly(isoprene). The SANS measurements of the networks under uniaxial extension could be performed up to extension ratio of 2. Finally, the results are precise enough to afford quantitative comparison with the theoretical predictions.

EXPERIMENTAL

1) Preparation of Prepolymers

Perdeuterated isoprene was synthesized following the procedure by Normant and Angelo.[20] The final product showed a boiling point of 32-34°C and higher than 99% purity by gas liquid chromatography. The isotopic purity was determined as 96.6% by mass spectrometry.

Anionic polymerization of isoprene was carried out using practically the same method as reported by Rubio[18] and by Morton et al.[19] The bifunctional initiator, 2,4-hexadienyl dilithium was prepared through the reaction of trans,trans-2,4-hexadiene with Li metal in triethylamine. The initiator mixture was extensively purified to remove monofunctional side-product. The initiator concentration was determined by potentiometric titration with HCl after removing triethylamine as much as possible. Isoprene and isoprene-d_8 and were dried over CaH_2 and distilled in vacuo. Following treatment with sodium mirror and distillation, the final purification was carried out by having the monomers dispersed in n-butyllithium at 0°C and subsequently flash distilling out of n-butyllithium right before any significant polymerization could take place. The other chemicals used, 2,4-hexadiene, α-methylstyrene, ethylene oxide, cyclohexane, and triethylamine, were also similarly purified.

The apparatus and procedures of the polymerization were similar to those reported by Kitano et al.[21] After polymerization for 12 hours at room temperature, a small amount of ethylene oxide was added in order to cap the living ends of the polymers with OH, and the resultant gel was kept at 4°C for at least 2 days to ensure complete capping. The concentrations of the initiator, monomer and triethylamine used for the polymerizations are summarized in Table I. The prepolymer was finally obtained by freeze drying the cyclohexane solution under vacuum ($\sim 10^{-6}$ Torr).

Table I

Anionic Polymerization Conditions
and Microstructures of H- and
D-Chains of Poly(isoprene)

Prepolymer	$[I]^a \cdot 10^3$ (Li mol/ℓ)	$[M]^b$ (mol/ℓ)	TEA^c (vol %)	Microstructures of Prepolymer[d]		
				(1.4)	(3.4)	(1.2)
H-2213	2.1	1.1	9.3	50	37	13
H-2214	2.1	1.0	3.7	56	33	11
H-2206	6.3	1.3	2.0	55	35	10
D-3201	1.4	0.74	11			
D-3202	1.7	0.73	6			
D-3203	3.9	0.66	13			

a) Concentration of initiator
b) Concentration of monomer
c) Triethylamine
d) Determined from IR spectrum. (1.4), (3.4), and (1,2) mean 1.4-, 3,4-, and 1,2-adduct units in mol %.

2) Characterization of Prepolymers

The molecular weight and its distribution of prepolymers were determined in THF by gel permeation chromatography (Waters Associates)[‡]. Four different pore size columns (10^3, 10^4, 10^5 and 10^6 Å μ-Styrogels) and a refractive index detector were used. The instrument was calibrated with polyisoprene samples which were prepared similarly in this laboratory and whose number-average molecular weights were determined with a vapor phase osmometer (Wescan, Model 232 A) in toluene at 50°C. The results of the characterization of prepolymers are summarized in Table II.

Table II

Characterization of Prepolymers

Prepolymer	V_e^a (ml)	$[\eta]^b$ (dl/g)	M_s^c	M_{GPC}^d	M_w/M_n^e
H-2213	30.0	0.732	7.2×10^4	7.5×10^4	1.07
H-2214	30.5	0.697	6.8	6.1	1.15
H-2206	31.6	0.421	2.9	3.2	1.19
D-3201	29.3	0.611	8.2	8.4	1.17
D-3202	30.3	0.513	6.4	6.4	1.15
D-3203	31.8	0.280	2.6	2.9	1.13

a) Elution volume of GPC in THF at 25°C
b) Intrinsic viscosity in THF at 25°C
c) Calculated from [I] and [M] in Table I
d) Calculated from the peak of GPC, $\log M_{GPC} = -0.235 \, V_e \text{(ml)} + 5.917$
e) Calculated from GPC curve without any correction for spreading effect

3) Network Formation

Crosslinking reagent, triphenylmethane triisocyanate (TTI), provided by Mobay Chemical Company, was first precipitated by adding n-heptane to a methylene chloride solution of TTI under N_2 atmosphere. Repeated vacuum distillation (2-4 times) produced a pale pink viscous liquid, which was the finally purified TTI. The functionality of TTI was determined by a modified ASTM method.[22]

After an appropriate mixture of poly(isoprene-d_8) and poly(isoprene) was dried under vacuum, it was dissoved in a small amount of THF and sealed into an ampoule. The stoichiometric amount of TTI solution was added to the prepolymer solution and mixed well with a glass rod in a glove bag under argon atmosphere. Upon pouring the mixture into a Teflon mold, it was transferred to a desiccator through which argon was flushed gently at room temperature to remove the solvent. When almost all the THF was evaporated, which took about 1 day, the mold was transferred to a vacuum oven and heated gradually under a gentle stream of argon for about 4 days; the slow heating was necessary to avoid the formation of bubbles in the sample. The final temperature, 110-115°C, was maintained for another 2 days to complete the crosslinking reaction. The crosslinked rubbers were soaked in n-hexane overnight in the dark to extract the sol fraction which resulted from incomplete crosslinking reactions. The extraction procedures were repeated until no solute could be detected in the n-hexane after several days of soaking. The rubbers, thus purified, were finally soaked overnight in n-hexane containing .05 wt % antioxidant, 2,6-di-tert-butyl-4-methylphenol, relative to the amount of rubbers and dried under vacuum to constant weight. For each prepolymer, a blank network and two network samples with different D-chain contents 0, 8 and 15% (by weight), were prepared. The details of the network samples are listed in Table III.

Table III

Preparation of Rubber Networks

Sample Code	Composition (wt%)		Sol Fraction[a] (wt%)
KR-3-0[b]	D-3202 (0.0) + H-2214 (100.0)		12
-8	(7.7) +	(92.3)	12
-15	(15.0) +	(85.0)	13
KR-4-0	D-3203 (0.0) + H-2206 (100.0)		12
-8	(8.0) +	(92.0)	12
-15	(15.3) +	(84.7)	11
KR-5-0	D-3201 (0.0) + H-2213 (100.0)		9
-8	(8.0) +	(92.0)	14
-15	(15.0) +	(85.0)	18

a) Calculated from the weights of dry rubber before and after extraction by n-hexane
b) The number shows the approximate deuterated chain concentration.

4) SANS Measurements

SANS measurements were carried out using a thermal neutron beam from a 10 MW reactor at NBS, Gaithersburg, MD. The detailed description of the SANS instrument has been provided elsewhere.[23] The average wavelength, λ_o, of the incident neutron beam was 4.85 Å with $\Delta\lambda/\lambda_o$ of about 25%. The scattering geometry employed is shown in Figure 1. The direction of the incident beam of neutrons is defined as the x-axis. A network sample in the shape of rectangular sheet (1.2 cm x 2.5 cm) is set perpendicular to the x-axis and stretched along the y-axis. The scattering angle and azimuthal angle are defined as θ and ϕ, respectively. The scattering from a given network sample was first measured at the unstretched state and then at increasing extension ratios. The distance between marks spotted on the rubber sheet and the sheet thickness were measured before and after each scattering run in order to calculate the extension ratio λ, and to monitor the sample volume. At the end of a stretching cycle, the strain was reduced to a lower λ or to $\lambda=1$ to check the reversibility. For the swollen rubber sample, a quartz cell (2 cm x 2 cm x .3 cm) was used and the CS_2 swollen sample was suspended in pure CS_2. All the data for SANS were corrected for the sample thickness change, scattering contributions from the cell and solvent, and the incoherent background. For the isotropic samples such as unstretched and swollen networks the scattered intensity was circularly averaged ($0° < \phi < 360°$), whereas for the anisotropic samples, the sector averaging was effected at $\phi = 0°$ and $90°$ within $\Delta\phi = \pm 10°$. Thus the signal counting statistics for anisotropic samples were worse than the isotropic ones and the data points are correspondingly more scattered.

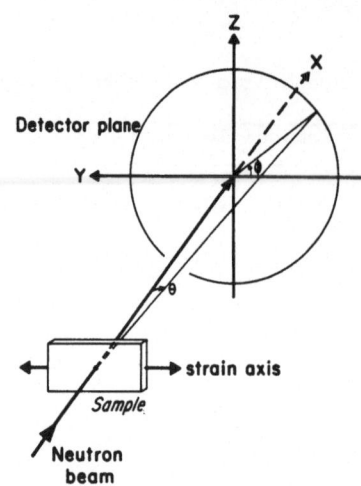

Fig. 1. Scattering geometry relative
to the strain axis of rubber
sample. In the detector
plane, $\phi = 0$ and $\phi = 90°$ refer
respectively to those parallel
(the y-axis) and perpendicular
(the z-axis) to the strain
axis.

RESULTS AND DISCUSSION

A set of representative neutron scattering intensity contour plots
is shown in Figure 2. The two dimensional scattering pattern becomes
progressively anisotropic as the extension ratio λ increases. Upon
stretching, the sample dimensions change from (ℓ_x^o, ℓ_y^o, ℓ_z^o) to (ℓ_x, ℓ_y, ℓ_z).
Since the network sample is stretched along the y-axis ($\phi = 0$), it is
being compressed along the z-axis ($\phi = 90°$) as well as the x-axis.
Assuming the homogeneous deformation, the extension ratios parallel and
perpendicular to the strain axis are defined as

$$\lambda_\parallel = \ell_y / \ell_y^o \tag{1}$$

$$\lambda_\perp = \ell_x / \ell_x^o = \ell_z / \ell_z^o \tag{2}$$

The volume change, deduced by measuring the thickness (ℓ_x) and the
distance between a pair of dot marks (ℓ_y), was found to be trivial such
that the volume conservation criterion

$$\lambda_\parallel = \lambda_\perp^{-2} \tag{3}$$

was confirmed within experimental error.

The scattering form fator $S_s(q)$ of an elastic strand in a stretched network is a function of θ and ϕ. In this study of uniaxial deformation, we focus on two directions, $\phi = 0°$ and $90°$. Still, $S_s(q)$ can be written in the usual form,

$$S_s(q)^{-1} = 1 + 1/3 \, R_\phi^2 q^2 + \cdots \cdots \qquad (4)$$

$$q = \left(\frac{4\pi}{\lambda_0}\right) \sin(\theta/2) \qquad (5)$$

where R_ϕ is a ϕ-dependent z-average radius of gyration. The two components of the radius of gyration, parallel and perpendicular to the strain axis, are given as follows:[17]

$$R_\parallel^2 = (1/2) \, R_g^{o2} + (1/4) \, \langle r_y^2 \rangle \; : \; \phi = 0° \qquad (6)$$

$$R_\perp^2 = (1/2) \, R_g^{o2} + (1/4) \, \langle r_z^2 \rangle \; : \; \phi = 90° \qquad (7)$$

R_g^o is the radius of gryation at unstretched state, and $\langle r_y^2 \rangle$ and $\langle r_z^2 \rangle$ are the y and z components of the mean square end-to-end distance in the stretched state. Thus we treat the SANS data of the anisotropic system as in the case of isotropic system in order to deduce the components of the radius of gyration parallel and perpendicular to the strain axis which will eventually be used to define the deformation at the molecular level.

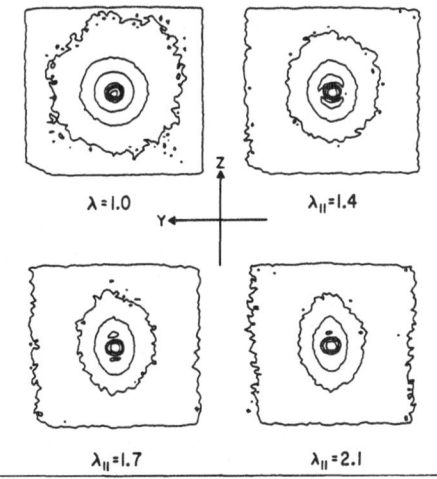

Fig. 2. Four examples of the contour maps of scattered neutron intensity patterns at different extension ratios of sample KR-3-15.

In order to arrive at the correct values of radius of gyration from the raw SANS data, we should taken into account the concentration effect of D-chains, since the network samples used had high contents of deuterated strands, 8 and 15 wt %. Two different methods were employed to take care of it. First, if we assume that the sample is incompressible or equivalently that the total structure factor $S_T(q)$ is negligible, then[24] the scattering intensity is directly proportional to the single chain scattering form factor $S_s(q)$;

$$I(q) = KMc(1-c)S_s(q) \qquad (8)$$

where K is a constant containing the scattering length and M and c are the molecular weight and the concentration of deuterated polymer, respectively. This approximation (such as eq 8) was also used by Beltzung et al.[15,16] for poly(dimethylsiloxane) networks. In this approximation, we can, therefore, use directly the intensity of 8 wt % sample, I_8, which is different from $S_s(q)$ by a constant factor. A more elaborate treatment is to deduce $S_s(q)$ from the scattering intensities of the samples with different contents of deuterated chain at the same λ. As reported earlier,[24]

$$S_s(q) = K' \left(\frac{I_8}{\ell_{x,8}|\overline{a}_8|^2} - \frac{I_{15}}{\ell_{x,15}|\overline{a}_{15}|^2} \right) \qquad (9)$$

where K' is a constant, I_8 and I_{15} are the intensities of 8 and 15 wt % sample, $\ell_{x,8}$ and $\ell_{x,15}$ are the thicknesses of the samples at a given λ, and a_8 and a_{15} are the average monomer scattering lengths for 8 and 15 wt % samples, respectively; a_8 and a_{15} were calculated using the values of the coherent neutron scattering amplitudes reported by Schoenborn and Nunes.[25] In eq 9, we take into account the thickness change of the sample relative to λ.

The values of $R_{\parallel}$ and $R_{\perp}$ are deduced by fitting $S_s^{-1/2}(q)$ to an expansion in x up to the quadratic term,

$$S_s^{-1/2}(q) = 1 + (1/6)\, x + bx^2 + \ldots \qquad (10)$$

where $x \equiv q^2 R_g^{o\,2}$ and b is a positive constant empirically determined by a simulation study similar to that by Ullman.[34] We call this QFIT scheme, meaning the inclusion of bx^2 in eq 10. The results of such analyses are displayed in Figure 3, where solid curves in (a) refer to QFIT analyses and dashed lines to the limiting slopes. In the bottom, we show the result in the undeformed case where a Debye function[26] fits the data consistently with $R_g^o = 83\,Å$.

We now turn to the problem of network imperfection. As indicated in Table III, our samples had significant sol fraction. Thus, the network samples ought to have some dangling chains which are not elastically active upon deformation. Since those R_ϕ values obtained through the above outlined fitting procedures are the z-averaged quantities of all D-chains, it is necessary to correct for their contribution to the scattering in order to monitor the net deformation of elastically active chains. From the gelation theory of Miller and Macosko,[27] we estimate

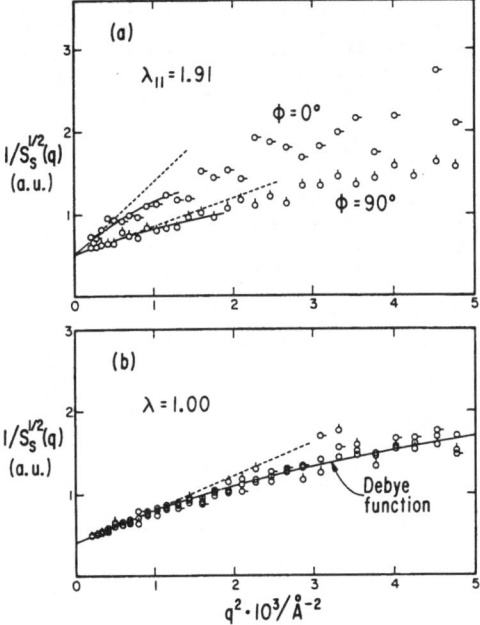

Fig. 3.(a)The square-root plot of the
        data of KR-3 sample at $\lambda_{\parallel}$ =
        1.91.
        (b)The square-root plot of the
        data of the sample at $\lambda = 1$.

the extent of reaction as about 83% and calculate the weight fraction of
dangling chains in a network as large as 51%. We regard this rather high
value to be the maximum limit of correction. If we assume, on the basis
of the high molecular weight of a network strand and entanglement
coupling molecular weight of polyisoprene,[28] that only a single unconnected
strand is elastically inactive, then the contribution turned out to be
26%. We regard this as the other limit of the correction. The true
radius of gyration of active chains, R_ϕ, can be calculated from the
observed z-average radius of gyration $R_{\phi,z}$ as follows

$$R_{\phi,z} = R_g^{\,o} \, w_d + R_\phi (1-w_d) \tag{11}$$

or

$$R_\phi = (R_{\phi,z} - R_g^{\,o} \, w_d)/(1-w_d) \tag{12}$$

where w_d is the weight fraction of dangling chains and $R_g^{\,o}$ is the radius
of gyration of a dangling chain which should be the same as that of
undeformed chain ($\lambda = 1.0$). The R_ϕ values obtained according to eq 17
are summarized in Table IV.

Table IV

The Longitudinal and Transverse Components of Radius of Gyration[a] (in units of Å) after Dangling Chain Contributions Corrections at Maximum & Minimum

| | | Longitudinal | | | | | | | Transverse | | | |
| | | Max.[b] | | | Min.[c] | | | | Max.[b] | | Min.[c] | |
Sample	$\lambda_\parallel^d$	$R_{\parallel,z}$	$R_\parallel$	$R_\parallel/R_g^\circ$	$R_\parallel$	$R_\parallel/R_g^\circ$	$\lambda_\perp^d$	$R_{\perp,z}$	$R_\perp$	$R_\perp/R_g^\circ$	$R_\perp$	$R_\perp/R_g^\circ$
KR-3	1.00	82[e]		1.00			1.00	82[e]		1.00		1.00
	1.21	86	90.2	1.1	87.4	1.07	0.91	78	73.8	0.90	76.6	0.93
	1.40	85	88.2	1.08	86.0	1.05	0.85	79	75.8	0.92	78.0	0.95
	1.50	91	101	1.23	94.1	1.15	0.82	73	63.5	0.77	69.9	0.85
	1.61	93	105	1.28	96.8	1.18	0.79	71	59.4	0.72	67.2	0.82
	1.71	96	111	1.05	101	1.23	0.77	66	49.1	0.60	60.5	0.74
	1.91	102	123	1.50	109	1.33	0.72	64	45.0	0.55	57.8	0.70
	2.12	110	140	1.70	120	1.46	0.69	68	53.2	0.65	63.2	0.77
	1.50[f]	90	98.5	1.20	92.8	1.13	0.82[f]	73	63.6	0.77	69.9	0.85
KR-4	1.00	58[e]		1.00		1.00	1.00	58[e]		1.00		1.00
	1.31	67	76.5	1.32	70.1	1.21	0.87	53	47.7	0.82	51.3	0.88
	1.60	68	78.6	1.35	71.4	1.23	0.79	48	37.4	0.95	44.6	0.77
	1.90	73	88.9	1.53	78.2	1.35	0.73	54	49.8	0.86	52.6	0.91

a) Calculated from $S_g(q)$
b) Corrected for all dangling chains
c) Corrected for single terminal dangling chains
d) Average value of 8% and 15% deuterated sample
e) Calculated from circular average intensity (R_g°)
f) Recovered value after reaching the maximum $\lambda_\parallel$ value and returning to $\lambda_\parallel = 1.50$.

We now compare our data with three different models. The first is the chain affine deformation, whereby every chain segment within an elastic strand deforms affinely with a macroscopic strain, and yields

$$\frac{R_\phi}{R_g^\circ} = \lambda_\phi.$$

(13)

Secondly, the junction affine model, where the junctions deform affinely and a strand between a junction pair remains Gaussian-like. This gives

$$\frac{R_\phi}{R_g^o} = (\frac{1 + \lambda_\phi^2}{2})^{1/2} \qquad (14)$$

Finally, the phantom network theory where junctions only deform affinely on the average but with strain independent and unconstrained fluctuations For a tri-functional network, this gives

$$\frac{R_\phi}{R_g^o} = (\frac{5 + \lambda_\phi^2}{6})^{1/2} \qquad (15)$$

The three model predictions are drawn in Figure 4 to compare with the data plotted in the same graph and collected in Table IV. Although data points are somewhat scattered, it is clear that none of these models can predict both chain deformation behaviors under extension in one axis and simultaneous compression in the two transverse axes; and this is true regardless of the extent of dangling chain corrections applied. With the minimal dangling chain correction (26%), the deformation behaviors are as shown in Figure 4. With the maximum correction (51% dangling chains), we find the chain deformation behavior to follow closely the junction affine model along the extensive direction whereas it is more sensitive than the chain affine model in the compressive direction (transverse to the strain axis), which is probably nonphysical. This is one of the reasons why we reject the maximum correction. Having thus put forth the minimum correction as more appropriate and plausible, we can now provide an explanation for the observed intermediate behavior between the junction affine and phantom network models. We do this by appealing to the recent efforts to revise the phantom network model[29-31] by taking into account anisotropic constraints on the junction fluctuation by chain interference, network unfolding and rearrangement. Once we accept that the magnitude of the junction fluctuation is constrained, the observed chain deformation behavior seems entirely expected. Quantitative comparison with any revised model however appears premature inasmuch as our data are still subject to a good deal of uncertainties. As for the chain deformation in the compressive directions, we cannot offer a plausible explanation at this juncture. Further theoretical and experimental efforts are obviously required to come to grips with this puzzling observation. Why should there be an asymmetry between the parallel and perpendicular directions of chain deformation? The observation clearly indicates that the chain segment distribution is not conserved under uniaxial deformation; the spherical distribution at $\lambda = 1$ deforms to an ellipsoidal distribution with the semi-minor axes much smaller than those expected under the volume conservation condition. However, we should point out that all three model calculations in Figure 4 are based on the Gaussian segment distribution with fixed end-to-end distance which is the deforming junction separation. It may be an indication that the segment distribution deviates from a Gaussian statistics at increasing strain since $S_s(q)$ follows Debye function at small $\lambda_\parallel$ but deviates from it at larger $\lambda_\parallel$.

In concluding the paper, we emphasize that our network samples showed excellent reversibility upon releasing the strain and opacity in the sample was never observed in the extension range studied. Some opacity was observed in very high extension ratio (result not shown),

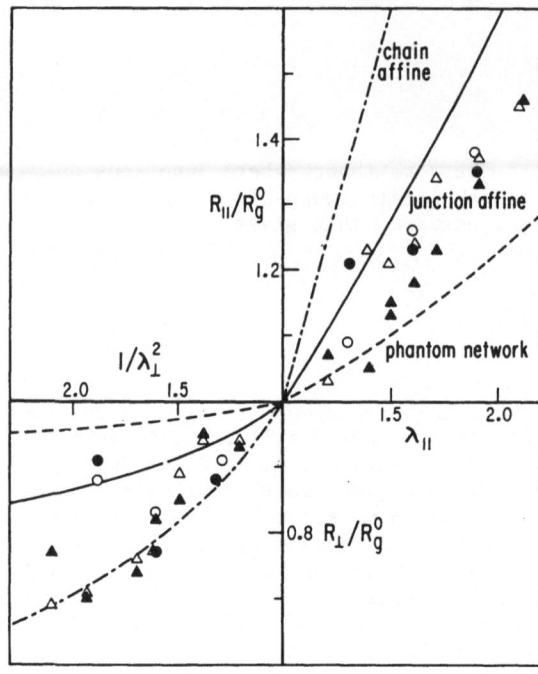

Fig. 4. Comparison of the experimental
radii of gyration listed in
Table V and three model
predictions. The two different
molecular weights of the
labeled chains are
distinguished by triangles
($M = 2.9 \times 10^4$, Sample KR-3)
and circles ($M = 6.4 \times 10^4$,
Sample KR-4). Filled data
points are obtained from $S_s(q)$
and unfilled ones are from $I_8(q)$.
Note that the extensive regime
(the upper right quadrant) is
plotted against $\lambda_\parallel$ (1.0-2.1) and
the compressive regime (the
lower left quadrant) is plotted
against $1/\lambda_\perp^2$, under the volume
conservation condition (see
text).

which might be due to partial crystallization in the sample. In the
isotropic expansion, the microscopic deformation seems to be much smaller
than that of uniaxial extension as shown in Table V. For $\lambda = 2.4$ the
predicted value for trifunctional network is 1.34 for phantom network and
1.84 for junction affine model compared to the measure value of 1.1.
Such a small expansion in swelling is also reported by Beltzung et al.
for a poly(dimethylsiloxane) networks.[16] Here again, we confirm a
puzzling observation without offering any sensible explanation beyond the
previous attempts to justify the observed insensitivity of chain
deformation with isotropic swelling $Q^{1/3}$.

Table V

Isotropic Expansion by Swelling in CS_2

Sample	Q	λ	R^o_g, (Å)	$R_{g,z}{}^d$(Å)	R_g(Å)	R_g/R^o_g,
KS-3-0 [a]	14	2.4	83 [c]	89	91	1.11
-8 [a]	12	2.3				
-15 [a]	15	2.5				
KS-3-S [b]		2.4 [c]	83 [c]	89	91	1.10

a) The same sample as KR-3-0, KR-3-8, and KR-3-15.
b) $S_s(q)$ calculated from eq. 17.
c) Average value of λ of KR-3-8 and KR-3-15.
d) Calculated from QFIT in square root plot.

ACKNOWLEDGEMENT

This work was in part supported by a research grant awarded to H.Y. by the Polymers Programs of NSF (DMR-7908652). We acknowledge the contributions of Drs. Carl R. Kessel and Peter M. Gannett for the synthesis of perdeuterated isoprene and of Dr. George B. Caflisch and his coworkers at Eastman Chemicals Division of Eastman Kodak Company for the measurements of the number average molecular weights of some of the poly(isoprene) samples. We thank Mobay Chemical Company for the generous gift of the crosslinker.

REFERENCES

1a) On leave from the School of Materials Science, Toyohashi 440, Japan.
 b) Permanent address: Laboratory for Polymeric Materials, Korea Advance Institute of Science and Technology, Seoul, Korea.
 c) Work was performed in part as a NRC postdoctoral fellow at N.B.S.; present address: Department of Chemistry, University of Southern California, Los Angeles, CA 90089-0482.
 d) Present address: Korea Research Institute of Chemical Technology, Daeduck Danji, Daejeon 300-32, Korea.
 e) Permanent address: Research Laboratories, Eastman Kodak Company, Rochester, NY 14650.
 f) Work was performed during the residence at N.B.S. as a NRC postdoctoral fellow; present address: Department of Chemistry, University of Minnesota Minneapolis, MN 55455.
2. Treloar, L.R.G. "Physics of Rubber Elasticity", 1975, Oxford University Press, London.
3a) Flory, P.J. "Principles of Polymer Chemistry", 1953, Cornell University Press, Ithaca, NY.
 b) Kuhn, W. Kolloid-Z. 1936, 76, 256.
 c) Wall, F.T. J. Chem. Phys. 1942, 10, 132; ibid. 1943, 11, 527.
 d) Flory, P.J.; Rehner, J. J. Chem. Phys. 1943, 11, 512.
 e) Hermans, J.J. J. Polym. Sci. 1952, 59, 191.
4a) James, H.M. J. Chem. Phys. 1947, 15, 651.
 b) James, H.M.; Guth, E. J. Chem. Phys. 1947, 15, 669.
5. Graessley, W.W. Macromolecules 1975, 8, 186; ibid., 1975, 8, 865.

6. Ronca, G.; Allegra, G. J. Chem. Phys. 1975, 63, 4990.
7. Flory, P.J. Proc. Roy. Soc. 1976, A351, 351; J. Chem. Phys. 1977, 12, 5720.
8. Deam, R.T.; Edwards, S.F. Philos. Trans. Roy Soc, 1976, A280, 1296.
9. Daoud, M.; Cotton, J.P.; Farnoux, B.; Jannink, G.; Sarma, G.; Benoit, H.; Duplessix, R.; Picot, C.; de Gennes, P.G. Macromolecules 1975, 8, 804.
10. Kirste, R.G.; Kruse, W.A.; Ibel, K. Polymer, 1975, 16, 120.
11. Cotton, J.P.; Decker, D.; Benoit, H.; Farnoux, B.; Higgins, J.; Jannink. G.; Ober, R.; Picot, C.; des Cloizeaux, J. Macromolecules, 1974, 7, 863.
12. Benoit, H.; Decker, D.; Duplessix, R.; Picot, C.; Rempp, P.; Cotton, J.P.; Farnoux, B.; Jannink, G.; Ober, R. J. Poly. Sci., Polym. Phys. Ed. 1976, 14, 2119.
13. Hinkley, J.A.; Han, C.C.; Mozer, B.; Yu, H. Macromolecules 1978, 11, 836.
14. Clough, S.; Maconnachie, A.; Allen, G. Macromolecules 1980, 13, 774.
15. Beltzung, M.; Picot, C.; Rempp, P.; Herz, J. Macromolecules 1982, 15, 1594.
16. Belzung, M.; Herz, J.; Picot, C. Macromolecules 1983, 16, 58.
17. Ullman, R. in "Elastomers and Rubber Elasticity", 1982, eds., Mark, J.E.; Lal, J. Am. Chem. Soc., D.C.
18. Rubio, D. Ph.D. Thesis, 1975, University of Arkron, Akron, OH.
19. Morton, M.; Fetters, L.J.; Inomata, J.; Rubio, D.C.; Young, R.N. Rubb. Chem Tech. 1976, 49 303.
20. Normant, H.; Angelo, B. Bull. Chim. Soc. Fr., 1960, 354.
21. Kitano, T.; Fujimoto, T.; Nagasawa, M. Macromolecules 1974, 7, 719.
22. ASTM D2572-80.
23. Glinka, C.J. AIP Conf. Proc. No. 89, Neutron Scattering, Argonne National Lab 1981, p. 395.
24. Akcasu, A.Z.; Summerfield, G.C.; Jahshan, S.N.; Han, C.C.; Kim, C.Y.; Yu, H. J. Polym. Sci., Polym. Phys. Ed. 1980, 18, 863.
25. Schoenborn, B.P.; Nunes, A.C. Annual Rev. Biophys. Bioeng. 1972, 1, 529.
26. Debye, P. J. Phys. Colloid. Chem. 1947, 51, 18.
27. Miller, D.R.; Macosko, C.W. Macromolecules 1976, 9, 206.
28. Flory, P.J. Chem. Rev. (1944) 35, 51.
29. Flory, P.J. J. Chem. Phys. 1977, 66, 5720.
30. Bastide, J.; Picott, C.; Candan, S. J. Macromol. Phys. 1981, B19, 13.
31. Ullman, R. Macromolecules 1982, 15, 582.

‡ Certain commercial equipment and materials are identified in this paper in order to adequately specify experimental procedures. In no case does such identification imply recommendation or endorsement by the National Bureau of Standards, nor does it imply that the material or equipment identified is necessarily the best available for the purpose.

NONLINEAR STRAIN MEASURES OF RUBBER NETWORKS AND POLYMER MELTS

Paul J.R. Leblans and Boudewijn J.R. Scholtens

DSM, Research and Patents
PO Box 18, 6160 MD Geleen
Netherlands

INTRODUCTION

Flexible polymer molecules can assume a tremendous number of different configurations. This characteristic is responsible for the remarkable dynamic and mechanical properties of molten polymers[1,2].

The difference between a rubber network and a polymer melt is that in the former type of material the chains are permanently interconnected to form a three-dimensional network, whereas in the latter they are not. As a consequence, crosslinked rubbers must be considered as solids with an equilibrium modulus, whereas polymer melts are basically fluids, which only behave as networks under special conditions[2].

In the field of rubber elasticity both experimentalists and theoreticians have mainly concentrated on the equilibrium stress-strain relation of these materials, i.e. on the stress as a function of strain at infinite time after the imposition of the strain[1,3]. This approach is obviously impossible for polymer melts[2]. Another complication which has thwarted the comparison of stress-strain relations for networks and melts is that crosslinked networks can be stretched uniaxially more easily, because of their high elasticity, than polymer melts. On the other hand, polymer melts can be subjected to large shear strains and networks cannot because of slippage at the shearing surface at relatively low strains. These seem to be the main reasons why up to some time ago no experimental results were available to compare the nonlinear viscoelastic behaviour of these two types of material. Yet, in the last decade, apparatuses have been built to measure the simple extension properties of polymer melts[4,5]. It has thus become possible to compare the stress-strain relation at large uniaxial extension of crosslinked rubbers and polymer melts.

The general factorable single integral constitutive equation is an equation that describes well the viscoelastic properties of a large class of crosslinked rubbers[6,7] and the elasticoviscous properties of many polymer melts[8-10] under various types of deformation. An appropriate way to compare the stress-strain relation for cured elastomers and polymer melts therefore is to calculate the strain-dependent function contained in this constitutive equation from experimental results and to compare the strain measures so obtained.

It is the aim of the present paper to compare the strain measures of a large number of polymeric materials, both crosslinked elastomers and polymer melts. In addition these will be examined in the light of a number of molecular theories, such as the affine theory,[1,3] the theory of James and Guth[11] and the more sophisticated model of Flory and Erman[12,13] for rubber networks, and the Doi-Edwards model[14] and its modified version by Marrucci[15] for concentrated polymer solutions and melts.

THE FACTORABLE SINGLE INTEGRAL EQUATION AND THE SIGNIFICANCE OF THE STRAIN MEASURE $S_E(\lambda)$

A general single integral constitutive equation results if the Boltzmann superposition principle is applied to a non-specified tensor functional, $\underline{S}(\underline{C}^{-1})$, of the macroscopic strain, represented by the Finger strain tensor $\underline{C}^{-1}$. If in addition the present state is chosen as the reference state, the total stress tensor at present time t, $\underline{p}(t)$, is related to the strain history – i.e. to the strain at previous times t' with respect to the present reference state, represented by $\underline{C}_t^{-1}(t')$ – by the equation:

$$\underline{p}(t) = - p_o \underline{1} - \int_{-\infty}^{t} G(t-t') \frac{d\underline{S}(\underline{C}^{-1})}{dt'} dt'$$

$$= - p_o \underline{1} + G_e \underline{S}[\underline{C}_t^{-1}(0)] + \int_{-\infty}^{t} \frac{dG(t-t')}{dt'} \underline{S}(\underline{C}^{-1}) dt' \tag{1}.$$

Here $p_o \underline{1}$ represents the isotropic pressure field, $G(t)$ is the stress relaxation shear modulus, G_e is the equilibrium shear modulus which has a finite value for crosslinked rubbers, but equals zero for melts, and $\underline{C}^{-1}$ under the inte-gral stands for $\underline{C}_t^{-1}(t')$. Often $dG(t-t')/dt'$ is identified with a memory function[8] $m(t-t')$. In that case the equation is transformed into:

$$\underline{p}(t) = - p_o \underline{1} + G_e \underline{S}[\underline{C}_t^{-1}(0)] + \int_{-\infty}^{t} m(t-t') \underline{S}(\underline{C}^{-1}) dt' \tag{2}.$$

The equilibrium modulus and the memory function $m(t-t')$ can be obtained from measurements in the linear viscoelastic region. Oscillatory shear data are most appropriate to determine the linear viscoelastic functions[2,16]. We will not go further into this matter here, since the present article is concerned with the comparison of the shape of the nonlinear tensor functionals $\underline{S}$ of different materials.

In simple extension the difference between the stress in the elongation direction and that in the direction perpendicular to the elongation direction is measured. For conciseness this stress difference will be denoted by σ_E. According to Eq. (2) σ_E is determined by the difference between the 11- and 22-components of the strain tensor, which will be denoted by S_E. $\underline{S}_t(t')$ represents a tensorial strain measure determined by the first and second invariants $I_t(t')$ and $II_t(t')$ of the relative Finger strain tensor. In the case of uniaxial extension these invariants can always be expressed in the ratio of the stretch ratios at times t and t', so that $\underline{S}_t(t') = \underline{S}[\lambda(t)/\lambda(t')]$. Equation (2) thus yields the following expression for σ_E:

$$\sigma_E(\lambda,t) = G_e S_E[\lambda(t)] + \int_{-\infty}^{t} m(t-t') S_E[\lambda(t)/\lambda(t')] dt' \tag{3}.$$

A measure which is frequently used in polymer melt constitutive equations is the Hencky strain measure, related to the stretch ratio by:

$$\varepsilon = \ln \lambda \tag{4}.$$

Hence, $\underline{S}_t(t')$ may also be expressed in the difference between the Hencky strains at times t and t': $\underline{S}_t(t') = \underline{S}[\varepsilon(t) - \varepsilon(t')]$.

If a sample is instantaneously extended by a factor λ at zero time, the relative strain history is given by:

$$\lambda(t)/\lambda(t') = 1 \text{ for } t' > 0$$

$$\lambda(t)/\lambda(t') = \lambda \text{ for } t' < 0 \tag{5}$$

and Eq. (3) reduces to:

$$\sigma_E(\lambda,t) = G_e S_E(\lambda) + S_E(\lambda) \int_{-\infty}^{0} m(t-t')dt' = S_E(\lambda)\,[G_e + G_r(t)] \tag{6}$$

where $G_r(t)$ is the relaxing part of the modulus. At equilibrium conditions, at infinite time after the imposition of the strain, $G_r(t)$ has completely decayed, and the familiar stress equation of rubber elasticity results:

$$\sigma_E(\lambda) = G_e S_E(\lambda) \tag{7}.$$

THEORETICAL STRESS-STRAIN RELATIONS

Rubber Networks

The early versions of the statistical theory of rubber elasticity assumed an affine displacement of the average positions of the network junctions with the macroscopic strain[1,3]. This is tantamount to the assertion that the network junctions are firmly embedded in the medium of which they are part. The elastic equation of state derived on this basis for simple extension at constant volume takes the familiar neo-Hookean form, i.e. Eq. (7), with

$$S_E(\lambda) = \lambda^2 - \lambda^{-1} \tag{8}$$

and

$$G_e = \nu kT \tag{9}.$$

Here ν represents the total number of elastically effective network chains per unit volume, k is the Boltzmann constant and T the absolute temperature.

James and Guth[11] dispensed with the premise of an affine displacement of all network junctions conceived of as fixed in space. Only those junctions which are located on the boundary surfaces are specified as fixed, and all other junctions are allowed complete statistical freedom, subject only to the restrictions imposed by their interconnectedness. This theory was later called the phantom network model[17] because the chains are devoid of material characteristics. Their only action is to exert forces on the junctions to which they are attached, but they can move freely through one another. This also leads to a stress-strain relation of the form of Eq. (7) with $S_E(\lambda)$ given by Eq. (8), but with an equilibrium modulus equal to

$$G_e = \xi kT \tag{10}$$

where ξ is the cycle rank per unit volume of the network. For perfect networks free of defects the following two relations hold[13,17]:

$$\xi = \nu - \mu \tag{11}$$

and

$$\nu = \mu\phi/2 \tag{12}$$

where μ is the concentration of junctions with functionality ϕ. Thus for a perfect network with $\phi = 4$ Eq. (9) yields a value for G_e which is twice as large as given by Eq. (10). James and Guth showed that in their theory the mean positions of the junctions are affine in the macroscopic strain, but the fluctuations from these mean positions are Gaussian and independent of strain.

Ronca and Allegra,[18] and independently Flory,[17] advanced the hypothesis that real rubber networks show departures from these theoretical equations as a result of a transition between the two extreme cases of behaviour. In subsequent papers Flory[13,19] and Flory and Erman[12] derived a theory based on this concept. At small deformations the fluctuations of the network junctions are constrained by the extensive interpenetration of neighbouring, but topologically remote chains. The severity of these constraints is characterized by the value of the parameter κ ($\kappa = 0$ corresponds to the phantom network, $\kappa = \infty$ to the affine network). With increasing deformation these constraints become less restrictive in the direction of the principal extension. The parameter ζ describes the departures from affine transformation of the shape of the domains of constraints. The resulting stress-strain relation also takes the form of Eq. (7) with

$$S_E(\lambda) = \left[\frac{1 + (f_c/f_{ph})_\lambda}{1 + (f_c/f_{ph})_{\lambda=1}} \right](\lambda^2 - \lambda^{-1}) \tag{13}$$

and

$$G_e = \xi[1 + (f_c/f_{ph})_{\lambda=1}]kT \tag{14}$$

where f_c/f_{ph} is a complicated function of λ containing the network characteristic μ/ξ, and the parameters κ and ζ.[12] For $\kappa = 0$ $f_c/f_{ph} = 0$ and Eqs. (13) and (14) degenerate to Eqs. (8) and (10), respectively. For $\kappa \to \infty$ Eq. (13) still degenerates to Eq. (8) but $\lim_{\kappa \to \infty} f_c/f_{ph} = \mu/\xi$, so that in that case Eq. (14) equals to Eq. (9).

Polymer Melts

In the Doi-Edwards theory[14] the environment of a chain is modeled as a tube with a diameter which is constant over the tube length. Each subchain, which is the part of the polymer chain between localized entanglements, resides in a tube segment. The subchains react to an instantaneous strain by affine deformation, as in the classical theory of rubber elasticity. Therefore, immediately after the imposition of a simple elongational strain at $t = 0$, the stress is given by Eq. (6), with $G_e = 0$; $S_E(\lambda)$ is given by Eq. (8), and, in analogy with Eq. (9), the relaxational part of the modulus at $t = 0^+$ equals:

$$G_r(0^+) = cNkT \tag{15}$$

cN being the number of subchains per unit volume. At this stage, the tensile forces acting on the neighbouring subchains are in general unequal. Since the connections between the subchains are not chemical crosslinks as in permanent networks, a part of the chain will slide over the temporary link – the entanglement – that separates the two subchains so as to balance these forces. During the first relaxation step, the chain retracts to its equilibrium length inside the affinely deformed tube by the described process, resulting in a decrease of the number of tube segments. The lengths of the tube segments remain constant whereas the diameter of the tube returns to its equilibrium value. Hence, after completion of this process with characteristic time τ_A, the number of subchains – i.e. parts of the chain contained in one tube segment – has decreased and the subchains are no longer strained. The remaining stress is uniquely due to the orientation of the chains and is given by Eq. (6), with $G_r(t \gg \tau_A)$ still equal to cNkT, but with a different strain measure[14]:

$$
S_E(\lambda) = Q_E(\lambda) = \frac{15}{4} \frac{\lambda^3 + 1/2}{\lambda^3 - 1} \left[\frac{1 - \dfrac{4\lambda^3 - 1}{2\lambda^3 + 1} \dfrac{\sinh^{-1}\sqrt{\lambda^3 - 1}}{\sqrt{\lambda^3}\sqrt{\lambda^3 - 1}}}{1 + \dfrac{\sinh^{-1}\sqrt{\lambda^3 - 1}}{\sqrt{\lambda^3}\sqrt{\lambda^3 - 1}}} \right] \qquad (16).
$$

In a second relaxation process the chain diffuses out of the tube, and, as the newly created subchains are randomly oriented, the stress gradually decays. Since the two relaxation processes are believed to be well separated in time and since the degree of relaxation during the first process only depends on the magnitude of the strain, whereas the second relaxation process is independent of the macroscopic strain, for times longer than τ_A and strain rates smaller than τ_A^{-1}, the stress is given by a factorable constitutive equation, Eq. (3), in which $G_e = 0$, m(t-t') has a definite shape which is of no concern to us here, and $S_E(\lambda)$ is given by Eq. (16).

Marrucci and de Cindio[15] proposed a different assumption to describe the retraction of the chains during the first relaxation step: the topological constant volume assumption. According to these authors not only the lengths of tube segments remain constant during the chain retraction, but also the tube segments' diameters. So, after the first relaxation step, the tube segments are deformed longitudinally, but not transversely in the Doi-Edwards model, whereas in the Marrucci model the tube segments' shapes are deformed affinely. This different assumption leads to a stress equation of the same form as that of Doi and Edwards, but with:

$$
S_E(\lambda) = Q_E'(\lambda) = \frac{15}{8} \lambda \frac{\lambda^3 + 1/2}{\lambda^3 - 1} \left[1 - \frac{4\lambda^3 - 1}{2\lambda^3 + 1} \frac{\sinh^{-1}\sqrt{\lambda^3 - 1}}{\sqrt{\lambda^3}\sqrt{\lambda^3 - 1}} \right] \qquad (17).
$$

EXPERIMENTAL RESULTS

Strain Measures for Various Crosslinked Rubbers

The most customary way to obtain the strain measure for permanent networks is to measure the equilibrium stress as a function of strain. However, it is also possible to calculate it from stress measurements at constant stretching rates, which at the same time yields information on the possibility of separation of time and strain effects[20]. The strain functions shown in Fig. 1 for the EPDM vulcanizates identified in Table I have been obtained with this type of measurement, using the approximate analytical method proposed by Scholtens, Leblans and Booij[20]. In addition the $S_E(\lambda)$-function for one of the EPDM copolymers prior to crosslinking (●) is shown, calculated from the measuring data by using the numerical method presented in the paper by Scholtens et al.[20]. To facilitate comparison with the strain measures for polymer melts (see below) the $S_E(\lambda)$-functions are transformed into $S_E(\varepsilon)$-functions making use of Eq. (4).

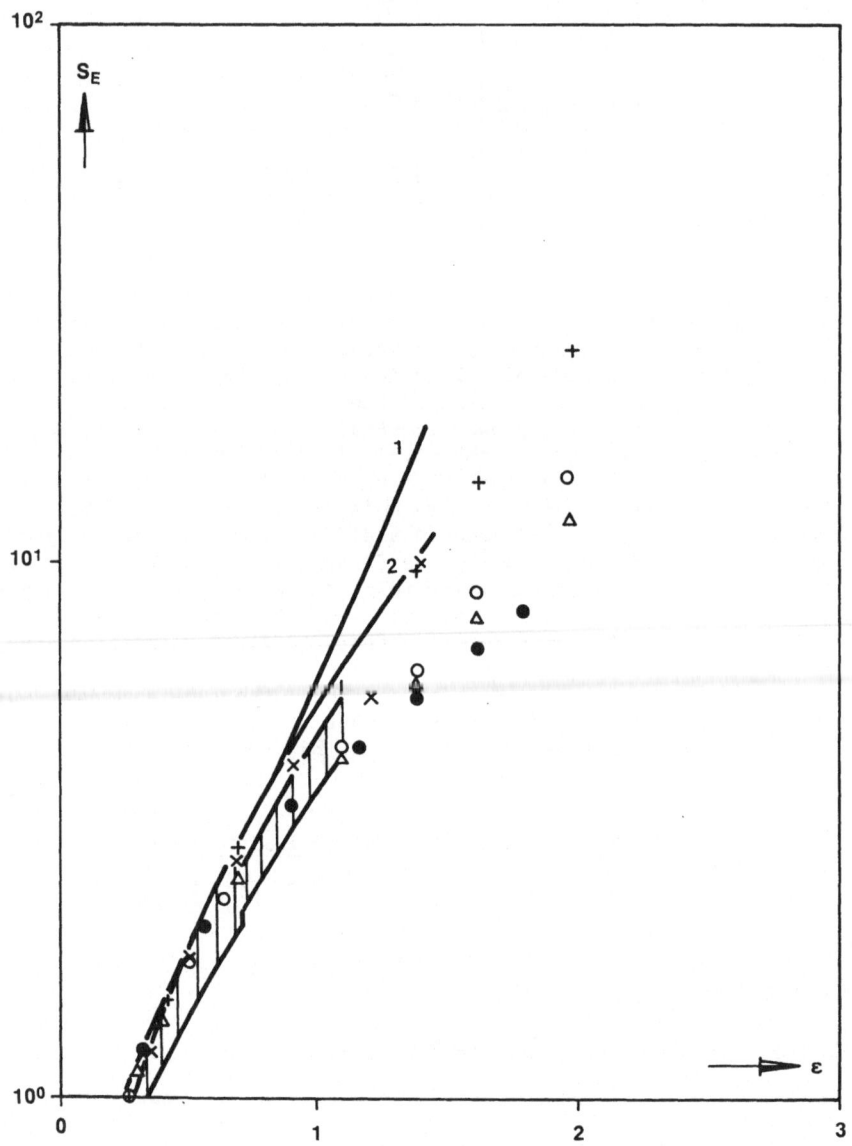

Fig. 1. Experimental strain measures for various rubber networks identified in Table I. The shaded area pertains to networks characterized in refs.[23-27].

Table I. Identification of the Rubber Networks of Fig. 1

Sample Characteristics	Type of Measurement	Reference	Curve Number or Symbol
styrene–butadiene rubber vulcanizate	constant stretching rate	21	1
styrene–butadiene rubber vulcanizate, vulcanized with 0.57 wt % dicumyl peroxide at 162 °C for 60 min	constant stretching rate	22	2
EPDM elastomer with 65 mol % ethylene and 3 mol % diene, vulcanized with 0.2 wt % dicumyl peroxide at 177 °C for 10 min (K 0.2)	constant stretching rate	7	+
EPDM elastomer with 65 mol % ethylene and 1 mol % diene, vulcanized with 0.4 wt % dicumyl peroxide at 177 °C for 10 min (N 0.4)	constant stretching rate	7	Δ
EPDM elastomer with 58 mol % ethylene and 1 mol % diene, vulcanized with 0.8 wt % dicumyl peroxide at 177 °C for 10 min (D 0.8)	constant stretching rate	7	O
unvulcanized EPDM elastomer with 58 mol % ethylene and 1 mol % diene (D 0.0)	constant stretching rate	7	●
PDMS elastomer	equilibrium stress-strain measurements	27	×

The function $g(\lambda)$, arrived at by Smith[21] for a crosslinked styrene-butadiene rubber is related to the strain measure in the sense of the present approach by the equation $S_E(\lambda) = 3(\lambda^2 - \lambda^{-1})/g(\lambda)$. $S_E(\lambda)$ calculated from Smith's data is also shown in Fig. 1. Using the approximate analytical method to analyze constant-stretching-rate data, Hong et al.[22] obtained a function $\Gamma(\lambda)$, which equals $S_E(\lambda)/3$, also for a styrene-butadiene rubber vulcanizate. Figure 1 also shows $3\Gamma(\lambda)$.

It must be emphasized that in the three investigations mentioned above the strain measures obtained are each independent of the stretching rate, although the ratio between the equilibrium contribution to the stress and the relaxational contribution to the stress is larger for the lower than for the higher rates. This implies that the same $S_E(\lambda)$ is appropriate for the relaxational part of the stress and for the equilibrium part, and thus that Eq. (3) is useful for all the investigated materials.

The strain measures for dry (unswollen) vulcanizates of a large number of natural rubbers, butadiene-styrene and butadiene-acrylonitrile copolymers, polydimethylsiloxanes, polymethylmethacrylates, polyethylacrylates and polybutadienes with different degrees of crosslinking and measured at various temperatures[23-27] are confined within the shaded area in Fig. 1. These measures were determined from the stress as a function of extension at (or near) equilibrium, i.e. by applying Eq. (7). Therefore they only reproduce the equilibrium stress-strain relation for the crosslinked rubbers. In all cases the strain dependence of the tensile force (and hence of the tensile stress) was expressed in terms of the well-known Mooney-Rivlin equation[3], equating the equilibrium tensile stress to:

$$\sigma_E = 2(\lambda^2 - \lambda^{-1})(C_1 + C_2\lambda^{-1}) \tag{18},$$

where C_1 and C_2 are parameters characterizing the vulcanizate. Given the values of C_1 and C_2, the strain measure in the sense meant here can be calculated as:

$$S_E(\lambda) = \left[\frac{1 + \lambda^{-1}C_2/C_1}{1 + C_2/C_1}\right](\lambda^2 - \lambda^{-1}) \tag{19}.$$

The data given in the literature on these crosslinked rubbers have thus been transformed into S_E-curves.

Strain Measures for Various Polymer Melts

Step-strain stress-relaxation measurements have been frequently used to determine $S_E(\lambda)$ for polymer melts[15,28-30]. Equation (6) shows that if separability of time and strain effects is possible for the melt under consideration, the stress after a step elongational strain can be factored into a time-dependent function, the linear shear relaxation modulus $G(t)$, and a strain-dependent function, the nonlinear strain measure $S_E(\lambda)$. Also other types of experiment may be performed to obtain $S_E(\lambda)$, such as constant-strain-rate experiments[30-32], creep under constant stress[30] and constant-stretching-rate experiments[20,30], but these methods require more involved analytical and/or numerical calculations.

Some authors[31,32] obtained a so-called damping function, $h(\lambda)$, from their measuring results. This function is related to the strain measure, $S_E(\lambda)$, by:

$$S_E(\lambda) = h(\lambda)(\lambda^2 - \lambda^{-1}) \tag{20}.$$

In Fig. 2 the resulting strain measures are collected from all these studies. The identification of these polymer melts is given in Table II.

DISCUSSION

Figures 1 and 2 reveal some surprising facts. First, the nonlinear strain measures obtained from equilibrium measurements on a large number of crosslinked rubbers lie in the same range as those obtained from transient measurements on the same types of material. This indicates that the nonlinear relation between stress and elastic strain is the same for permanent and temporary structures in the network. The fact that time and strain effects are separable over a large time scale for a number of crosslinked rubbers, which is revealed by the stretching rate independence of the strain measure calculated for these networks, is in accordance with this conclusion.

Second, crosslinked rubbers and polymer melts seem to differ very little in nonlinearity. Comparison of Figs. 1 and 2 shows that the strain measures obtained for crosslinked networks lie in exactly the same range as those for polymer melts. Moreover, Fig. 1 shows that the strain measure of the elastomeric melt D 0.0 is not highly different from that of the permanent network D 0.8, obtained by crosslinking D 0.0. It seems, therefore,

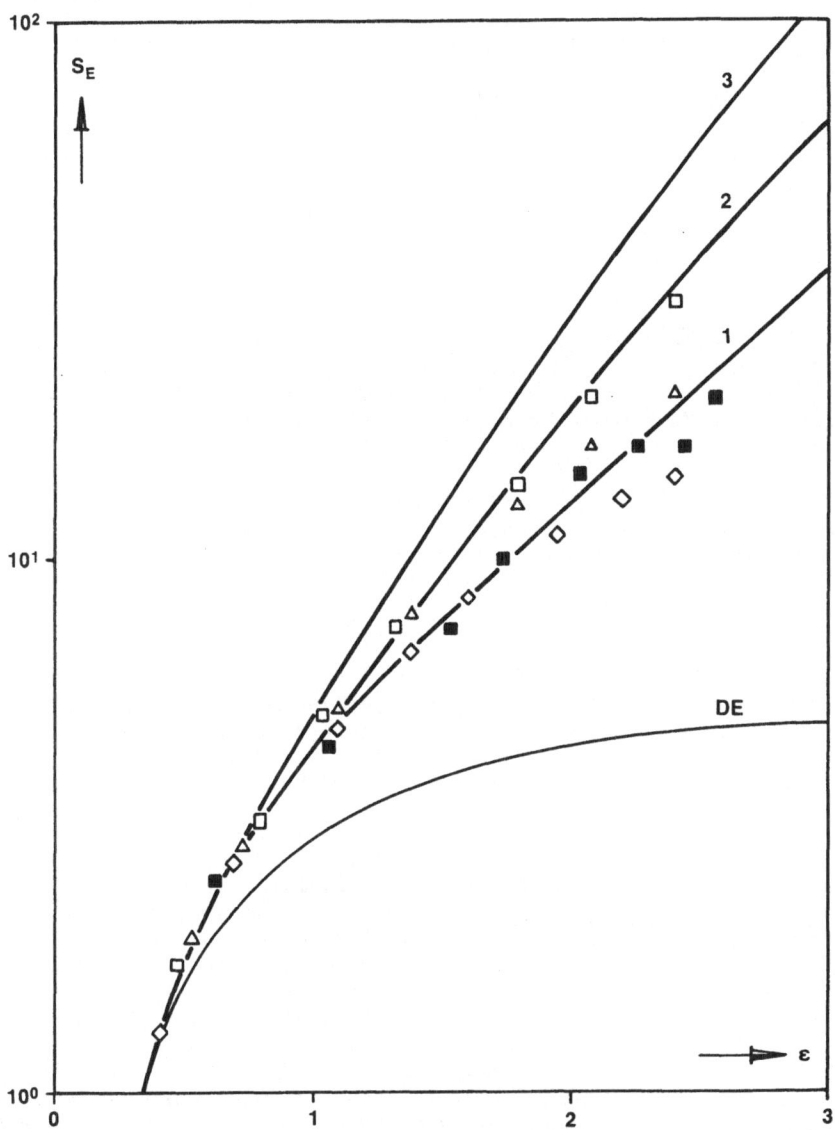

Fig. 2. Experimental and theoretical strain measures for various polymer melts identified in Table II. Curve DE pertains to Eq. (16) of the Doi-Edwards theory[14]; curve 1 is indistinguishable from Eq. (17) resulting from the modification of this theory by Marrucci and de Cindio.

Table II. Identification of the Polymer Melts of Fig. 2

Sample Characteristics	Type of Measurement	Reference	Curve Number or Symbol
PMMA	step strain stress relaxation	28	1
PS	step strain stress relaxation	28	1
PIB $\bar{M}_n = 8 \ 10^4$; $\bar{M}_\eta = 3.8 \ 10^5$ $\eta_{0,20} \ °C = 1.5 \ 10^8$ Pa·s $\eta_{0,100} \ °C = 8 \ 10^5$ Pa·s	strep strain stress relaxation	29	1
PS	constant strain rate	32	◇
HDPE	constant strain rate	32	△
LDPE	constant strain rate	32	□
LDPE $\bar{M}_w = 1.5 \ 10^5$; $\bar{M}_w/\bar{M}_n = 6.2$	step strain	30	■
LDPE $\bar{M}_w = 1.5 \ 10^5$; $\bar{M}_w/\bar{M}_n = 6.2$	constant strain rate constant stretching rate creep at constant stress	30	2
LDPE $\bar{M}_w = 4.8 \ 10^5$; $\bar{M}_w/\bar{M}_n = 2.8$	constant strain rate creep at constant stress	8,31	3

that the origin of nonlinearity is the same for crosslinked rubbers and for polymer melts. In other words, the deformation and orientation on molecular level that take place immediately or at least a very short time after the imposition of a macroscopic strain appear to be the same for chains connected to a network at both ends and for free chains.

This conclusion drawn from the experimental data is obviously incompatible with the molecular models reviewed above. The Flory theory is based on the presence of permanent junctions which are absent in polymer melts. As a consequence it evidently cannot explain why strain measures for crosslinked rubbers and polymer melts lie in the same range. On the other hand, the experimental $S_E(\lambda)$-curves for the EPDM materials of Table I (including the non-crosslinked D 0.0) can be excellently fitted with Eq. (13) using κ and μ/ξ as adjustable parameters, as is shown in Fig. 3. It is noteworthy that the values of μ/ξ are, however, in general much higher than 2, which is the value of μ/ξ for real networks with tetrafunctional junctions just beyond the gel point. A possible explanation for this finding in terms of the Flory theory is that also many two-functional junctions, i.e. chain segments, are actively constrained. This is equivalent to the hypothesis, that the restriction of possible chain configurations must play a very important role as well. This assumption explains why $S_E(\lambda)$-curves for crosslinked and non-crosslinked networks are almost identical.

The strain measure predicted by the Doi-Edwards theory is shown in Fig. 2 as well. This theory is obviously inappropriate to describe the tensile stress-strain relation for polymeric materials. Also, the fact that the nonlinearity is about the same for rubbers and melts is completely inconsistent with the Doi-Edwards model, because the relaxation process which is responsible for the nonlinearity, the chain retraction, cannot possibly take place in crosslinked rubbers where the chains are connected to the network at both ends.

The strain measure which follows from the model by Marrucci is indistinguishable from curve 1 in Fig. 2. It is in good agreement with expe-

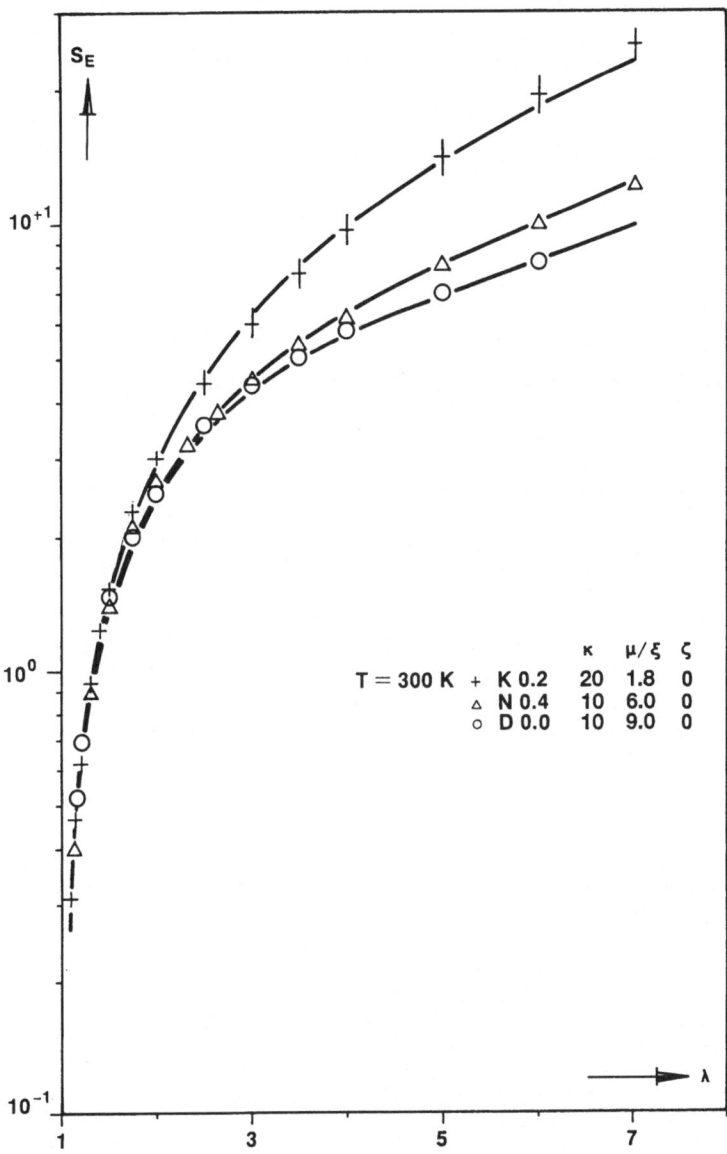

Fig. 3. Experimental strain measures for some EPDM elastomers (symbols) compared with theoretical predictions of the theory of Flory and Erman[12], using Eq. (13) and the parameter values indicated in the figure.

rimental strain measures of a considerable number of polymeric materials, but the molecular-kinetic picture underlying the model is incompatible with the fact that the strain measures of permanent networks and melts are almost identical. Also, Eq. (17) contains no adjustable parameter, which seems to be necessary to give a quantitative description of all experimental data.

CONCLUDING REMARKS

Experimental data seem to indicate that the origin of nonlinearity is the same for crosslinked rubbers and polymeric melts. Existing molecular models are inconsistent with this finding because they are designed to explain the nonlinear viscoelastic behaviour of one of the two types of material only. This situation calls for the development of a unified model which deals with the nonlinearity of both crosslinked and non-crosslinked materials.

The nonlinear viscoelastic behaviour in simple shear of crosslinked materials should be investigated to find out whether also in this deformation geometry similar strain measures are found for elastomers and melts.

REFERENCES

1. P.J. Flory, 'Principles of Polymer Chemistry', Cornell University Press, Ithaca, NY, 1953.
2. J.D. Ferry, 'Viscoelastic Properties of Polymers', 3rd ed., Wiley, New York, 1980.
3. L.R.G. Treloar, 'The Physics of Rubber Elasticity', 3rd ed., Clarendon, Oxford, 1975.
4. J. Meissner, Rheol. Acta, 8, 78 (1969).
5. H. Münstedt, Rheol. Acta, 14, 1077 (1975).
6. W.V. Chang, R. Bloch and N.W. Tschoegl, Rheol. Acta, 15, 367 (1976).
7. B.J.R. Scholtens and P.J.R. Leblans, J. Rheol., in press.
8. M.H. Wagner, Rheol. Acta, 18, 681 (1979).
9. C.J.S. Petrie, J. Non-Newtonian Fluid Mech., 5, 147 (1979).
10. H.C. Booij and J.H.M. Palmen, Rheol. Acta, 21, 376 (1982).
11. H.J. James and E. Guth, J. Chem. Phys., 11, 455 (1943); ibid., 15, 669 (1947); J. Polym. Sci., 4, 153 (1949).
12. P.J. Flory and B. Erman, Macromolecules, 15, 800 (1982).
13. P.J. Flory, British Polym. J., 17, 96 (1985).
14. M.S. Doi and J. Edwards, J. Chem. Soc., Faraday Trans. II, 74, 1789, 1802, 1818 (1978).
15. G. Marruci and B. de Cindio, Rheol. Acta, 19, 68 (1980).
16. B.J.R. Scholtens and H.C. Booij, in 'Elastomers and Rubber Elasticity', ACS Symp. Ser. 193, J.E. Mark and J. Lal, Eds., American Chemical Society, Washington, DC, 1982, chap. 28, p. 517.
17. P.J. Flory, Proc. Roy. Soc. London, A 351, 351 (1976).
18. G. Ronca and G. Allegra, J. Chem. Phys., 63, 4990 (1975).
19. P.J. Flory, J. Chem. Phys., 66, 5720 (1977).
20. B.J.R. Scholtens, P.J.R. Leblans and H.C. Booij, J. Rheol., in press.
21. T.L. Smith, Trans. Soc. Rheol., 6, 61 (1962); T.L. Smith and R.A. Dickie, J. Polym. Sci. Part A-2, 7, 635 (1969).
22. S.D. Hong, R.F. Fedors, F. Schwarzl, J. Moacanin and R.F. Landel, Polym. Eng. Sci., 21, 688 (1981).
23. S.M. Gumbrell, L. Mullins and R.S. Rivlin, Trans. Faraday Soc., 49, 1495 (1953).
24. A. Ciferri and P.J. Flory, J. Appl. Phys., 30, 1498 (1959).
25. L. Mullins, J. Appl. Polym. Sci. 2, 1 (1959).
26. G. Kraus and G.A. Moczvgemba, J. Polym. Sci., Part A, 2, 277 (1964).

27. B. Erman and P.J. Flory, Macromolecules, $\underline{15}$, 806 (1982).
28. B. de Cindio, Polymer, $\underline{25}$, 1049 (1984).
29. G. Titomanlio, G. Spadaro and F.P. La Mantia, Rheol. Acta, $\underline{19}$, 477 (1980).
30. P.J.R. Leblans, J. Sampers and H.C. Booij, J. Non-Newtonian Fluid Mech., in press.
31. M.H. Wagner, T. Raible and J. Meissner, Rheol. Acta, $\underline{18}$, 427 (1979).
32. H.M. Laun, Colloid Polymer Sci., $\underline{259}$, 97 (1981).

INDEX